MICROGRAVITY TWO-PHASE FLOW AND HEAT TRANSFER

THE SPACE TECHNOLOGY LIBRARY
Published jointly by Microcosm Press and Springer

An Introduction to Mission Design for Geostationary Satellites, J. J. Pocha
Space Mission Analysis and Design, 1st edition, James R. Wertz and Wiley J. Larson
**Space Mission Analysis and Design*, 2nd edition, Wiley J. Larson and James R. Wertz
**Space Mission Analysis and Design Workbook*, Wiley J. Larson and James R. Wertz
Handbook of Geostationary Orbits, E. M. Soop
**Spacecraft Structures and Mechanisms, From Concept to Launch*, Thomas P. Sarafin
Spaceflight Life Support and Biospherics, Peter Eckart
**Reducing Space Mission Cost*, James R. Wertz and Wiley J. Larson
The Logic of Microspace, Rick Fleeter
Space Marketing: A European Perspective, Walter A. R. Peeters
Fundamentals of Astrodynamics and Applications, 2nd edition, David A. Vallado
Mission Geometry: Orbit and Constellation Design and Management, James R. Wertz
Influence of Psychological Factors on Product Development, Eginaldo Shizuo Kamata
Essential Spaceflight Dynamics and Magnetospherics, Boris Rauschenbakh, Michael Ovchinnikov, and Susan McKenna-Lawlor
Space Psychology and Psychiatry, Nick Kanas and Dietrich Manzey
Fundamentals of Space Medicine, Gilles Clément
Fundamentals of Space Biology, Gilles Clément and Klaus Slenzka
*Also in the DoD/NASA Space Technology Series (Managing Editor Wiley J. Larson)

Microgravity Two-Phase Flow and Heat Transfer

by

Kamiel S. Gabriel

University of Ontario Institute of Technology (UOIT)
Oshawa, Canada

Published jointly by
Microcosm Press
El Segundo, California

Springer

A C.I.P. Catalogue record for this book is available from the Library of Congress.

ISBN-10 1-4020-5142-5 (HB)
ISBN-13 978-1-4020-5142-5 (HB)
ISBN-10 1-4020-5143-3 (e-book)
ISBN-13 978-1-4020-5143-2 (e-book)

Published jointly by
Microcosm Press
401 Coral Circle, El Segundo, CA 90245-4622 U.S.A.
and
Springer,
P.O. Box 17, 3300 AA Dordrecht, The Netherlands.

www.springer.com

Sold and distributed in North, Central and South America
by Microcosm.

In all other countries, sold and distributed
by Springer.

Cover Photo of Earth from Space: View of Africa and the Indian Ocean taken in Dec. 1972, by Apollo 17, the last of the Apollo missions to explore the Moon. Photo Courtesy of NASA.

Cover design by Jeanine Newcomb and Joy Sakaguchi.

Printed on acid-free paper

This book is dedicated to all my graduate students and research personnel of the Microgravity Research Group (1988–2004). Over the years, they made it possible to achieve the impossible.

TABLE OF CONTENTS

PREFACE

Multiphase thermal systems (involving more than one phase or one component) have numerous applications in aerospace, heat-exchangers, transport of contaminants in environmental systems, and energy transport and energy conversion systems. Advances in understanding the behavior of multiphase thermal systems could lead to, among other applications, higher efficiency energy production systems, improved heat-exchanger design, safer and enhanced treatment of hazardous waste. But such advances have been greatly hindered by the strong and masking effects of gravitational acceleration on the flow. Depending on the flow orientation and the phase velocities, gravitational forces could significantly alter the flow regime, and hence the momentum and heat transport mechanisms associated with the flow. A reduced gravity environment (or "microgravity"), provided an excellent tool to study the flow without the masking effects of gravity. This book presents for the first time a comprehensive coverage of all aspects of two-phase flow behavior in the virtual absence of gravity.

The chapters of the book introduce in depth coverage of the hydrodynamics and heat-transfer aspects of two-phase flow in reduced gravity including: flow regimes, void-fraction, pressure drop, and heat-transfer rates. Much of the material included in the book is based on extensive experimental research work which was conducted aboard specially equipped aircraft flying parabolic flights (e.g., NASA KC-135, ESA Zero-G Airbus 300, Learjet, etc.). The experimental data covered a wide range of liquid and gas flow rates and flow regimes ranging from bubbly to annular flow. Flow observations were made using high speed video cameras and non-intrusive sensors.

This book is written with the intent to provide the reader with a comprehensive coverage of all aspects of microgravity two-phase flow. Data analysis is presented with a view to give the reader a basic understanding of the two-phase flow behavior in a reduced-gravity environment. This serves well as a starting point for further studies and analysis.

I do hope that the reader will find the material interesting and inspiring to continue research in this important field.

Kamiel S. Gabriel
April 2006

ABOUT THE AUTHOR

Dr. Kamiel Gabriel is the Associate Provost of Research & Graduate Programs at UOIT; one of Canada's newest technical universities. He holds a Masters and Doctorate degrees in engineering, an MBA in international studies and a diploma in Space Science from the International Space University. Dr. Gabriel established the Microgravity Research Group in 1988. Since then, the group logged over 35 hours of near-zero gravity experiments aboard parabolic flights, space shuttles and in drop towers. He is internationally renowned for his pioneering work on multi-phase flow in near zero-gravity conditions. Dr. Gabriel resides in Whitby, Ontario, Canada.

ACKNOWLEDGEMENTS

I am indebted to all my graduate students, post-doctoral fellows and research associates who participated in this research through the Microgravity Research Group. I also wish to thank all the visiting professors and international researchers from Japan, Germany, France, USA and others who took part in our research activities during their stay in Canada. My sincere appreciation goes to Nanette Roberton, Gwen Jones, and Vaska Krabbe for their diligent work in editing the material included in this book.

NOMENCLATURE

A	Annular
B	Bubble
cv	sample coefficient of variation
C	Churn
C_o	distribution coefficient
D	diameter [m]
D	inner tube diameter, mm
E	electric power to heated test section, W
e	deviation between the experimental data and correlations
F	volume fraction function
Fr	Froude number, $V/(gD)^{1/2}$
g	gravity [ms^{-2}]
g	gravitational acceleration, m/s^2
g_0	normal earth gravity, $9.81\,m/s^2$
Gr	Grashof number, $g\beta D^4 q''/(\nu^2 k)$
Gz	Graetz number, $RePr$D/L
h	average convective heat transfer coefficient, $W/m^2\,K$
ID	inside diameter [m]
i	summation index
j	volumetric flux [ms^{-1}]
J	number of variables
k	thermal conductivity, W/m-K
L	length [m]
L	length of heated test section, cm
L	length of solution domain
$\overline{L}$	mean length
L_{gas}	length of gas bubble
L_{liquid}	length of liquid slug
Mo	Morton number, $\rho\sigma^3/(g\mu^4)$
N	number of data points
n	number of occurrences
N	total number of samples
n	time level
$\hat{n}$	unit normal vector
Nu	Nusselt number, hD/k
N_V	normalized voltage
OD	outside diameter [m]
p	pitch [m]
p	pressure

$\hat{p}$	pressure deviation
P_{gas}	gas pressure
p_{liquid}	liquid pressure
Pr	Prandtl number, ν/γ
q''	heat flux at the surface of the heated test section, W/m^2
r	experimental result
r	radial direction
Re	Reynolds number, $\rho VD/\mu$
s	sample standard deviation
S	Slug
S-A	slug-annular
$\hat{s}$	unit tangential vector
t	time
T	temperature, °C
Tail	location of bubble tail
U	velocity vector
u	radial component of velocity
$U_{i,j}$	discrete velocity component at location (i,j)
$\tilde{u}_{i,j}$	provisional discrete velocity component at (i,j)
$\bar{u}_{i,j}+1/2$	average discrete velocity at $(i,j+1/2)$
U_{x_i}	uncertainty of X_i
U_r	results of the experimental uncertainty
U_n	velocity in (n)-direction
u_s	velocity in {s}-direction
V	velocity, m/s
V	velocity vector
$\tilde{V}$	provisional velocity vector
v_{sg}, v_{sg}	superficial gas velocity
v_{sl}, v_{sl}	superficial liquid velocity
v	axial component of velocity
$\overline{v}$	mean velocity
V_b	bubble velocity
v_{face}	velocity at control volume face
v_g	actual gas velocity $[ms^{-1}]$
v_{gj}	drift velocity of the gas relative to the mean fluid velocity $[ms^{-1}]$
$v_{i,j}$	discrete velocity component at location (i, j)
$\tilde{v}_{i,j}$	provisional discrete velocity component at (i, j)
$\overline{v}_{i+1/2,j}$	average discrete velocity at $(i+1/2, j)$
v_{imag}, v_{real}	velocities used for application of boundary conditions
V_{nose}	velocity of bubble nose
V_{sg}	superficial gas velocity $[ms^{-1}]$
V_{sl}	superficial liquid velocity $[ms^{-1}]$
w	width [m]
We	Weber number, $\rho V^2 D/\sigma$

x gas mass quality
x gas quality
x co-ordinate direction
X_i Variable
$\bar{X}$ sample mean
y co-ordinate direction
y_a Variable
y_b Variable
z co-ordinate direction

GREEK SYMBOLS

α void fraction
β gas quality
β volumetric thermal expansion, 1/K
Δp pressure correction
κ local surface curvature
κ_i, κ_j local interface curvature
μ Mean
μ dynamic viscosity, Ns/m^2
ν kinematic viscosity, m^2/s
ρ density, kg/m^3
ρ_g density of the gas [kgm^{-3}]
ρ_l density of the liquid [kgm^{-3}]
ψ^2 normalized heat transfer coefficients, (h_{TP}/h_L)
α' pseudo-void fraction, V_{SG}/V_{SL}
ε_H eddy diffusivity for heat, m^2/s
σ surface tension, N/m^{-1}

SUBSCRIPTS

b bulk fluid
i interface subscript / column index
j row index
G single-phase gas
L single-phase liquid
SG superficial gas
SL superficial liquid
TP two-phase, two-component

w	inner tube wall
avg	mean
corr.	correlation
exp.	experimental
rms	root-mean-square

LIST OF FIGURES

LIST OF TABLES

CHAPTER 1

INTRODUCTION

Gas–liquid (or vapor–liquid) two-phase flows are commonly found in many industrial processes, engineering applications, and in ordinary life. Due to their important applications in chemical engineering processes, nuclear reactors, air conditioning and refrigeration systems, and heat exchangers, two-phase flows have been actively searched for several decades. Valuable experimental data were collected on-ground for different flow orientations and flow passage geometries. Theories and correlations were developed to predict the engineering parameters of gas–liquid flows in conduits in terms of flow pattern transitions, pressure drops, void fraction, heat-transfer rates, etc. However, due to the complexity of the flow, predictions were largely in terms of empirical or semi-empirical correlations, which were for the most part based on specific test conditions; e.g., conduit size and shape, adiabatic or boiling flows, method of heating or gas injection, etc. Extrapolation of these correlations to other conditions may not be valid, particularly when gravity is significantly reduced.

The two phases in gas–liquid flows are distributed in several configurations in pipe flows. The flow distribution itself is a function of – among other parameters – the liquid and gas flow rates, the fluid properties, the conduit shape, and the gravity level and direction relative to the flow direction. Furthermore, the position and shape of the interface and the interfacial velocity are very difficult (if not impossible) to determine analytically or experimentally. As a consequence, application of conservation equations to two-phase pipe flow in order to predict which fluid occupies a specific portion of the tube at a specific time has been unsuccessful. Because the gas–liquid flows are so chaotic and sensitive to the test conditions (which are, in many cases, beyond the control of researchers), the experimental data are widely scattered and predictions are somewhat elusive.

Most previous research on gas–liquid flows was conducted for terrestrial conditions where gravity plays a very important role. Mainly due to the demand for active heat transport systems in future communications and earth observation satellites and space stations, two-phase gas–liquid flows at microgravity (μ-g) conditions have emerged as an active research area in the last decade. When gravity is reduced, buoyancy forces are minimized, and secondary forces which are usually very small at terrestrial conditions (such as surface tension forces) are expected to play an important role. The behavior of gas–liquid flow may be different under such conditions. Theories and correlations which were derived at terrestrial conditions must be tested and possibly modified (or rejected) before they are used with confidence for conditions of reduced gravity. New theories and correlations have to be developed for microgravity conditions, where necessary. At reduced gravity, the slip velocity between the two phases is minimized, and in some cases is very small or totally diminished (bubble and slug flows). This allows for better measurements and predictions of actual phase velocity and interfacial areas (since the interface is much more stable in μ-g bubble and slug flows). Better understanding of the actual contribution of each force acting on the flow can then be gained from the analysis of two-phase flow measurements at μ-g.

1. APPLICATIONS OF TWO-PHASE FLOW AT MICROGRAVITY CONDITIONS

One of the major applications of two-phase flow at microgravity conditions is the design and maintenance of active thermal control systems for future space stations and high-power communications satellites. Past thermal management requirements for satellites and orbiting spacecraft have been characterized by low power ($< 25\,\text{kW}$) and short lifetime (< 3 years). These modest requirements can usually be satisfied by passive devices, such as heat pipes, or by pumped single-phase fluid cooling systems. The performance of passive devices is mostly independent of gravity. Knowledge gained of the performance of such systems on-ground can be directly used in the design of the space systems. Two-phase flow systems (capillary or mechanically driven) have been chosen as potential candidates for the design of future thermal subsystems. There are several characteristics that make two-phase flows more desirable than single-phase flow for heat transfer purposes. First, the heat transfer coefficient in two-phase flow with phase change can be several orders of magnitude higher than that in single-phase flow due to the large latent heat of vaporization of a liquid. This results in a physically smaller system that can carry as much heat as a single-phase system with much larger size. Secondly, heat can be transferred to the fluid while maintaining a constant surface temperature. This is a highly desirable feature since many advanced instruments onboard payloads require an isothermal environment (e.g., material processing, heat treatment, thermal control subsystems, etc.).

Another important application of two-phase flow is in the design of space nuclear power systems. Such systems have been proposed to meet escalating future power needs particularly in the area of surveillance and space transportation and logistics. A liquid-metal working fluid has been selected for these energy conversion systems. The details of such nuclear-powered devices proposed by General Electric can be found in Kirpich et al. (1990).

Two-phase flow phenomena also occur in many life-support systems in space stations and space labs. For example, the environmental conditions inside a space station (e.g., level of O_2, humidity, etc.) have to be maintained at comfortable levels. Water used for personal hygiene or other purposes, in many instances, has to be collected and processed for reuse. Two-phase flow is also prevalent in material processing in space, cryogenic transfer and storage, and many other enabling technologies (mining, water processing, etc.).

Studies of two-phase flow under microgravity conditions are also of great interest to better understand the behavior of terrestrial flows. Since gravity force plays an overwhelming role on-ground, the microgravity conditions provide an ideal environment to study the influence of other less dominant forces such as those due to surface tension, lift force, and turbulence.

2. SIMULATION OF MICROGRAVITY CONDITIONS

Orbiting spacecraft and free-falling objects experience near weightless conditions. For a spacecraft orbiting Earth, the gravitational attraction force is balanced by the centrifugal force at the location of the center of mass. Objects inside the spacecraft will therefore experience weightlessness. However, it should be noted that weightlessness is an ideal state, which cannot be practically achieved. Due to the tidal effects and oscillatory accelerations, the gravity level an orbiting object may experience is not exactly zero. Typical gravity levels for objects inside the Space Shuttle range from 10^{-2} to 10^{-5} g_0. For the International Space Station the gravity level would range between 10^{-3} and 10^{-6} g_0. Detailed discussions of this topic may be found in Hamacher et al. (1987).

The best place to conduct microgravity research at this time would be aboard the Space Shuttle, where longtime duration of reduced gravity and reasonable micro-gravity levels can be attained. However, conducting research in space tends to be very expensive, complex, and time-consuming. It is generally desirable to acquire as much knowledge and experience as possible through ground-based research before a flight is warranted. Ground-based facilities can provide limited micro-gravity duration ranging from a fraction of a second in drop tubes to about 7–15 min on sounding rockets.

Drop tubes and drop towers are facilities that allow for simulation of microgravity conditions on Earth. Drop tubes are capable of accommodating only small experi-mental packages. Drop towers differ from drop tubes in their ability to accommodate larger experimental packages and can provide longer durations of microgravity conditions. Depending on the drop distance, drop tubes may provide 1.7–4.6 sec of microgravity, and drop towers usually provide 4.2–10 sec of test time.

Another alternative to simulate a reduced-gravity environment is offered by aircraft flying parabolic trajectories. These include the KC-135 aircraft at NASA Johnson Space Center, the Learjet and DC-9 aircraft at NASA Glenn Research Center, the French Caravelle "Zero-G" aircraft, and the Japanese Learjet MU-300. During a typical flight parabola, the airplane first climbs rapidly at a 45° angle in a pull-up maneuver, slows down at the top, and then descends at a 45° angle in the pullout maneuver. During the pull-up and pullout legs of the flight, the gravity level inside the aircraft cabin is about 1.8 g_0. While at the top of the parabola, the flight produces a "quasi-steady" gravity level of about 10^{-2} g_0 for approximately 20 sec.

Another alternative to orbital flight that provides an excellent reduced-gravity environment is a sounding rocket. A payload can be launched to altitudes ranging from 90 to 290 km with a single- or multiple-stage sounding rockets. The duration of the free fall is about 4–15 min depending on the type of rocket used. The residual gravity level is usually below 10^{-4} g_0, and is mainly determined by atmospheric drag, residual spin, and operational disturbances.

3. SCOPE OF THIS BOOK

The most recent results of heat transfer, pressure drop, gas void fraction, and flow regimes at microgravity conditions are presented in this book. Definitions of the various gas–liquid flow patterns observed at microgravity conditions are given in Chapter 2. Actual photographs of the observed two-phase flow pattern images are presented to convey some of the characteristics of gas–liquid flow at microgravity conditions. Previous models for the prediction of flow pattern transitions are reviewed in Chapter 3. Ground flow models are extrapolated to microgravity conditions and compared with experimental data. New models that were specifically developed for microgravity conditions are introduced and compared with existing experimental data. In Chapter 4, experimental results for pressure drop at microgravity conditions are presented. Various methods for the estimation of pressure drop are tested against the microgravity experimental data. Chapter 5 introduces gas void fraction measurements using non-intrusive capacitance sensors. Signal analysis and Probability Density Functions (PDFs) are also presented in that chapter. In Chapter 6, experimental results for the local and length-averaged heat transfer coefficients are presented and discussed. Transient effects due to short durations of μ-g, as well as a change in the flow regime as the flow proceeds from hypergravity to microgravity, are presented in some detail. The effect of free convection on the 1-g results compared with microgravity results is also discussed. New empirical correlations are presented for the prediction of the average heat transfer coefficients. In Chapter 7, numerical simulation results are presented for bubbly flow using an interface-tracking model. The model is based on the Volume of Fluid (VOF) method. Finally, the summary and conclusion are presented in Chapter 8. Appendix A includes comprehensive data sets for the microgravity two-phase (adiabatic) experiments conducted in Canada, the United States, and Europe. Appendix B includes a large set of photographs showing the different flow regimes observed at microgravity conditions.

CHAPTER 2

CLASSIFICATION OF GAS–LIQUID FLOW PATTERNS

The liquid–gas distribution pattern inside a conduit varies considerably depending on, among other factors, the mass flow rates of both phases, the conduit size, the phase properties, and the gravity vector with respect to the flow direction. Groups with similar phase distribution characteristics are classified into flow patterns (or flow regimes). Tremendous efforts have been made to study the flow patterns and their transitions under normal and microgravity conditions. Since the identification of a particular phase distribution is somewhat subjective, the literature contains a host of flow pattern definitions and descriptions. In recent years, there has been some effort to reach a "standard" for flow pattern definitions (Chisholm 1983; Dukler and Taitel 1986; Barnea and Taitel 1986; Whalley 1987).

1. HORIZONTAL FLOWS ON-GROUND

In horizontal pipe flow, gravity is acting on the two-phase mixture in a direction perpendicular to the flow direction. Typical flow patterns for horizontal flow in a circular tube are illustrated schematically in Fig. 1.

The flow pattern definitions adopted by Dukler and Taitel (1986) and Barnea and Taitel (1986) are described as follows:

(i) *Bubble flow*, in which the gas-phase is distributed as discrete bubbles in an axially continuous liquid-phase. The gas bubbles tend to flow near the top of the tube, but as the liquid flow rate increases, the bubbles are dispersed more uniformly in the liquid-phase.

(ii) *Intermittent flow*, in which the small gas bubbles have coalesced to produce large gas bubbles. Plugs or slugs of liquid, which fill the whole pipe cross-section, are separated by gas zones that overlay a stratified liquid layer flowing along the bottom section of the pipe. The intermittent pattern is sometimes subdivided into slug and elongated bubble patterns, but the distinction between them has not been clearly defined.

(iii) *Stratified flow*, in which liquid flows along the bottom section of the pipe with the gas-phase flowing along the top. Both phases are continuous in the axial direction. The interface may be smooth or wavy.

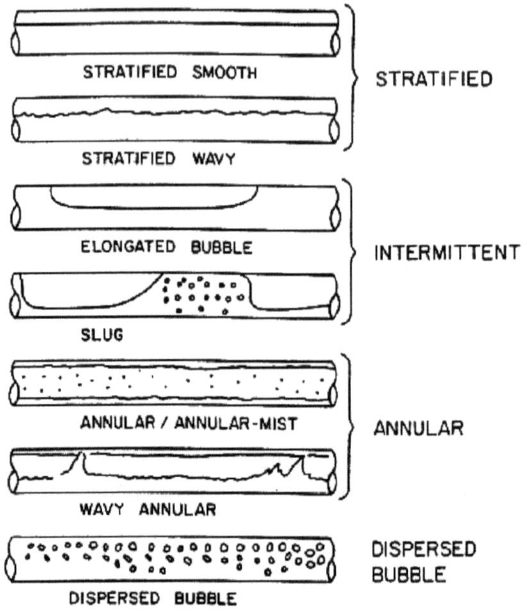

Figure 1. Two-phase flow in horizontal pipe on-ground. (Barnea and Taitel, 1986)

(iv) *Annular flow*, in which the liquid travels partly as a continuous film around the perimeter of the pipe, and partly as small drops distributed in the gas-phase which flows in the center of the tube. The liquid film is usually thicker at the bottom section of the tube than at the top.

2. VERTICAL FLOWS ON-GROUND

In this context, we are mainly concerned with co-current vertical flows. The flow may be upward or downward. The flow pattern definitions are less diversified in vertical upward flow than in horizontal flow. As depicted in Fig. 2, the flow patterns in vertical upward flows can be defined as follows:

(i) *Bubble flow*, in which the gas-phase is somewhat uniformly distributed in the form of discrete bubbles in a continuous liquid-phase.

(ii) *Slug flow*, in which most of the gas-phase is located in large "bullet-shaped" bubbles, generally called "Taylor bubbles," which occupy most of the pipe cross-sectional area and can vary in length from one to several tube diameters. The "Taylor bubbles" are separated by slugs of continuous liquid that bridge the pipe and usually contain small gas bubbles.

(iii) *Churn flow*, is a highly unstable flow of oscillatory nature. It is similar to slug flow in that "pockets" of gas are observed to be separated by liquid slugs, but it is much more chaotic, frothy, and disordered. The "Taylor bubbles" are

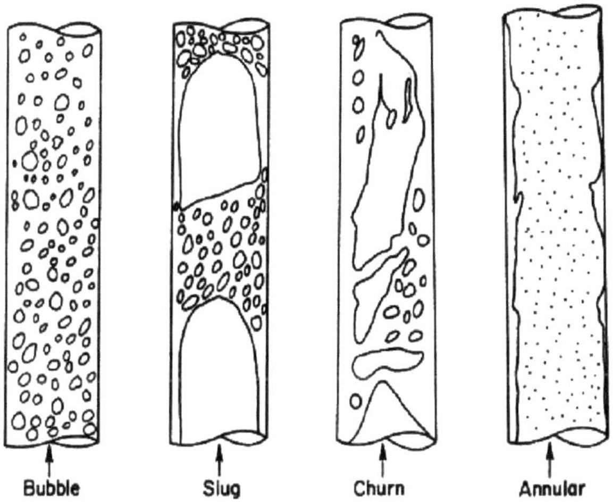

Figure 2. Two-phase flow patterns in vertical pipe on-ground. (Barnea and Taitel, 1986)

bridged with liquid and their shape is severely distorted. The continuity of the liquid in the slug region is destroyed by a high gas concentration. A typical characteristic of churn flow is the oscillatory (or alternating) motion of the liquid-phase as the liquid hold up is repeatedly overcome by gravitational pull.

(iv) *Annular flow*, in which the liquid travels partly in the form of an annulus at the wall and partly in the form of small droplets that are distributed in the gas-phase. The latter usually flows at the center of the tube forming the core region.

3. FLOW PATTERNS AT REDUCED-GRAVITY CONDITIONS

Studies of forced convective two-phase pipe flow under microgravity conditions are still in the formative stage. Due to the limited access to such an environment, and the high cost associated with it, only limited experimental data sets have been collected to date. Most of the available data were collected aboard aircraft flying parabolic trajectories or in drop towers. These ground facilities offer only short durations of microgravity (up to 10 sec in drop tower/drop tubes, and approximately 22 sec aboard aircraft). Despite their shortcomings, valuable information has been obtained using such facilities.

A pioneering work on gas–liquid flow patterns under microgravity conditions was conducted by Heppner et al. (1975) aboard the KC-135 zero-gravity aircraft using a water–air system. Flow pattern observations indicated that, qualitatively speaking, the behavior of the two-phase system differed from that at 1-g. Earth-based theories were used to classify the reduced-gravity flows; e.g., Quandt's criteria for flow transition.

A further work on the microgravity flow patterns and their transitions was presented by Dukler et al. (1988). Experiments were carried out using water–air in 12.7 mm and 9.525-mm i.d. tubes in the NASA Glenn30-m drop tower and aboard a Learjet flying trajectories. Flow pattern observations were presented for the range of $0.07\,\mathrm{m/s} < V_{SL} < 1\,\mathrm{m/s}$ and $0.09\,\mathrm{m/s} < V_{SG} < 25\,\mathrm{m/s}$. Models were proposed to predict the transitions between bubble and slug flows, as well as between slug and annular flows. Colin et al. (1991) reported observations from a flight experiment using a larger tube size (40 mm i.d.). Only bubble and slug flows were found to exist over the narrow range of the liquid and gas flow rates tested. More recently, flight data were reported by Huckerby and Rezkallah (1992) using a 9.525-mm i.d. water–air system. A model for the prediction of the flow transitions in microgravity two-phase flow was proposed by Zhao and Rezkallah (1993a), and later modified by Rezkallah (1996).

3.1 Description of Gas–Liquid Flow Patterns at Microgravity

In order to convey some of the basic flow pattern observations, a series of prints of frames at different gas and liquid velocities obtained aboard NASA's KC-135 aircraft are presented in Figs 3–7. These figures are taken from the work by Zhao and Rezkallah (1993a). The tube orientation is vertical with respect to the aircraft floor. The details of the flight hardware are given elsewhere in this document.

Figure 3 shows the flow patterns at $V_{SL} = 0.2\,\mathrm{m/s}$ and $V_{SG} = 0.11\,\mathrm{m/s}$. Hydrodynamically stable "bullet-shaped" bubbles with spherical nose and tail move along the pipe, separated by liquid slugs, which may contain several small gas bubbles (frame 4). In some cases (frames 2, 5, and 6), there are very thin membranes in the "Taylor bubbles" that bridge the gas bubble (or give the appearance of two

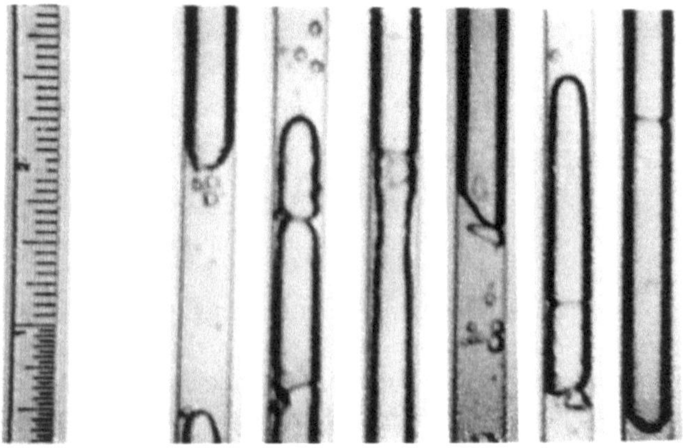

Figure 3. μ-g Water–air flow patterns at $V_{SL} = 0.2\,\mathrm{m/s}$, $V_{SG} = 0.11\,\mathrm{m/s}$

or more connected bubbles). It was found that the small gas bubbles in the liquid slug and the "Taylor bubbles" move at the same speed. At this flow condition, the shape of the bubbles is decided mainly by the surface tension force, as well as the turbulent force in the liquid. This is perhaps the reason for the spherically nosed "Taylor bubbles," and almost spherical small gas bubbles in the liquid slugs (frames 1, 2, and 5). The liquid Reynolds number, based on the liquid superficial velocity (Re_{SL}), is approximately 2000 in this case, which is about the critical value for transition from laminar to turbulent flow in single-phase flow. The disturbances in the liquid-phase are responsible for the distortion of the gas bubbles.

Flow patterns at $V_{SL} = 0.2\,m/s$ and $V_{SG} = 6.97\,m/s$ are given in Fig. 4. It is found that the liquid slug in the previous flow pattern has been gradually replaced by a frothy slug (frames 3 and 5). The liquid-phase has been broken into droplets and mixed with the gas-phase, giving it the appearance of a frothy mixture. The liquid droplets continuously deposit onto the liquid film at the wall. As the thickness of the film increases, the liquid is entrained back into the core flow again. Outside the frothy-slug region, there is a typical annular flow; gas is flowing at the center core, and liquid is flowing as a film at the wall. The transition from slug to annular flow is a slow, gradual process. After a slug flow pattern is formed, and with further increase in the gas flow rate, the gas in the "Taylor bubbles" breaks into the liquid slugs forming many small gas bubbles. As the gas flow rate further increases, the density of the gas bubbles increases, and eventually the gas bubbles form a continuous gas-phase filling what was previously occupied by the liquid slugs. At these flow rates, the gas inertial force is comparable with that due to surface tension. The inertial force gradually overcomes surface tension. Similar phenomena were also reported by Dukler et al. (1988) where they described it as a "locally thick" annular film. The frothy slugs become thinner as the gas flow rate increases until eventually a fully annular flow pattern is established.

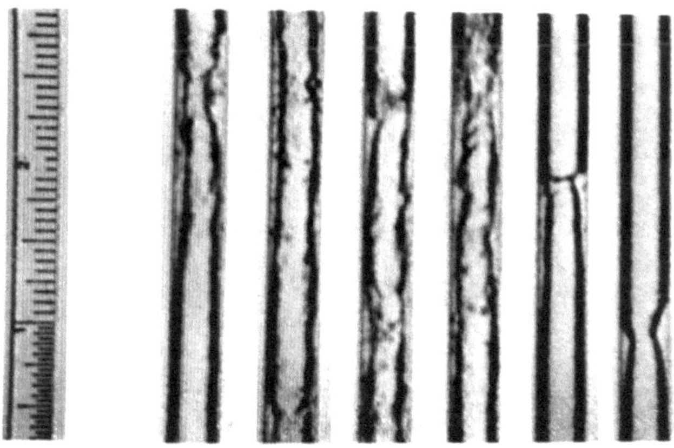

Figure 4. μ-g Water–air flow patterns at $V_{SL} = 0.2\,m/s$, $V_{SG} = 6.97\,m/s$

At a higher liquid flow rate and a relatively low gas flow rate, $V_{SL} = 2.3\,m/s$ and $V_{SG} = 0.20\,m/s$, a bubble flow (or perhaps a transitional flow from bubble to slug) is observed in the tube, as shown in Fig. 5. The size, shape, and length of the gas bubbles vary considerably, but still with a diameter that is smaller than the tube diameter. The shape of the bubbles is quite irregular due to the high turbulence in the liquid-phase ($Re_{SL} = 22,000$).

At the same liquid superficial velocity ($V_{SL} = 2.3\,m/s$) and a higher gas superficial velocity (0.79 m/s), irregular "Taylor-shaped" bubbles are formed in the center of the tube with some fine gas bubbles dispersed in them. These large bubbles are separated by liquid slugs containing some fine gas bubbles in them as well, as shown in Fig. 6.

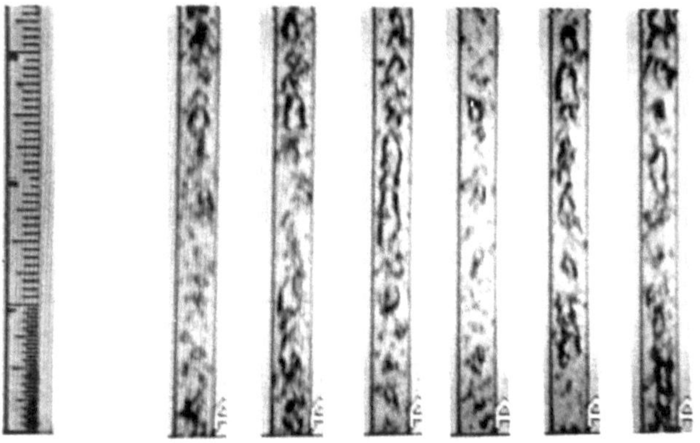

Figure 5. μ-*g* Water–air flow patterns at $V_{SL} = 2.3\,m/s$, $V_{SG} = 0.20\,m/s$

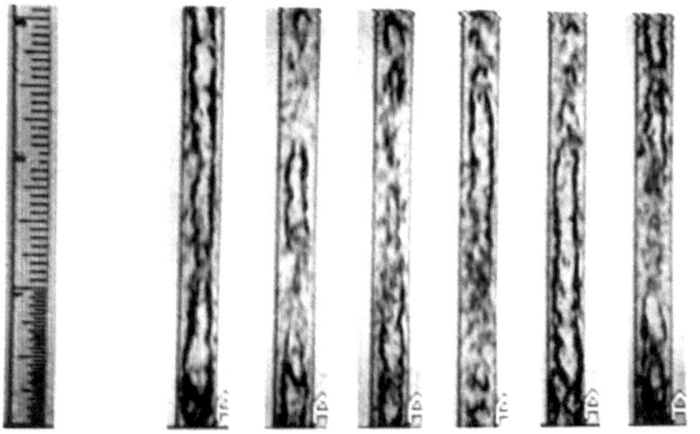

Figure 6. μ-*g* Water–air flow patterns at $V_{SL} = 2.3\,m/s$, $V_{SG} = 0.79\,m/s$

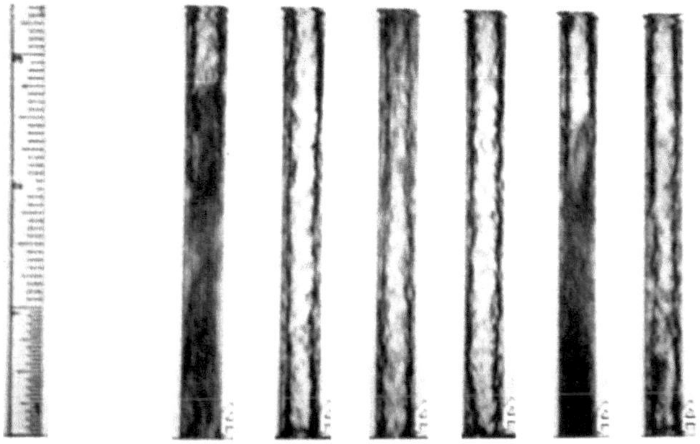

Figure 7. μ-g Water–air flow patterns at $V_{SL} = 2.3\,m/s$, $V_{SG} = 3.98\,m/s$

At a yet higher gas velocity ($V_{SG} = 4\,m/s$), frothy slug-annular flow occurs. For such high gas concentrations, the frothy liquid portions are more frequent, and also quite packed, as shown in Fig. 7.

3.2 Definition of Flow Patterns at Microgravity

Basically four flow patterns are observed to exist under microgravity conditions. The flow patterns can be categorized as follows:

(i) *Bubble flow*, in which the gas bubbles, distributed in a liquid continuum, are of a size less than or equal to the tube diameter, such as in Fig. 8a.

(ii) *Slug flow*, in which the length of the gas bubbles is greater than the tube diameter, and its diameter is close to the tube diameter. The liquid slugs that separate the large "Taylor bubbles" may or may not contain small gas bubbles (Fig. 8b).

(iii) *Transitional* or *frothy slug-annular flow*, in which case the liquid is flowing in the form of a film at the tube wall, and the gas-phase is flowing in the center with frequent appearances of frothy slugs (Fig. 8c). Because the details of the frothy slugs cannot be clearly seen, it is speculated that they contain densely packed fine gas bubbles in a liquid continuum at relatively low gas flow rates; or small liquid droplets in a gas continuum at relatively high gas flow rates.

(iv) *Annular flow*, which is observed when the liquid-phase flows at the tube wall and the gas-phase flows uninterruptedly at the center of the tube. The gas core may contain in it dispersed liquid droplets (Fig. 8d).

The experimental air–water data of Huckerby and Rezkallah (1992) and Rite and Rezkallah (1994) are plotted on a V_{SL} vs. V_{SG} flow pattern map according to the above definitions. This is shown in Fig. 9. As can be seen from this map, bubble flow was observed to exist at high liquid superficial velocities and low

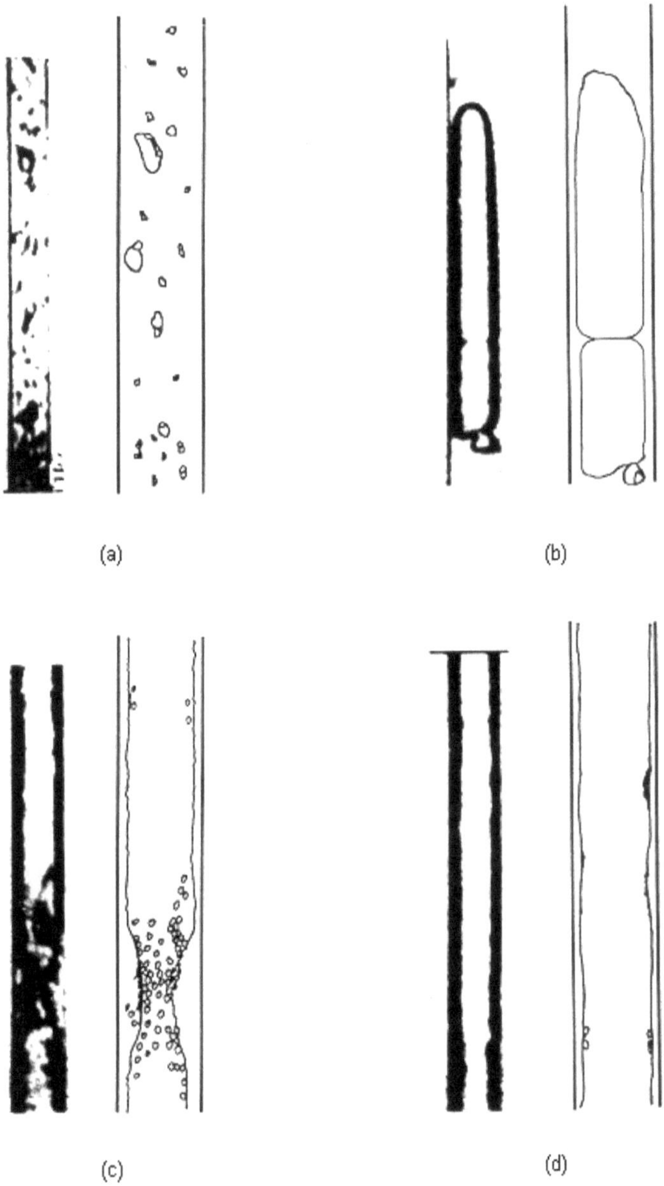

Figure 8. Typical two-phase flow patterns at μ-g conditions. (**a**) Bubbly flow. (**b**) Slug flow. (**c**) Frothy slug-annular flow. (**d**) Annular flow

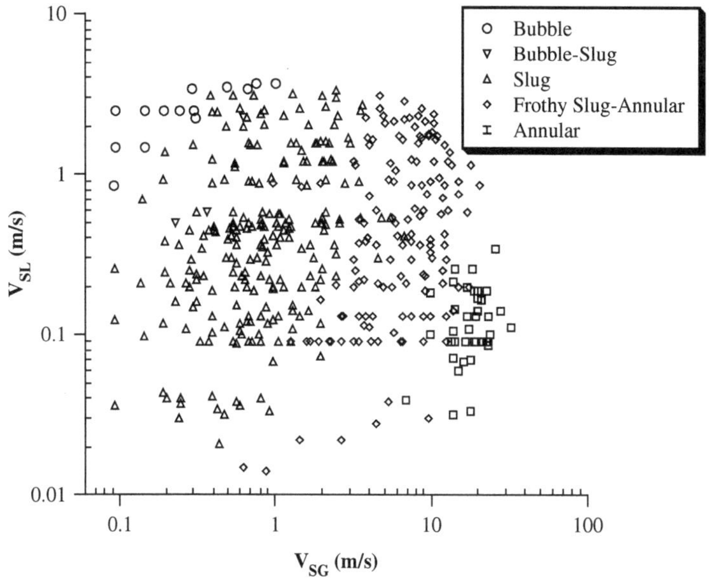

Figure 9. μ g Air water flow pattern observations. (Huckerby and Rezkallah 1992; Rite and Rezkallah 1994)

gas superficial velocities ($V_{SL} > 1.0\,\mathrm{m/s}$ and $V_{SG} < 1.0\,\mathrm{m/s}$). Annular flow was observed at high gas superficial velocities ($V_{SG} > 10\,\mathrm{m/s}$). Slug flow and frothy slug annular flow were found to occupy a wide range of gas and liquid flow rates between the previous two regimes. The onset to annular flow seems to occur at a constant gas superficial velocity of $V_{SG} = 10.0\,\mathrm{m/s}$.

CHAPTER 3

FLOW PATTERN TRANSITION MODELS

Flow pattern transition models were developed to provide methods to predict which type of flow will exist in the tube/conduit for some given flow conditions. A typical approach to generate those predictions is purely empirical, that is, based on experiments at certain flow conditions. Flow pattern observation results are plotted on a two-dimensional map, typically in terms of the superficial liquid and gas velocities (Fig. 18 in Chapter 2). Other co-ordinate systems have been used for the purpose of mapping the flow, including both dimensional and dimensionless groups. Extension of such flow pattern maps to other flow conditions and conduit geometries and sizes is of uncertain reliability, since the maps strongly depend on the particular flow conditions at which they were generated with no rigorous theoretical basis. Lately, models based on physical explanation of the flow behavior have been proposed for the prediction of flow pattern transitions in ground-based horizontal and vertical flows. These models provide a good step toward more quantitative prediction methods, even though the phenomenon itself is far from completely understood.

1. MODELS FOR HORIZONTAL FLOWS ON-GROUND

Two-phase gas–liquid flows in horizontal pipes have been classified as bubble flow, intermittent flow, stratified flow, and annular flow. The most widely used models for the prediction of transitions between these flow patterns are those by Taitel and Dukler (1976), and Weisman et al. (1979).

Taitel and Dukler (1976) proposed models for the various transitions based on some theoretical and empirical theories. For a stratified flow with waves, it was argued that as the gas accelerates, the pressure in the gas-phase over a wave decreases owing to Bernoulli's effect. This tends to cause the wave to grow. On the other hand, gravity force acting on the wave tends to eventually cause it to decay. The Kelvin-Helmholtz theory provides stability criteria for waves of infinitesimal amplitude formed on a flat sheet of liquid flowing between horizontal parallel plates. According to that theory, waves will grow when

$$(1) \qquad V_G > \left[\frac{g(\rho_L - \rho_G)h_G}{\rho_G}\right]^{1/2}$$

where V_G is the actual gas velocity and h_G is the distance between the upper plate and the equilibrium liquid level. This analysis was extended to the round pipe geometry to yield the following transition criteria:

$$V_G > C_2 \left[\frac{(\rho_L - \rho_G)\, g A_G}{\rho_G\, (\mathrm{d}A_L/\mathrm{d}h_L)} \right]^{1/2}, \tag{2}$$

where

$$C_2 = 1 - \frac{h_L}{D},$$

and A_G and A_L are the areas occupied by the gas and liquid phases, respectively, and h_L is the height of the liquid-phase. Equation (2) describes the conditions for the transition in pipes from stratified to intermittent and to annular flow. As the finite waves grow, two events can take place: slug flow or annular flow. It was suggested that if the equilibrium liquid level in the pipe was above the pipe centerline, intermittent flow would develop, and if $h_L/D < 0.5$, annular flow would result. The A value of h_L/D equal to 0.5 corresponds to a constant value of X, which is the Martinelli parameter; the latter is defined as:

$$X^2 = \frac{\left|\frac{\mathrm{d}p}{\mathrm{d}x}\right|_{LS}}{\left|\frac{\mathrm{d}p}{\mathrm{d}x}\right|_{GS}}. \tag{3}$$

For horizontal tubes, the value of X is equal to 1.6. Therefore, if the value of X is smaller than 1.6 it is a slug flow, otherwise it is annular flow.

For the transition between intermittent and dispersed bubble, it was suggested that the transition takes place when the turbulent fluctuations are strong enough to overcome the buoyant forces tending to keep the gas near the top of the pipe. This yields

$$V_L \geq \left[\frac{4A_G g}{S_i f_L} \left(1 - \frac{\rho_G}{\rho_L} \right) \right]^{1/2} \tag{4}$$

as the transition criteria from intermittent to dispersed bubble regimes, where S_i is the perimeter at the gas–liquid interface and f_L is the Fanning friction factor for the liquid-phase. A flow pattern map according to the above transition is plotted in Fig. 10 on V_{SL} vs. V_{SG} co-ordinates for water–air flow in 2.5 cm i.d. tube.

Weisman et al. (1979) conducted numerous experiments to determine the effect of various fluid properties on the flow pattern transitions. Using empirical correlations, they proposed transition criteria between different flow patterns based on their experimental data. The transition correlation between stratified and intermittent regimes was given by:

$$\frac{V_{SG}}{(gD)^{1/2}} = 0.25 \left(\frac{V_{SG}}{V_{SL}} \right)^{1.1} \tag{5}$$

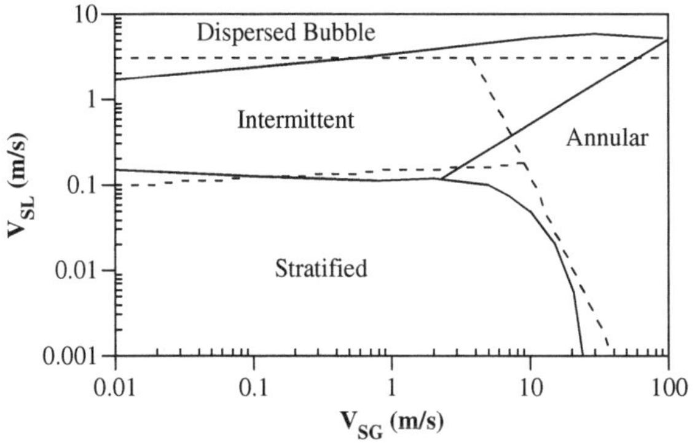

Figure 10. Flow pattern map in 1-g horizontal tube. (Air–water flow in 2.5 cm i.d. tube—Taitel–Dukler model and Weisman et al. model)

while the correlation for the transition to annular flow was given as:

$$1.9\left(\frac{V_{SG}}{V_{SL}}\right)^{1/8} = Ku^{0.2}Fr^{0.18} = \left(\frac{V_{SG}\rho_G^{1/2}}{[g(\rho_L \quad \rho_G)\sigma]^{1/4}}\right)^{0.2}\left(\frac{V_{SG}^2}{gD}\right)^{0.18}, \tag{6}$$

and the transition to dispersed bubble flow was correlated by:

$$\left[\frac{\left(\frac{dp}{dx}\right)_L}{(\rho_L-\rho_G)g}\right]^{1/2}\left(\frac{\sigma}{(\rho_L-\rho_G)gD^2}\right)^{-0.25} = 9.7. \tag{7}$$

The transition lines according to the above empirical correlations are superimposed on the Taitel–Dukler lines as shown in Fig. 10 (broken-lines). The transitions were generated for water–air flow in 2.5 cm i.d. pipe. It is clear that the same flow patterns are predicted by both models in roughly the same regions of the map.

2. MODELS FOR VERTICAL UPWARD FLOWS ON-GROUND

Gas–liquid flows in vertical tubes have been classified as bubble flow, slug flow, churn flow, and annular flow. Many two-phase flow pattern maps have been proposed for the prediction of the flow pattern transitions in vertical, upward flows. Among them, the most widely used models are the ones proposed by Taitel et al. (1980) and Weisman and Kang (1981).

In order to develop general applicable transition models for vertical flow, Taitel et al. (1980) attempted to suggest physically based mechanisms and to model the transitions based on these mechanisms. The transition from bubble flow to slug flow requires a process of agglomeration or coalescence of the gas bubbles. At low liquid flow rates, where turbulent fluctuations are not high enough to break large gas bubbles, the bubble density increases, as the gas flow rate is increased, to a point where the dispersed bubbles become so closely packed that many collisions occur and the rate of agglomeration to large bubbles increases sharply. This results in a transition from bubble flow to slug flow. Experiments have shown that this transition takes place at a bubble void fraction ranging from 0.25 to 0.3. Thus, the equation for the transition from bubble flow to slug flow is of the form:

$$V_{SL} = 3.0V_{SG} - 1.15\left[\frac{g(\rho_L - \rho_G)\sigma}{\rho_L^2}\right]^{1/4}. \tag{8}$$

At higher liquid flow rates, turbulent forces act to break the large gas bubbles and disperse them into small bubbles. From a balance between surface tension forces and those due to turbulent fluctuations, a relationship between V_{SG} and V_{SL}, above which slug flow cannot exist, was obtained:

$$V_{SL} + V_{SG} = 4.0\left\{\frac{D^{0.429}(\sigma/\rho_L)^{0.089}}{\nu_L^{0.072}}\left[\frac{g(\rho_L - \rho_G)}{\rho_L}\right]^{0.446}\right\}. \tag{9}$$

However, regardless of how much turbulent energy is available to disperse the mixture, bubble flow cannot exist at packing densities above 0.52. This is the limit of dispersed bubble flow.

There is another mechanism in the case of tubes of small diameter. The rise velocity of small gas bubbles depends only on the properties of the fluids. However, the rise velocity of the "Taylor bubbles" is independent of fluid properties, but it depends on the tube diameter and gravity. Whenever the small bubble velocity is larger than the "Taylor bubble" velocity, the rising small bubbles approach the back of the large bubble and coalesce with it. Bubble flow cannot exist under these conditions. This region is occupied by slug flow. The criterion for small diameter tube is

$$\left[\frac{\rho_L^2 g D^2}{(\rho_L - \rho_G)\sigma}\right]^{1/4} \leq 4.36. \tag{10}$$

For water–air system at atmospheric pressure, the critical diameter is $D \approx 5.0\,\text{cm}$.

It should be noted that it is difficult to identify accurately the slug–churn transition because there is confusion as to the description of the churn flow itself. Based on observations, Taitel et al. (1980) suggested that the churn flow might be an entry region phenomenon associated with the existence of slug flow further down the pipe. A correlation for the transition from slug to churn flow from those ideas was given as:

$$\frac{l_E}{D} = 40.6\left(\frac{V_{SL} + V_{SG}}{\sqrt{gD}} + 0.22\right), \tag{11}$$

where l_E is the entry length of the tube.

For the transition to annular flow, it was suggested that annular flow cannot exist unless the gas velocity in the gas core is sufficiently high to lift the entrained droplets. The transition boundary was then calculated using the expression:

$$\frac{V_{SG}\rho_G^{1/2}}{[\sigma g(\rho_L-\rho_G)]^{1/4}}=3.1. \tag{12}$$

A flow pattern map according to the above transitions is plotted in Fig. 11 for air–water upward flow in a 2.5-cm i.d. vertical tube.

Weisman and Kang (1981) extended their early transition correlations for horizontal flow to vertical flow. Based on experimental observations, they suggested that at high mass velocities the inertial forces will greatly exceed gravitational forces, and hence the orientation of the test section should have little effect. The correlation for transition to dispersed bubble flow for horizontal flow can be used then for vertical flow. The transition to dispersed bubble flow was given by:

$$\left[\frac{\left(\frac{dp}{dx}\right)_L}{(\rho_L-\rho_G)g}\right]^{1/2}\left[\frac{\sigma}{(\rho_L-\rho_G)gD^2}\right]^{-1/4}=9.7. \tag{13}$$

For the transition to annular flow in vertical tubes, a simplified version of the correlation for horizontal flow was found to work well; this was given as:

$$Ku\ Fr=\left\{\frac{V_{SG}\rho_G^{1/2}}{[g(\rho_L-\rho_G)\sigma]^{1/4}}\right\}\left(\frac{V_{SG}^2}{gD}\right)=25\left(\frac{V_{SG}}{V_{SL}}\right)^{5/8}. \tag{14}$$

Finally, the transition between bubble and slug flow was given by:

$$\frac{V_{SG}}{\sqrt{gD}}=0.45\left(\frac{V_{SG}+V_{SL}}{\sqrt{gD}}\right)^{0.78}. \tag{15}$$

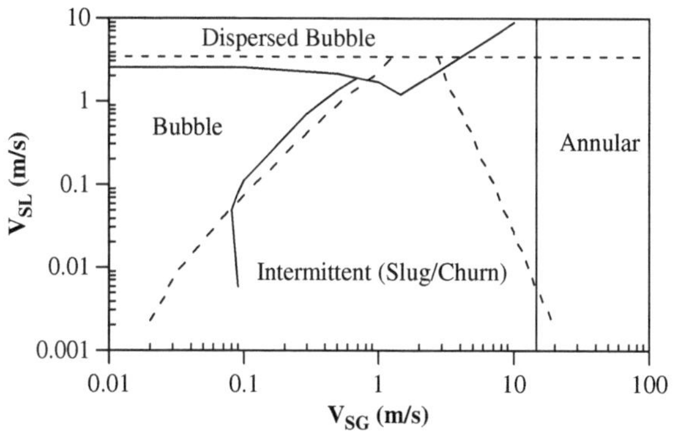

Figure 11. Flow pattern map in 1-*g* vertical tube. (Air–water flow in 2.5 cm i.d. tube—Taitel et al. model—Weisman and Kang model)

A flow pattern map according to these correlations is also plotted on Fig. 11 for water–air system. A comparison of the two models reveals that they agree reasonably well at 1-g conditions.

3. EXTENSION OF GROUND MODELS TO CONDITIONS AT REDUCED GRAVITY

Karri and Mathur (1988) extrapolated the above flow pattern models for ground flow to microgravity conditions. When the models of Taitel and Dukler (1976) and Weisman et al. (1979) for horizontal flow were extrapolated to a gravity level of 0.01-g_0, both models predicted a similar directional shift in the transition boundaries. A flow pattern map for a water–air system at a reduced gravity level of 0.01-g_0 is plotted in Fig. 12. The boundaries between the various flow regimes all move toward lower superficial velocities of both phases as gravity is reduced. The transition to dispersed bubble flow starts at a lower superficial liquid velocity. Transition from stratified flow to intermittent flow shifts to a lower superficial liquid velocity. Both models predict that annular flow would start at a lower gas superficial velocity (for the same liquid superficial velocity).

For vertical upward flow, the models proposed by Taitel et al. (1980) and Weisman and Kang (1981) were also extrapolated to microgravity conditions (Fig. 13). These two models also predict that the boundaries would shift to lower superficial velocities. The transition to dispersed bubble flow starts at a lower liquid superficial velocity, the intermittent flow (including slug flow and churn flow) starts at a lower superficial gas velocity, and the onset of annular flow, once again, is shifted to a lower gas superficial velocity ($V_{SG} \approx 5.0\,\mathrm{m/s}$).

A large set of two-phase flow pattern data has been collected by the Microgravity Research Group at the University of Saskatchewan aboard the NASA KC-135 aircraft. Observations were recorded on a high-speed video camera (1000 fps), and were later analyzed using an image-processing system. These data were collected

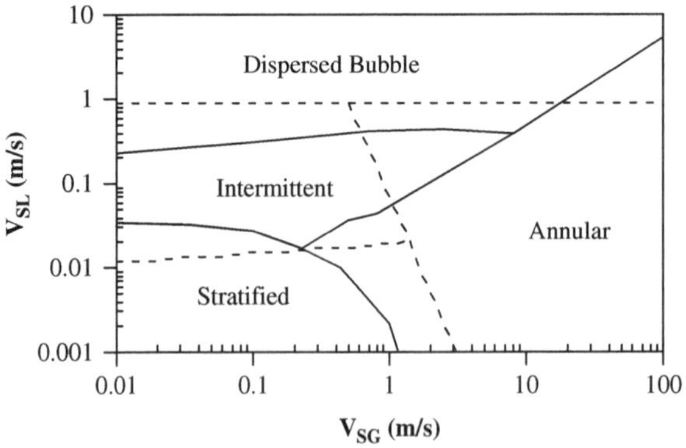

Figure 12. Flow pattern map in 0.01-g_0 horizontal tube. (Air–water flow in 2.5 cm i.d. tube—Taitel–Dukler model—Weisman et al. model)

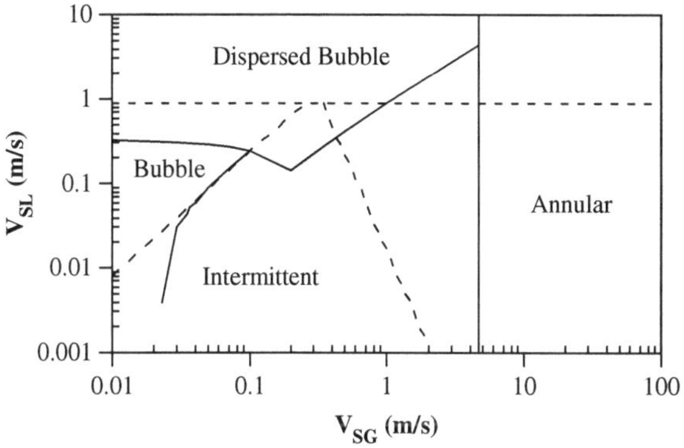

Figure 13. Flow pattern map in $0.01\text{-}g_0$ vertical tube. (Air–water flow in 2.5 cm i.d. tube—Taitel et al. model—Weisman and Kang model)

at gravity levels varying from 10^{-6} to $0.04\,g_0$, but the average was approximately $0.01\text{-}g_0$. The data can be used to test the above models.

The horizontal models by Taitel and Dukler (1976), and Weisman et al. (1979) were compared with the experimental data, and the results are shown in Fig. 14. Generally speaking, the predictions are poor. Slug flow was found to exist in a much larger region than that predicted by both models. On the other hand, annular flow occupies a smaller area than the region predicted by both models. The negative slope transition line between intermittent flow and annular flow for the extrapolated model of Weisman et al. (1979) gives especially poor predictions when compared with the experimental data. In the latter, a zero, or slightly positive slope or zero slope was empirically obtained instead. Bubble flow was observed within a very limited area ($V_{SL} > 0.9\,\text{m/s}$, $V_{SG} < 1\,\text{m/s}$), much smaller than the region predicted by both models. Finally, no data was available in the region predicted by both models as stratified flow. More data are needed in that region to verify, if at all possible, the existence of such flow at microgravity.

The models of Taitel et al. (1980), and Weisman and Kang (1981) for vertical flow, extrapolated to a gravity level of $0.01\text{-}g_0$, were compared with the present microgravity experimental data, and the results are shown in Fig. 15. The model of Taitel et al. (1980), extrapolated to $0.01\text{-}g_0$, gave better predictions than in the previous case. However, the onset of annular flow was underpredicted while slug flow (according to the model) would begin at a lower V_{SG} than experimentally determined. In other words, slug flow and transitional flow occupy a much larger area than the one predicted by the model. The extrapolated model of Weisman and Kang (1981), similar to the horizontal case, gave poor predictions. Once again, first the negative slope transition line between intermittent flow and annular flow predicts a large portion of slug flow and transitional flow in what is actually occupied by annular flow. Secondly, a dispersed bubble flow is predicted by that model above a constant superficial liquid velocity value. In reality, slug flow, as well as frothy slug-annular or transitional flow were also observed in that region.

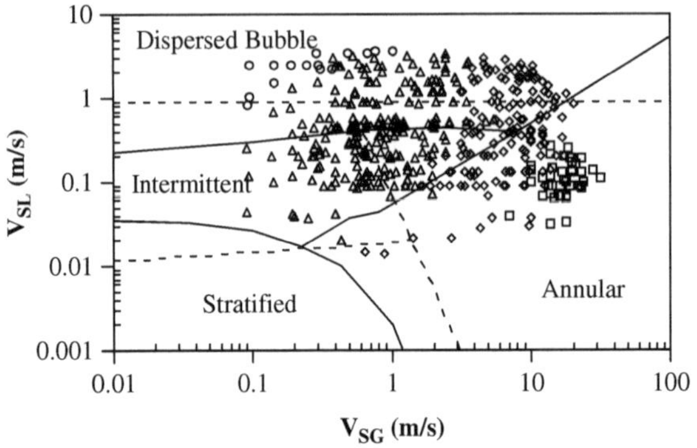

Figure 14. Comparison of experimental data with horizontal flow models (Air–water flow in 0.9525 cm i.d. tube at 0.01-g_0; o – bubble flow, Δ – slug flow, ◇ – frothy slug-annular flow, □ – annular flow, — Taitel–Dukler model, — Weisman et al. model)

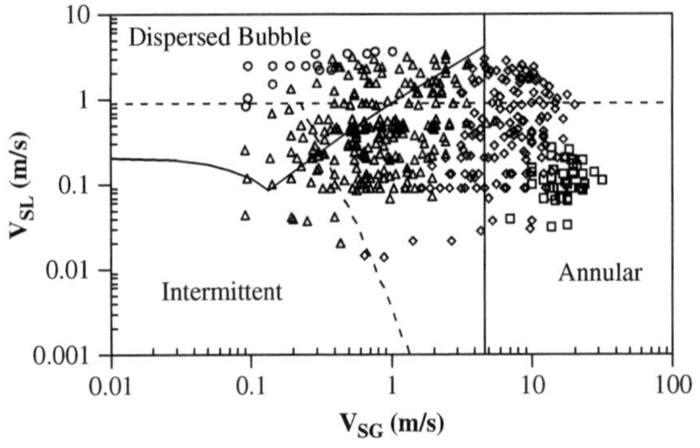

Figure 15. Comparison of experimental data with vertical flow models (Air–water flow in 0.9525 cm i.d. tube at 0.01-g_0; o – bubble flow, Δ – slug flow, ◇ – frothy slug-annular flow, □ – annular flow, — Taitel et al. model, — Weisman and Kang model)

4. MODELING AT REDUCED GRAVITY

Dukler et al. (1988) reported a set of experimental data with models to predict flow pattern transitions at microgravity. They classified gas–liquid flow into three flow patterns: bubble, slug, and annular. Assuming that the local relative velocity between liquid and gas is negligible, a reasonable assumption in bubbly and slug flows at μ-g, they suggested that the critical void fraction for the transition from bubble to slug flows is $\alpha = 0.45$. Based on these assumptions, the transition correlation is

$$V_{SL} = 1.22 V_{SG}. \quad (16)$$

For the transition from slug to annular flow, it was speculated that it takes place when the void fractions, as predicted by two separate models for each flow, become equal. The average void fraction in slug flow was given as:

$$\alpha = \frac{1}{C_0} \frac{V_{SG}}{(V_{SL} + V_{SG})}, \tag{17}$$

where C_0 is the void fraction distribution coefficient, and it ranges between 1.15 and 1.30. For annular flow, the void fraction is given by:

$$\frac{\alpha^{5/2}}{(1-\alpha)^2} = \left(\frac{f_i}{f_w}\right)\left(\frac{\rho_G}{\rho_L}\right)\left(\frac{V_{SG}}{V_{SL}}\right)^2. \tag{18}$$

The above model was tested using the experimental data collected by the Microgravity Research Group at the University of Saskatchewan, and the results are given in Fig. 16. It should be noted that only three flow patterns were classified by Dukler et al. (1988). The transition line from slug to annular flows predicted by the model under-predicts the actual transition. According to their prediction, the annular flow region includes much of the transitional flows (frothy slug-annular). Part of the discrepancy is due to the subjectivity in the definitions of flow pattern and classification based on visual observations. For the other transition line (between bubble and slug flows), several points identified as slug flow in the experiments fall into the bubble flow region predicted by the model. As discussed earlier, this transition line was based on a hypothetical critical void fraction of $\alpha = 0.45$

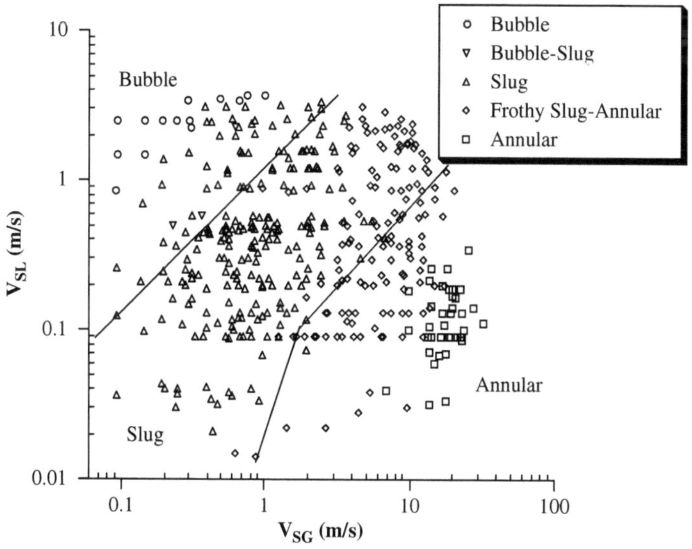

Figure 16. Comparison of experimental data with the model of Dukler et al. (1988)

(which is the maximum packing of bubbles in a conduit) and has no theoretical or empirical basis.

More recently, Zhao and Rezkallah (1993a) reported a large set of experimental data on gas–liquid flows at microgravity and suggested a model for the prediction of flow pattern transitions. The model is based on a mechanistic approach, in which the forces acting on the flows are considered.

In a two-phase flow system, there are several forces acting on the mixture, each of which has some impact on the overall flow configuration. These forces include those due to inertia, buoyancy, surface tension, and turbulent eddies. The flow pattern is determined by a delicate balance of these forces. The influence of buoyancy force due to gravity can be evaluated by Froude number and Eötvos number. The Froude number, *Fr*, is defined as:

$$Fr = \frac{V_{\mathrm{m}}^2}{gD} = \frac{\text{inertial force}}{\text{buoyancy force}}, \tag{19}$$

where the velocity, V_{m}, is the sum of the gas and liquid superficial velocities, V_{SG} and V_{SL}, $V_{\mathrm{m}} = V_{\mathrm{SG}} + V_{\mathrm{SL}}$, g is the apparent gravity, and D is the diameter of the tube. The average gravity level for the aircraft flight conditions was $0.01\text{-}g_0$ for mixture velocities as low as $V_{\mathrm{m}} = 0.2$, $Fr = 21.4$, and for $V_{\mathrm{m}} = 0.4$, $Fr = 85.6$. Thus, inertial forces are much larger than buoyancy forces for most of the flow conditions.

The Eötvos number, *Eo*, is represented by:

$$Eo = \frac{(\rho_{\mathrm{L}} - \rho_{\mathrm{G}})\, gD^2}{\sigma} = \frac{\text{buoyancy force}}{\text{surface force}}, \tag{20}$$

where ρ is phase density and σ is the surface tension. For a water–air flow in a 0.9525-cm i.d. tube at atmospheric conditions, the value of the Eötvos number is estimated to be 0.25. This suggests that even if the buoyancy force is still playing a role, that role is minor in comparison to other forces under microgravity conditions, namely those due to inertia and surface tension (especially when the inertial force is large).

The Weber number, *We*, is defined as:

$$We = \frac{\rho V^2 D}{\sigma} = \frac{\text{inertial force}}{\text{surface tension}}, \tag{21}$$

which represents the balance between inertial force and surface tension. This dimensionless group must be an important correlating parameter at microgravity conditions.

At low gas velocity, and hence for a low We_{G}, surface tension is dominant. The flow is a bubble flow, and the bubble shape is determined by surface tension. According to Colin et al. (1991), the bubble velocity at microgravity can be reasonably represented by:

$$V_{\mathrm{G}} = 1.2\,(V_{\mathrm{SG}} + V_{\mathrm{SL}})\,. \tag{22}$$

Since $V_G = V_{SG}/\alpha$, the above equation can be written as:

$$V_{SL} = CV_{SG} \tag{23}$$

or

$$C = \left(\frac{1}{1.2\alpha} - 1\right).$$

The value of C depends on the critical void fraction, which in turn is a function of the tube diameter and the length needed to reach fully developed conditions. Based on the experimental data, C was suggested to be 4.56. The corresponding critical void fraction is 0.15.

When the gas velocity is increased, the inertial force becomes large enough to overcome surface tension. The gas-phase breaks through the liquid slug, and forms tiny packed gas bubbles. This is the beginning of the transition from slug flow to annular flow. It was mentioned above that this transition is a slow and a gradual process, and that it occupies a wide range of liquid and gas flow rates. This transition region is called "frothy slug-annular" due to the continuous appearance of frothy mixtures in the liquid slugs which travel at a velocity that is relatively higher than that of the liquid-phase at the wall. Based on experimental data, the transition from slug flow to frothy slug-annular flow appears to take place at

$$We_G = \frac{\rho_G V_G^2 D}{\sigma} \approx 1. \tag{24}$$

As the gas flow rate further increases, the density of the liquid droplets in the frothy slugs decreases, and the frothy slugs become thinner until eventually annular flow is reached. In this region, the flow pattern is mainly dominated by forces due to inertia. From the experimental data, it was found that annular flow occurs at

$$We_G \approx 20, \tag{25}$$

that is, when the surface tension force is as low as 5% of the inertial force. Equation (25) presents the criterion for the transition from frothy slug-annular to annular flow.

Thus, in general, two-phase gas–liquid flow under microgravity conditions can be divided into three main flow regions. These are surface tension controlled, intermediate, and inertial force controlled. The first region is where the forces due to surface tension are significantly higher than those due to inertia ($We_G < 1$), which includes bubble and slug flows. The second region is where the two forces are comparable ($1 < We_G < 20$), which is occupied by transitional flows (frothy slug-annular flow). The third region is where the forces due to inertia are dominant ($We_G > 20$), which is occupied by annular flow. A flow pattern map based on these criteria is given in Fig. 17.

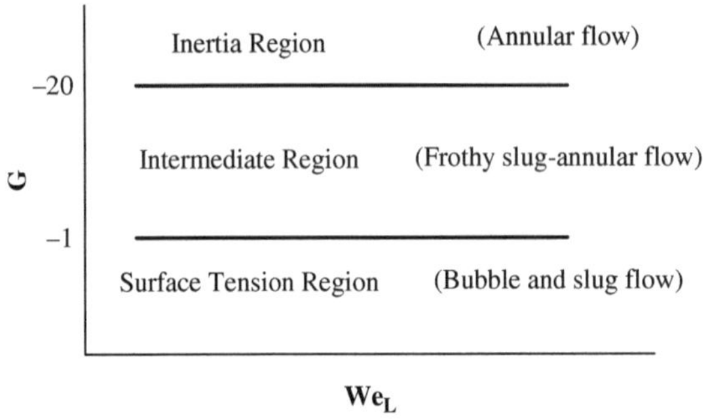

Figure 17. Two-phase flow regions at microgravity conditions

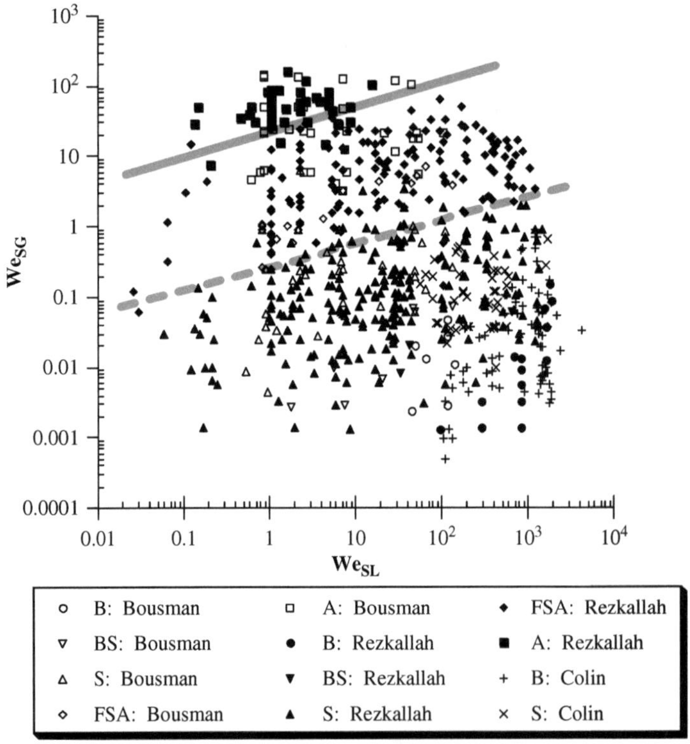

Figure 18. Comparison of present data with Bousman's (1995), air–water data for a 12.5-mm i.d. tube and Colin et al. (1991), air–water data for a 40-mm i.d. tube using Weber numbers based on superficial velocities

The model proposed above is compared with the experimental data collected by this research group, as well as with air–water data from Bousman (1995) and Colin et al. (1991) in Fig. 18. In this figure, superficial liquid and gas velocities have been used in the definitions of Weber number. Flow patterns are abbreviated as follows: B – bubble, BS – bubble-slug transition, S – slug, FSA – frothy slug-annular transition, A – annular. As can be seen in the figure, the model provides fairly good prediction for most data points. However, the transition lines are not horizontal as predicted, but rather have a definite positive slope as shown in the figure. At very low We_{SL}, where buoyancy force may not be negligible compared with inertial force, the slug flow ceases to exist at a lower We_{SG} than predicted by (24). At high We_{SL}, it seems that the frothy slug-annular flow prevails over the We_{SG} value predicted by (25). The transition to annular flow would occur at a higher We_{SG}.

If one uses actual liquid and gas velocities determined from void fraction measurements, the sloped lines that were empirically determined using the large data set in Fig. 18 become very close to the idealized constant Weber number lines that were first theorized (Fig. 17). The results of a comparison with experimental data where actual values of the volumetric gas void fractions were measured and reported in the literature are shown in Fig. 19. In the figure, the Rezkallah data for water–air is plotted along with data from Bousman (1995) for air–water and air–50% glycerin/water flows. It can be seen that a gas Weber number of 20 delimits annular flow from frothy slug annular flow very well for all values of liquid Weber number.

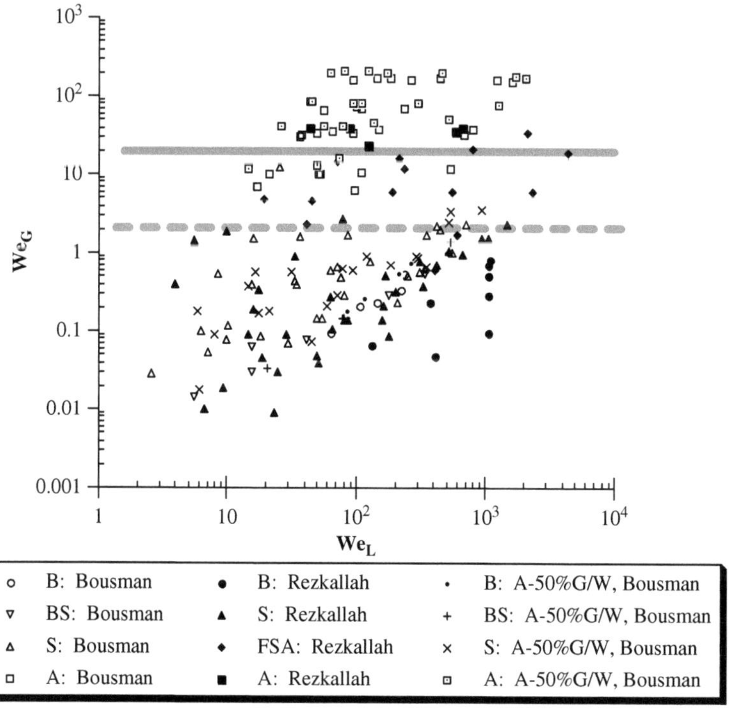

Figure 19. Comparison of present air–water data with Bousman's (1995) air–water and air–50% glycerin/water data for a 12.5-mm i.d. tube using Weber numbers based on the actual velocities

The transition from frothy slug annular to bubble and slug flows, based on this data set, is best represented by $We_G = 2$ rather than 1. This value for the transition line has also been suggested by other researchers, for example, Colin and Fabré (1995).

The present data agree quite well with these transitions lines for inertia-dominated to intermediate and intermediate to surface tension dominated (which are marked as solid and dashed-lines in the figure, respectively). There is some discrepancy with Bousman's data, especially for the annular flow pattern points. This discrepancy is most likely due to the subjective nature of the flow pattern identification, especially between annular and frothy slug-annular flows. A very conservative definition of annular flow was used for the Rezkallah data. It is possible that a more liberal definition was used by Bousman, thus leading to annular flows being designated at lower gas Weber numbers.

CHAPTER 4

GAS–LIQUID FLOW PRESSURE DROP

1. MOMENTUM EQUATIONS

The determination of pressure losses in two-phase flow systems is essential to the design of a variety of industrial processes. The pressure loss in a two-phase flow system includes pressure losses due to friction, gravitational force, and momentum change. The total pressure losses are therefore estimated by:

$$\left(\frac{\mathrm{d}p}{\mathrm{d}z}\right)_{\mathrm{tot}}=\left(\frac{\mathrm{d}p}{\mathrm{d}z}\right)_{\mathrm{F}}+\left(\frac{\mathrm{d}p}{\mathrm{d}z}\right)_{\mathrm{g}}+\left(\frac{\mathrm{d}p}{\mathrm{d}z}\right)_{\mathrm{m}}. \quad (1)$$

The calculations of the last two terms – gravitational and momentum terms – require the knowledge of the void fraction, and can be written as:

$$\left(\frac{\mathrm{d}p}{\mathrm{d}z}\right)_{\mathrm{g}}=[\alpha\rho_{\mathrm{G}}+(1-\alpha)\,\rho_{\mathrm{L}}]g, \quad (2)$$

and

$$\left(\frac{\mathrm{d}p}{\mathrm{d}z}\right)_{\mathrm{m}}=\mathrm{G}^2\frac{\mathrm{d}}{\mathrm{d}z}\left[\frac{x^2}{\alpha\rho_{\mathrm{G}}}+\frac{(1-x)^2}{(1-\alpha)\,\rho_{\mathrm{L}}}\right]. \quad (3)$$

When gravity is changed, the methods to calculate the two terms remain the same, only the void fraction might be different. In this chapter, we will concentrate our attention on the estimation of the frictional pressure drop component.

2. EMPIRICAL METHODS

A large number of investigations were carried out on-ground in the past several decades. Many empirical or semi-empirical correlations, based on experimental data collected at various conditions on-ground, were proposed for the prediction of the two-phase flow pressure drop at Earth's gravity (e.g., Lockhart and Martinelli, 1949; Dukler et al., 1964a, b; Chisholm, 1967; Friedel, 1979). Next, we will summarize some of the widely used methods.

2.1 The Homogeneous Model

The homogeneous model is a simple model that could be most suitable for two-phase flows at microgravity. The basic assumption of this model is that the two phases are well mixed, and that the velocities of the two phases are equal. These assumptions are closer to what is actually experienced in bubbly and slug flows (short slugs) at microgravity than at 1 g. The mixture density is given in terms of the gas quality, x:

$$\frac{1}{\rho_{\mathrm{m}}} = \frac{x}{\rho_{\mathrm{G}}} + \frac{1-x}{\rho_{\mathrm{L}}}. \tag{4}$$

The calculation of the mixture viscosity could be done using one of several methods. One of them is to use the liquid viscosity (let $\mu_{\mathrm{m}} = \mu_{\mathrm{L}}$) in the calculation of the Reynolds number. Another is to use a mixture viscosity instead of the liquid viscosity. Several correlations were recommended for μ_{m}, among which is the one suggested by Dukler et al. (1964b); the latter is given by:

$$\mu_{\mathrm{m}} = \mu_{\mathrm{G}}\frac{x\rho_{\mathrm{m}}}{\rho_{\mathrm{G}}} + \mu_{\mathrm{m}}\frac{(1-x)\,\rho_{\mathrm{m}}}{\rho_{\mathrm{L}}}. \tag{5}$$

The friction factor C_{f} can be calculated from the Blasius equation, in which the Reynolds number is given by:

$$Re_{\mathrm{m}} = \frac{\rho_{\mathrm{m}} V_{\mathrm{m}} D}{\mu_{\mathrm{m}}}. \tag{6}$$

The pressure drop can then be obtained from:

$$\left(\frac{\mathrm{d}p}{\mathrm{d}z}\right)_{\mathrm{F}} = \frac{2}{D} C_{\mathrm{f}} \rho_{\mathrm{m}} V_{\mathrm{m}}^2, \tag{7}$$

where V_{m} is the mixture velocity, i.e., $V_{\mathrm{m}} = V_{\mathrm{SL}} + V_{\mathrm{SG}}$.

2.2 The Lockhart-Martinelli Correlation

Another widely used method for the estimation of two-phase flow pressure drop is the Lockhart-Martinelli Correlation. The Martinelli parameter, X^2, is defined as:

$$X^2 = \frac{\left(\frac{\mathrm{d}p}{\mathrm{d}z}\right)_{\mathrm{L}}}{\left(\frac{\mathrm{d}p}{\mathrm{d}z}\right)_{\mathrm{G}}}, \tag{8}$$

where $(\mathrm{d}p/\mathrm{d}z)_{\mathrm{G}}$ and $(\mathrm{d}p/\mathrm{d}z)_{\mathrm{L}}$ are the single-phase gas and liquid frictional pressure gradients, calculated using the gas and liquid phase flow rates alone, respectively.

The original correlation is given in a graphical form (Lockhart and Martinelli, 1949). Later, Chisholm (1967) approximated these relationships by the expression:

$$\text{(9)} \qquad \phi_L^2 = 1 + \frac{C}{X} + \frac{1}{X^2},$$

where ϕ_L^2 is the two-phase multiplier, and is defined as:

$$\text{(10)} \qquad \phi_L^2 = \frac{\left(\frac{dp}{dz}\right)_F}{\left(\frac{dp}{dz}\right)_L}.$$

The coefficient C is a parameter and its value (for 1-g conditions) is given by:
$C = 20$, for turbulent liquid, turbulent gas flows,
$C = 12$, for laminar liquid, turbulent gas flows,
$C = 10$, for turbulent liquid, laminar gas flows,
$C = 5$, for laminar liquid, laminar gas flows.

2.3 Friedel's Model

A more sophisticated empirical correlation for two-phase pressure drop was proposed by Friedel (1979). The equation was given in terms of a multiplier, ϕ_{lo}^2, which is defined by:

$$\text{(11)} \qquad \phi_{lo}^2 = \frac{\left(\frac{dp}{dz}\right)_F}{\left(\frac{dp}{dz}\right)_{lo}},$$

where $(dp/dz)_{lo}$ is the single-phase frictional pressure gradient, assuming the liquid is flowing with the same mass flow rate as the total two-phase flow rate. Friedel's equation is of special interest since it contains both the Froude number and the Weber number. These two groups are obviously relevant to the μ-g application. The equation is given by:

$$\text{(12)} \qquad \phi_{lo}^2 = E + \frac{3.24FH}{(Fr^{0.045} We^{0.035})},$$

where

$$E = (1-x)^2 + x^2 \frac{\rho_L C_{f_{go}}}{\rho_G C_{f_{lo}}},$$

$$F = x^{0.78}(1-x)^{0.224},$$

$$H = \left(\frac{\rho_L}{\rho_G}\right)^{0.91} \left(\frac{\mu_G}{\mu_L}\right)^{0.19} \left(1 - \frac{\mu_G}{\mu_L}\right)^{0.7},$$

$$Fr = \frac{G^2}{gD\rho_m^2},$$

and

$$We = \frac{G^2 D}{\sigma \rho_m},$$

in which $C_{f_{go}}$ and $C_{f_{lo}}$ are the friction factors assuming that the gas and liquid are flowing separately with the total mass flux. (In each case, the gas or the liquid properties are used to determine these parameters, as appropriate.) It has been considered an accurate general correlation for the frictional two-phase flow pressure gradient when $\mu_L/\mu_G < 1000$ (Whalley 1987). The correlation is the only one that attempts to include the effects of surface tension and gravity.

3. EXPERIMENTAL RESULTS AT REDUCED-GRAVITY CONDITIONS

The pressure drop obtained from experimental measurement is the total pressure loss, which is the sum of frictional pressure loss, gravitational pressure loss, and accelerational pressure loss. Under the conditions of fully developed, adiabatic, and steady-state flow in a tube with a uniform cross section, the pressure loss due to acceleration can be ignored. The frictional pressure gradient can then be obtained from

$$\left(\frac{dp}{dz}\right)_F = \left(\frac{dp}{dz}\right)_{tot} - \left(\frac{dp}{dz}\right)_g . \quad (13)$$

At reduced-gravity conditions, the gravitational pressure loss is usually a small portion of the total pressure loss, and can be calculated using the actual gravity level measurements on-board the KC-135 zero-gravity aircraft. Such measurements were continuously recorded at a central unit in the aircraft, provided by the Canadian Space Agency.

The tables in Appendix A summarize the experimental frictional pressure drop data that were collected during three flight campaigns in 1992, 1993, and 1994 for air–water mixtures and three air–glycerin/water mixtures and the corresponding ground data collected using the same apparatus. In these experiments, the liquid superficial velocity, V_{SL}, ranged from 0.1 to 2.5 m/s, and the gas superficial velocity, V_{SG}, ranged from 0.1 to 18 m/s, thus covering a wide range of flow regimes from bubbly to annular flows.

During the experiments, gauge pressure was measured at a distance of 25.7 cm from the mixer outlet. Two additional gauge pressure measurements were taken at 30.5 and 69 cm downstream from the bottom transducer. All of the pressure readings were taken using Validyne pressure transducers with accuracy of 0.25% of full scale. The range for the middle transducer is 0–14 kPa (0–2 psi), and the ranges for the top and bottom transducers are 0–21 kPa (0–3 psi). The other sides of the three-gauge pressure transducers were connected to a Null Matic pressure regulator. The regulator was used to set the reading of the middle transducer to zero at all times. The uncertainties associated with the measurement of pressure gradient are 106 Pa/m for values obtained over the 69.0 cm length. Pressure drop values less than 2 kPa/m were discarded due to the poor transducer accuracy below that range. For the remaining data, the uncertainty in the measurement of pressure gradients due to the pressure transducers is 5% for bottom to top reading (69.0 cm length). A sampling rate of about 70 points per second was used in collecting the data. More details on the pressure drop measurements and the hardware used may be found in Zhao and Rezkallah (1995a).

Comparisons of the experimental frictional pressure drop at microgravity and normal gravity conditions are shown in Fig. 20 for three liquid velocities (superficial). The frictional pressure gradient is plotted against the gas-phase volume fraction

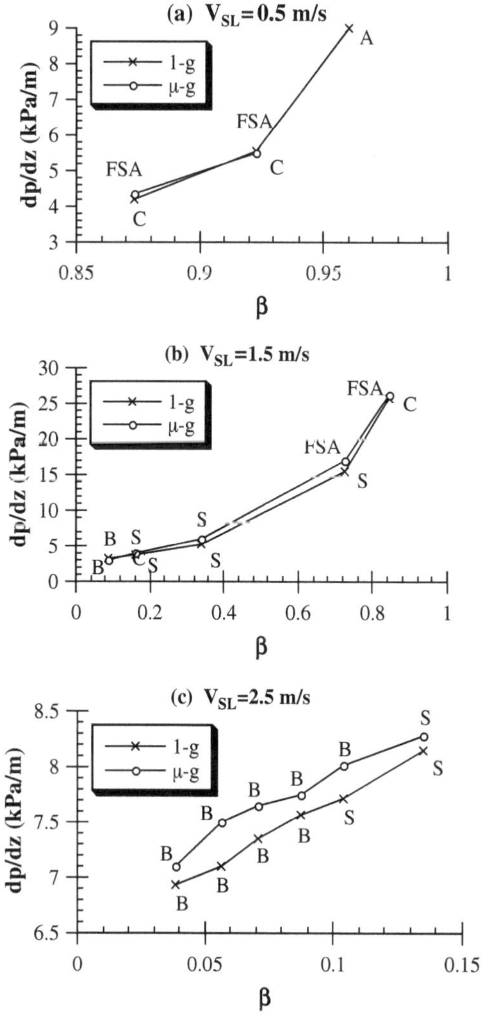

Figure 20. Comparison of two-phase pressure drops at μ-*g* and 1-*g* conditions (A: annular flow, B: bubble flow, C: churn flow, S: slug flow, FSA: frothy slug-annular flow)

(defined as $\beta = V_{SG}/(V_{SL} + V_{SG})$) as the independent variable for $V_{SL} = 0.5, 1.5$, and 2.5 m/s, respectively. Corresponding flow patterns are also indicated besides each data point. Among these, pressure drop data at $V_{SL} = 1.5\,\mathrm{m/s}$ cover the widest range, with β ranging from less than 0.1 to over 0.8. Generally, at a constant V_{SL}, the pressure drop gradually increases as the volume fraction increases. The frictional pressure drop at 1-g and μ-g are very comparable, with those at μ-g being slightly higher than (or equal to) dp/dz at 1-g. The difference is very small, ranging from 1 to 14%. For most of the cases, the differences are within 10%, being slightly higher for the μ-g data.

The two-phase flow frictional pressure gradient is often correlated by two-phase multipliers. Using the definition of the liquid two-phase multiplier given in Eq. (10), ϕ_L^2 is calculated for the experimental data and is plotted as a function of the gas quality, x, in Fig. 21 at both μ-g and 1-g conditions. Clearly, the two-phase multiplier values at 1-g and μ-g conditions are almost identical at the same gas quality.

For a two-phase gas–liquid mixture flowing upwardly and concurrently in a vertical tube, when gravity is reduced (while keeping the liquid and gas flow rates unchanged), the major effect on the flow hydrodynamics results from the reduction of the buoyancy forces on the gas-phase. For annular flow, a large change in the flow dynamics is not expected when gravity is changed since the flow in that region is inertia dominated. The behavior of the liquid film as well as its thickness may be altered under reduced-gravity conditions.

For bubble and slug flows in an upward vertical co-current system, the flow dynamics may change due to the change of bubble movement when gravity is reduced. The slip ratio between the two phases decreases due to the decrease in the gas-phase velocity at microgravity. This leads to a higher gas void fraction in upward flow systems. The liquid-phase flowing in the reduced flow area is then accelerated, and the velocity gradient becomes larger near the wall. This would cause the pressure drop to increase. On the other hand, it has been shown (Lance and

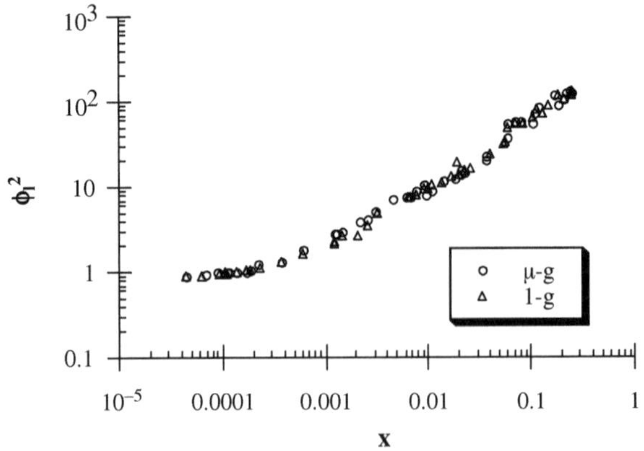

Figure 21. Two-phase multiplier vs. quality x at both μ-g and 1-g conditions

Bataille, 1991) that for bubble flow at 1-*g* conditions, the turbulence in the liquid-phase is amplified owing to the hydrodynamic interactions between the bubbles and the wakes of the bubbles (since the bubbles travel at higher speed than their surrounding liquid). The high-energy dissipation associated with the turbulence in the flow leads to high frictional pressure drop values. At microgravity, the gas bubbles are moving with virtually the same speed as the surrounding liquid. The turbulence amplification induced by the bubble movement is therefore much more reduced under those conditions. Thus, it could be argued that there are two competing effects at microgravity. One is to increase the liquid velocity due to the increase in void fraction, and consequently the reduction in the liquid flow area. The other is a decrease in the turbulence dissipation under microgravity conditions due to the reduction of the turbulence amplification because of a substantial reduction in the bubble movement at microgravity. The change in pressure drop is determined by the balance of the two effects and their relative magnitude compared with the flow inertia.

When the gas-phase density is much smaller than the liquid-phase density, the bubble velocity in a gas–liquid flow is expressed by (Wallis, 1969):

$$V_b = 1.2V_m + k_1\sqrt{gD}. \quad (14)$$

For "Taylor bubbles," $k_1 = 0.35$; for smaller bubbles, k_1 is a function of the bubble size, with a value less than 1. The second term in Eq. (14) has a value between 0.1 and 0.3 for the tube size used in the above experiment (0.009525 m i.d.). The value of the first term ranges from 1.8 to 3.4 under the experimental conditions when slug or bubble flows were observed. It is evident that, on-ground, less than 10% of the bubble velocity is due to buoyancy effect. The Reynolds number of the mixture, Re_m, is in the order of 10^4. This indicates that the flow is well into the turbulent region. A change of less than 10% in the bubble velocity would not make significant changes to the flow dynamics. Perhaps this is the reason that large differences in the pressure drop were not observed when gravity was reduced under the above experimental conditions.

At very low liquid and gas flow rates, the velocity component due to buoyancy would be a significant part of the total bubble velocity. A relatively larger pressure drop change may be observed when gravity is reduced.

The above experimental results are somewhat contradictory to the findings by Chen et al. (1991), where pressure drop at microgravity was reported to be about 40% higher than that at normal gravity. In that experiment, data were collected at very low liquid velocities (0.02–0.16 m/s), and the data at microgravity were compared with those at normal gravity, taken in a horizontal tube section. Since flow stratification takes place in the latter case (which does not occur at μ-*g*), comparisons of the μ-*g* data with horizontal flows is inappropriate and could also be misleading. In addition, the accuracy of the pressure transducers they used was 345 Pa (0.05 psi), and the pressure differences were from 62 to 1792 Pa (0.009–0.26 psi) with about half of the data around 345 Pa (0.05 psi). Also, the sampling rate was very low (about one point per 2 sec), which makes their conclusions even more questionable.

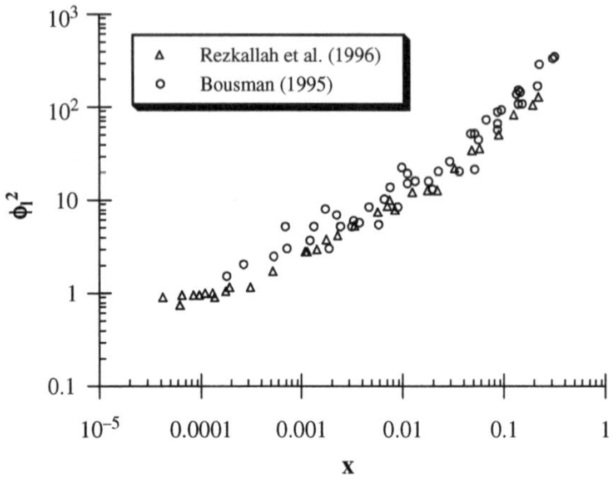

Figure 22. Comparison of experimental μ-*g* data for air–water with Bousman data (1995)

A comparison of the experimental results of the present study with recent μ-*g* data from Bousman (1995) shows better agreement than was found for the Chen study. The Bousman (1995) pressure gradient data for an air–water mixture flowing through a vertical 12.7-mm i.d. circular tube is plotted in Fig. 22 with the present data. It can be seen in the plot that Bousman's two-phase pressure multiplier is consistently higher than the present data for the range of gas qualities tested. The root-mean-square deviation between the two data sets is approximately 28% with the largest differences occurring between qualities of 0.001 and 0.01 (slug and frothy slug-annular flows).

4. COMPARISON OF EXPERIMENTAL DATA WITH EMPIRICAL METHODS

4.1 Comparison with the Homogeneous Model

Pressure drop calculations using the homogeneous model are compared with the experimental data at μ-*g* in Fig. 23. In the calculation, liquid viscosity and mixture viscosity were used separately. The overall root-mean-square deviations between the experimental and calculated pressure drop values are 28% when a mixture viscosity is used, and 40% when a liquid viscosity is used. The relatively poor performance of the latter is mainly due to the large overestimation (100%) of pressure drop at annular flow. Generally speaking, when the liquid velocity is much higher than the gas velocity (bubble flow), both models give equally good predictions. When the liquid velocity is much lower than the gas velocity (annular flow), using the liquid viscosity tends to largely overestimate the pressure drop: while using the mixture viscosity tends to slightly underestimate the pressure drop. When the two

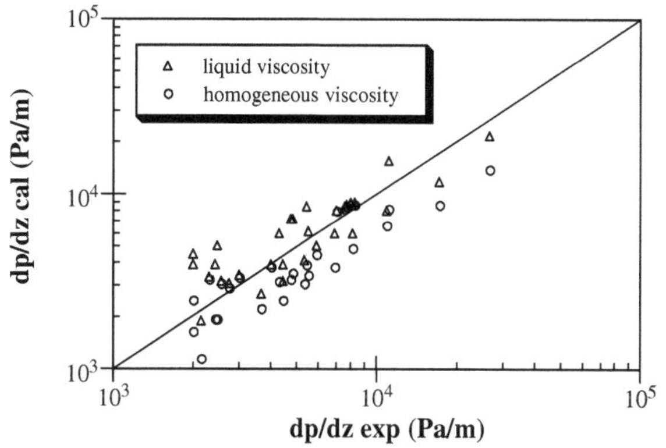

Figure 23. Comparison of two-phase experimental pressure drop at μ-g with homogeneous model

velocities are comparable (slug and transitional flows), both methods underpredict the pressure drop, with the liquid viscosity yielding better results.

It also has been reported by Colin et al. (1991) and Sridhar et al. (1992) that using the liquid viscosity in the homogeneous model would give good results. However, comparisons with the calculation using a mixture viscosity were not given. Chen et al. (1991) reported that the homogeneous model using a mixture viscosity gave good pressure drop prediction for bubble and slug flows, and poor prediction for annular flow.

4.2 Comparison with the Lockhart-Martinelli Correlation

The calculated two-phase multiplier, $\phi_{L\ cal}^2$, using Chisholm's Eq. (9) is compared with the experimental multiplier, $\phi_{L\ expl}^2$, in Fig. 24. Generally, the prediction was

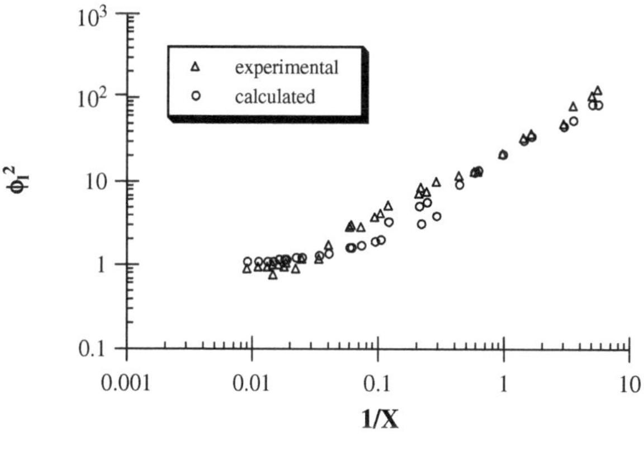

Figure 24. Comparison of two-phase experimental pressure drop at μ-g with Chrisholm's correlation

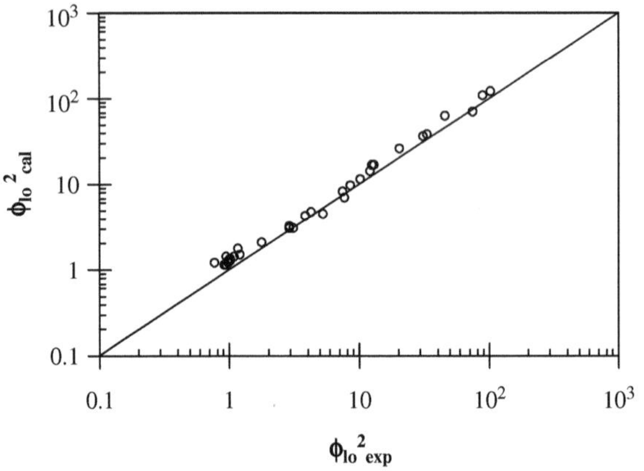

Figure 25. Comparison of two-phase experimental pressure drop at μ-g with Friedel's model

good with a root-mean-square deviation of 28%. It is found that when $1/X$ is about 0.01 or larger than 1.0, the predictions are very good; when $1/X$ is about 0.1, the correlation tends to underpredict the pressure drop.

4.3 Comparison with Friedel's Model

The calculated two-phase multiplier, ϕ^2_{lo}, using Eq. (11) is compared with the experimental data in Fig. 25. The root-mean-square deviation for the whole set of data is 29%. Generally speaking, the correlation overestimates the pressure drop. Under the above experimental conditions, the term E has a value of about 1 for bubble, slug, and transitional flows, and it contributes up to 85% for bubble flow, and 20–50% for slug and transitional flows, depending on the liquid and gas flow rates. For annular flow, the term E may be as high as 20, but it contributes only 10–20% of the ϕ^2_{lo} value. When gravity is reduced to 0.01 g_0 (which is the average gravity level for the flight data) the second term in Eq. (12) is increased by 20%. This means that the effect of changing gravity on the pressure drop is larger for annular flow and smaller for bubble flow. This is in contradiction with the analysis given earlier in this chapter, and could perhaps explain the reason why the correlation consistently overpredicts the pressure gradient. In general, the overall prediction is good.

CHAPTER 5

VOID FRACTION

1. INTRODUCTION

Measurement techniques of void fraction in two-phase gas–liquid flow continues to evolve owing to the transient nature of the flow that gives rise to difficulties in instrumentation and measurement. Among the various methods available for measuring void fraction in gas–liquid flows, non-intrusive means are most desirable. Capacitance sensors have been widely investigated in recent years; for example, Merilo et al. (1977), Masuda et al. (1980), and Heerens (1986) among others. However, it is not uncommon for those measurements to be in the range of 0.1 to 10 pF. Thus, proper shielding against stray capacitance and a good signal-to-noise ratio are needed.

Due to the transient nature of two-phase flows, most analysis techniques involve an ensemble average that is obtained from various data at the same flow conditions, or by a time average from a single recording. Some of the common techniques for obtaining void fraction measurements include the following: gamma-ray attenuation, conductance probe, quick-closing valves, and capacitance methods. The selection of any of these methods depends on its application, and whether a volumetric average or a local void fraction measurement is desired. The volume average void fraction, α, in a two-phase mixture is defined as:

$$\alpha = \frac{\text{Volume of gas in the mixture}}{\text{Total volume of the gas and liquid}} \tag{1}$$

In addition to obtaining a void fraction measurement, the statistical analysis of the signal can also be used to determine the flow regimes associated with the flow and their transitions.

Quick-closing valves provide an exact void fraction measurement and are useful for calibrating or comparing against other methods. However, it is not a practical means to determine void fraction for continuous processes since it disrupts the flow. Under adiabatic conditions it is important that the valves close simultaneously. However, if the quality, x, is increasing as in the case when heat is added to the two-phase flow mixture, the closing time must be shorter. For this condition, the experimental errors are small if the closing time is less than 1/100 per second

(Dounan et al., 1985). In this study, quick-closing valves were used to calibrate a helical wound and a concave plate capacitance sensor.

A capacitance sensor provides a non-intrusive way to measure void fraction. It provides time-averaged void fraction measurements, and its time-varying output signal can also be used for flow pattern identification. Since the area and distance between the capacitor's electrodes are constant, the only contribution to a change in capacitance is due to a change in the dielectric. For two-phase flow, the dielectric consists of the liquid and gas phases. The measured capacitance represents the amount of the phases and configuration of the phases within the tube for a pre-set flow rate.

Electrical impedance measurements are commonly used for gas–water mixtures to determine void fraction. One important factor is that when impedance gauges are used, drift will occur if the liquid conductivity changes. This was noted by Geraets and Borst (1988). If water temperature, for example, increases from 25 to 50 °C, conductivity doubles while the relative permittivity decreases by approximately 15%. Geraets and Borst (1988) found that drift can be reduced by operating at a frequency high enough to give dominance to capacitance. They used a Boonton 72BD capacitance meter, capable of phase detection, operating at a frequency of 1 MHz. Fluids having a specific conductivity of less than $0.5 \times 10^{-2}[\Omega\text{m}]^{-1}$ were reported to have a measurement accuracy within 1.5%.

A variety of electrode configurations have been designed by a number of researchers; ranging from flat plate, concave, helical, and multiple helical wound, where the electrodes were either in contact or isolated from the fluid. Two capacitance sensors were used by Geraets and Borst (1988), where one tube had an inside diameter of 50 mm and the other was 5 mm. The sensors used by them were of the helical wound type having 2 thin brass strips wound around a thin acrylic tube. The brass strips were placed such that they are always opposite to each other. A guard electrode was used to minimize edge effects and stray capacitance. Stray capacitance is essentially any undesirable capacitance which can occur between circuit wires, wires and the chassis, or components and the chassis of electronic equipment. Geraets and Borst (1988) used the sensor for void fraction measurement as well as to determine flow pattern information in horizontal pipe flow.

Tomographic imaging using capacitance sensors is also possible. Huang et al. (1989) reported results using eight electrodes mounted on the outside of an insulated pipe. By measuring the capacitance of different pairs of electrodes, image reconstruction was possible. The capacitance of the two-phase flow changed corresponding to a change in the dielectric within the pipe. The change in the amplitude of the signal was then processed by a computer where a linear back-projection algorithm was used to reproduce a cross-sectional image. This method is being further developed in the present time.

Other articles by Huang et al. (1988), and Huang et al. (1989) reviewed electrode guard methods and electronic measurement techniques. Additional research efforts with capacitance measurements include those by Albouelwafa and Kendall (1979); Li et al. (1992); and Shu et al. (1982).

Some of the more recent work which involved the measurement of void fraction at reduced gravity was done by Bousman (1995). He used air–water, air–water/glycerin, and air–water/zonyl FSP mixtures flowing in 12.7 and 25.4-mm i.d. tubes. The air–water mixture results are later compared to those obtained in this study.

An experimental study by Grossetête (1995), at 1-*g*, involved an investigation into the development of void fraction and velocity profiles for bubble flow in a 38-mm i.d. vertical tube. Water–air mixtures were used and local measurements were taken with a dual-fiber optical probe and a hot film-sensing probe at three lift-to-drag (L/D) locations. It was noted that as the L/D ratio increased, the high void fraction values at the wall diminished with an increasing gas concentration near the center of the tube. These results are discussed later in this chapter.

2. INSTRUMENTATION

A full discussion of the sensor design can be found in Elkow and Rezkallah (1996). In summary, a design was selected consisting of two helical wound electrodes with electronics operating on a charge/discharge principle. The helical design was selected based on the work of Gregory and Mattar (1973) and Geraets and Borst (1988). To determine the best electrode configuration, tests with flat plate, concave, to multiple helix electrodes were done by Gregory and Mattar (1973). They found that the best results were obtained with a two-electrode helical wound sensor. In addition, a number of helical wound electrode sensors were constructed (by trial and error), where the pitch and width of electrodes were varied (Gregory and Mattar, 1973). Criteria were developed relating the width to pitch and tube diameter; these arc

$$\frac{w}{p} = 0.100 \text{ to } 0.136, \tag{2}$$

and,

$$\frac{p}{OD} = 1.83 \text{ to } 3.66, \tag{3}$$

where w is the width of the electrode, p is the pitch, and $o.d.$ is the outside diameter of the tube.

A schematic of the sensor can be seen in Fig. 26. The sensor provided a non-intrusive means of obtaining the void fraction measurement. The two-phase flow was isolated from the electrodes by the 9.53 mm i.d. and 15.88-mm o.d. acrylic tube. To stay within the limits of Eqs. (2) and (3), the electrodes were wound around the acrylic tube according to:

$$\frac{w}{p} = 0.122, \tag{4}$$

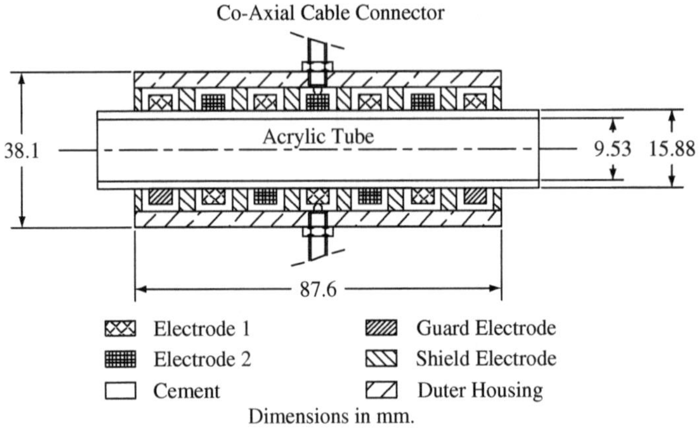

Figure 26. Electrode configuration for the helical plate capacitance sensor

and

$$\frac{p}{OD} = 1.84. \tag{5}$$

As a result, the pitch was 29.21 mm and the electrode width was 3.58 mm.

An outer cylinder – made of brass, tapped for two BNC connectors – was positioned over the electrodes. The outer cylinder, running the full length of the sensor, acts as a guard and was electrically in contact to the shield electrode. The shield electrode was grounded to the guard electrode to eliminate electric field lines running along the wall of the tube. Thus, the void fraction measurement is based on a measurement due to the electric field lines running across the tube diameter rather along the tube wall.

Two BNC connectors, diametrically opposed, were then threaded into the outer cylinder with the center electrode of each of the BNC connectors in contact with each of the two active electrodes. This design resulted in a total change in capacitance of approximately 0.5 pF when the sensor was filled with all water compared to when it was filled with all air.

2.1 Calibration

A calibration was needed which could be applied to ground and flight data over a wide range of liquid and gas flow rates and void fractions. De-ionized and distilled water was used during the calibration tests. Quick-closing, electrically triggered valves were used for calibrating the void fraction sensors. The desired water and air flow settings were entered into a data file which is then read by the computer. First, the void fraction sensor was zeroed, and its output was recorded during single-phase air and water flows. Pre-set gas–liquid flow settings were then read from the data file by the computer. The gas flow was controlled by the computer and the liquid flow setting was achieved by manually adjusting the control valves.

Once the flow reached a steady state, the DAS records V_{sl}, V_{sg}, ambient air and liquid temperatures, and the absolute pressure for approximately 20 sec. All three quick-closing valves were simultaneously triggered causing the two valves across the acrylic tube test section to close, hence trapping a volume of the two-phase flow. Meanwhile, the third valve opens, allowing the flow to bypass the test section and return to the tank. This allows the two-phase flow to continuously run without having any sudden changes imposed on the flow settings. The distance from the top quick-closing valve to the meniscus of the water column trapped into the test section was then measured to the nearest millimeter. For each of the flow settings, three void fraction measurements between the valves were recorded. The average value from the three measurements was then used to represent the void fraction for that particular flow setting. This procedure was followed for all flow settings.

In each test, the liquid flow was initially held at a constant value. The gas flow was then increased to cover the widest possible range of void fraction readings. After the upper limit for the gas flow was obtained, and the void fraction was recorded for that range, measurements were taken with single-phase water before the liquid flow was increased to its next setting. The single-phase recording of the sensor's output was used to determine if drift in the sensor's electronics had occurred.

The plots of the void fraction from the quick closing valves as a function of the normalized voltage output from the capacitance sensor can be seen in Fig. 27. The liquid superficial velocities at those tests were 0.1, 0.33, 0.6, 1.0, 1.7, 2.4, and 3.3 m/s. The gas flow rates varied depending on the liquid flow with the intent to cover the widest possible range of void fraction. The solid lines were obtained from the following equation (which relates the voltage output to the liquid flow rate):

$$\alpha = y_A + y_B N_V^3, \tag{6}$$

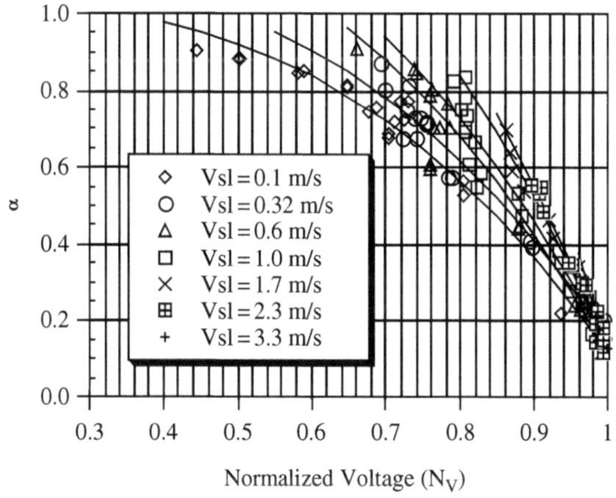

Figure 27. Void fraction measured by quick-closing valves as a function of normalized voltages

where α is the void fraction, y_A and y_B are variables, and N_V is the normalized voltage. If a flow rate other than those tested is used, the values for y_A and y_B are interpolated by relating those obtained in the test to the superficial liquid velocity V_{sl}.

The calculated void fractions were compared to the measured values, which were actually trapped in the test section between the two quick-closing valves. The results of this comparison are shown in Fig. 28. In that figure, it can be seen that most of the measured data are within ±10% of the actual values. Using linear regression, the standard error was found to be 0.046.

At a void fraction of approximately 0.7, there is a slightly larger scatter compared to other void fraction values. The scatter in this region is associated with the churn flow regime. This can be attributed to the oscillatory motion of the flow, which is the result of a reduced liquid "hold-up", interrupted by occasional high gas inertia intervals that tend to hold the liquid up. As a result, the void fraction in this region varies significantly depending on the flow conditions in the tube.

Measurement uncertainties (Coleman and Steele, 1989) were determined from:

$$(7) \qquad \mathrm{U}_r = \left[\left(\frac{\partial r}{\partial X_1} \mathrm{U}_{X_1} \right)^2 + \left(\frac{\partial r}{\partial X_2} \mathrm{U}_{X_2} \right)^2 + \mathrm{L} + \left(\frac{\partial r}{\partial X_J} \mathrm{U}_{X_J} \right)^2 \right]^{\frac{1}{2}},$$

where r, the experimental result to be determined, is a function of J variables X_i, and U_{X_i} are the uncertainties in the measured variables X_i. The measurement uncertainties for the quick-closing valves were determined to be 0.5%. Measurement uncertainties for the helical wound capacitance sensor were more difficult to determine since the electronics were built in-house. They are estimated to be approximately 5%. Some additional tests were conducted with the capacitance

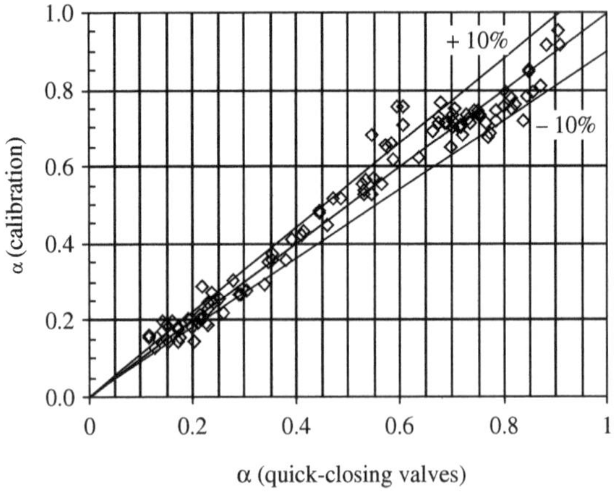

Figure 28. Calculated void fraction as a function of the void fraction trapped between two quick-closing valves

sensor to determine the effects of temperature, fluid properties, and operating frequencies. The results of these tests can be found in Elkow and Rezkallah (1996). A signal analysis was also conducted to ensure a good signal-to-noise ratio and that a non-aliased signal was being recorded (Elkow and Rezkallah (1996)).

3. EXPERIMENTAL RESULTS AND COMPARISONS

In February 1994, data for two-phase water–air flow was collected during low gravity periods on-board NASA's KC-135 aircraft. Five separate flights were conducted where pressure drop, heat transfer, and void fraction data were measured simultaneously. A total of 61 data points were used in the void fraction analysis. Superficial liquid flow settings from 0.07 to 2.5 m/s, and superficial gas velocities in the range of 0.1–18 m/s were tested during those flights. Void fractions from 0.1 to 0.9, covering the total range from bubble to annular flow regimes, were obtained. Ground data at approximately the same flow settings of the flight tests were obtained prior to and after the flights for the sake of comparisons. A helical plate capacitance sensor was used for both the flight and ground tests.

Figure 29 shows the average void fraction data sets obtained during both flight and ground tests as a function of the "pseudo" void fraction ratio, V_{sg}/V_{sl}. Further comparisons for each flow regime were made between the ground and flight data by matching the ratios of V_{sg}/V_{sl}. Only the data points which had similar values for V_{sg}/V_{sl} were used in the comparisons here. In general, the percent difference in the ratio of V_{sg}/V_{sl} for the flight and ground data is approximately within 3–4%. However, due to the limited number of data points in the annular flow regime, the percent difference in V_{sg}/V_{sl} is higher. It was found that the average void fraction values for bubble, transitional, and annular flows were comparable for the flight and

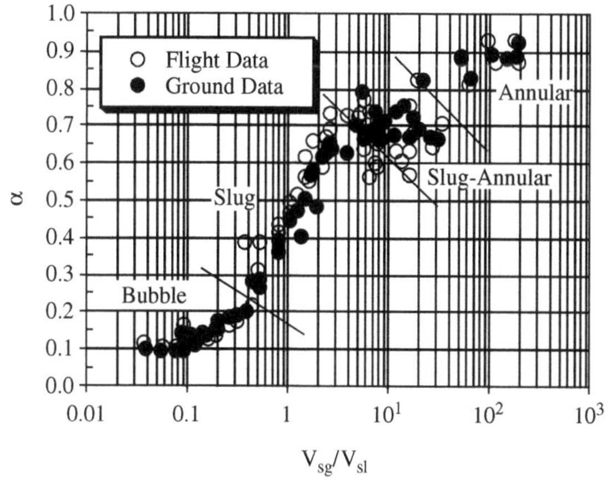

Figure 29. Flight and ground void fraction data

ground data. On the other hand, slug flow showed the highest difference between the two values of void fractions at μ-g being approximately 9–10% higher.

Since two-phase flow measurements are strongly flow regime dependent, more informative comparisons should be made for each flow regime individually. The results of these comparisons are shown in Figs. 30–33 for bubble, slug, transitional, and annular flows, respectively.

Data for the bubble flow regime is shown in Fig. 30. For a pseudo void fraction approximately below 0.1, the results show a tendency for the flight data to be approximately 8–25% higher than those taken at 1-g. Above $V_{sg}/V_{sl} = 0.1$, the difference is reduced until the trend is reversed with the 1-g data slightly higher than the μ-g counterpart (approximately 7–16% higher). The higher void fraction

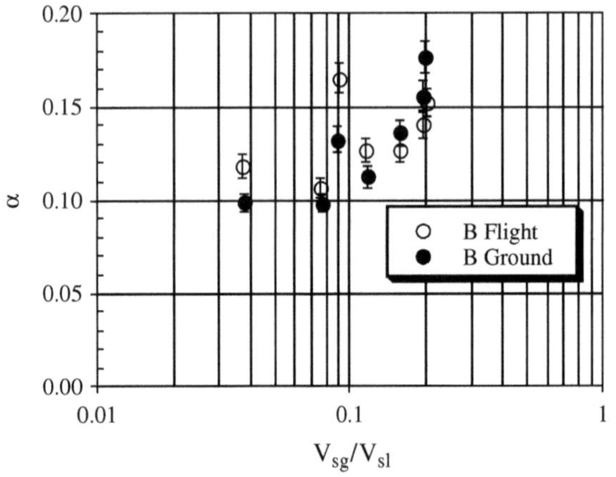

Figure 30. Comparison of the bubble flow regime for the flight and ground data

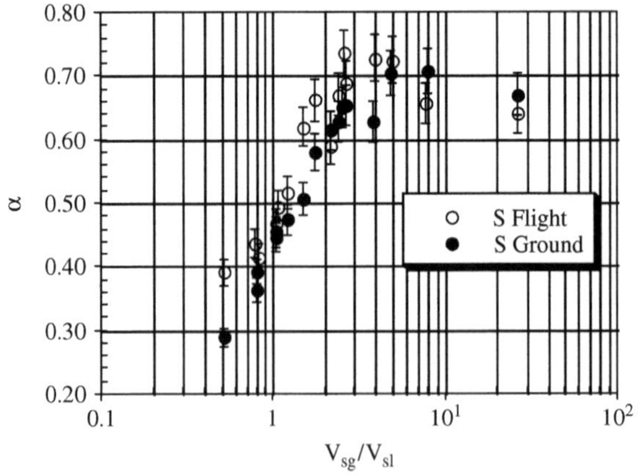

Figure 31. Comparison of the slug flow regime for the flight and ground data

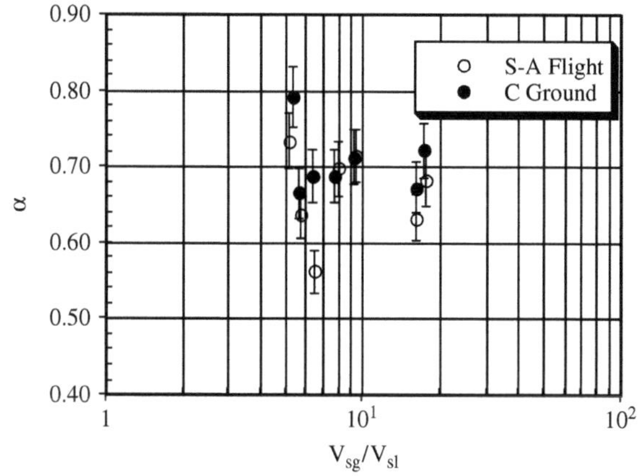

Figure 32. Comparison of the slug-annular flow regime at μ-*g* and the churn flow regime at 1-*g*

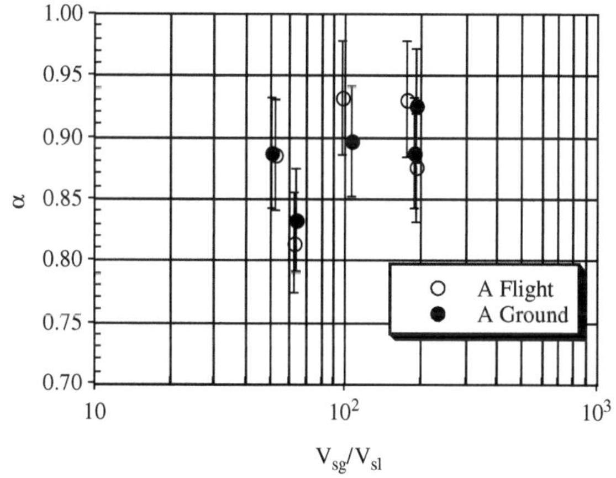

Figure 33. Comparison of the annular flow regime for the flight and ground data

at μ-*g* for $\alpha < 0.2$ can be attributed to the significant reduction in slip between the phases, combined with an increased influence of surface tension under reduced gravity. These changes lead to the bubbles being able to sustain their shape even at high gas content, while the rate of collision and coalescence of the bubbles is significantly reduced.

The slug flow results are shown in Fig. 31. For this regime, the void fraction appears to be consistently higher for the μ-*g* data, with a difference that ranges from 3 to 35%, with an overall average of approximately 10%. This trend seems to be consistent up to a void fraction of approximately 0.70 beyond which the percentage difference in void fraction decreases. This occurs near the transition to slug-annular flow at μ-*g*, and churn flow at 1-*g*. In the latter, flow reversal is observed and could greatly influence the void fraction readings in that region.

The results for transitional flows (slug-annular flow at μ-*g* and churn flow at 1-*g*) are shown in Fig. 32. The void fractions in this highly transitional region are comparable for both the μ-*g* and 1-*g* cases. As mentioned earlier, the slug-annular flow is in a transition where inertia forces are becoming more dominant when compared to surface tension forces. The liquid slug at these conditions contains a highly aerated frothy mixture, which could cause large fluctuations in the void fraction measurements.

Finally, the results for annular flow are shown in Fig. 33. The annular flow results show that the μ-*g* and 1-*g* data are very similar. This is expected since, under both 1-*g* and μ-*g* conditions, the flow is highly turbulent and inertia forces are dominant in both cases. The difference between the 1-*g* and μ-*g* void fraction values is within 5%.

Another comparison was made with the μ-*g* data collected by Bousman (1995). The μ-*g* void fraction data collected during the present study and those of Bousman (1995) are plotted in Fig. 34 in terms of the volumetric gas quality, β, as a function of the measured void fraction, α, where β is defined as:

$$\beta = \frac{Q_g}{(Q_g + Q_l)} = \frac{V_{sg}}{(V_{sg} + V_{sl})}. \tag{8}$$

This is a convenient parameter to use in situations where the reported data in the literature are in terms of V_{sl} and V_{sg}.

Bousman's μ-*g* data was collected for horizontal flow in 12.7 and 25.4 mm i.d. tubes aboard NASA's KC-135 aircraft using an impedance method (two parallel wires were located in the flow path, spanning the cross section of the tube). The wires were separated by 2.5 mm, and the void fraction was determined by measuring

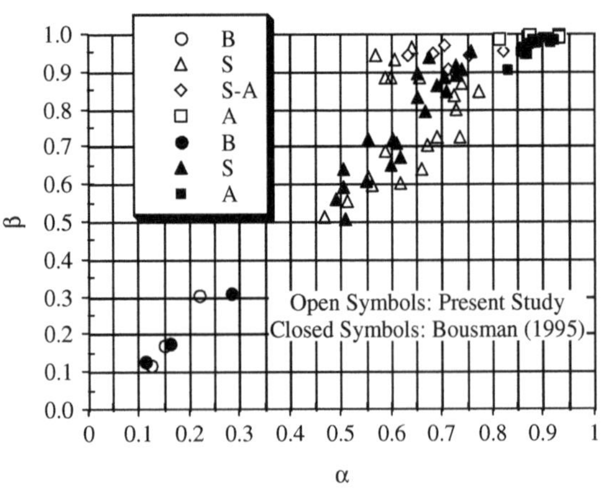

Figure 34. Comparison between the present study and Bousman's (1995) at μ-*g* conditions

Table 1. Comparison between the present study and data from Bousman (1995)

Flow Regime	Ave. % difference in β	RMS of α (%)
Bubble	±8	3.8
Slug	3	8.1
Slug-Annular	1	15.3
Annular	1.5	4.3
Average	3.4	7.9

the electrical conductance between the wires. Uncertainty in the void fraction measurements made by Bousman (1995) was reported to be ±0.65% in the void fraction range of $0 < \alpha < 0.75$. Void fraction data was reported for only the 12.7 mm i.d. tube.

The comparison between the present data points and those of Bousman (1995) was done by matching β for each flow regime. As seen in Table 1, β was matched within 3%, except for bubble flow (8%). It should be noted that in Bousman's (1995) reported void fraction data, none were classified as slug-annular flow. Therefore, the comparison of slug-annular flow shown in Table 1 is based on similar values of β that were classified as slug-annular flow in the present study. The classification of the slug-annular transitional flow, particularly at μ-*g* where this transition occurs over a much wider range of liquid and gas velocities, is very difficult resulting in some discrepancies between researchers. This could perhaps explain the large difference between the two sets of data in this region (15.3%), compared to the average of 7.9%. In the present study, however, visual images of the flow were substantiated with signal analysis from the void fraction sensor.

4. VOID FRACTION DISTRIBUTION COEFFICIENT

Since under most practical conditions the two phases may not travel at the same velocity, consideration must be given to the relative velocity (or slip velocity) between the two phases. Starting with the continuity equations for both phases, Zuber and Findlay (1965) developed a general expression for predicting the average void fraction, which takes into account the relative velocity between the two phases and the void fraction profile across the tube. The expression is given as:

$$\frac{\langle\beta\rangle}{\langle\alpha\rangle} = \mathrm{C_o} + \frac{\langle\alpha \mathrm{v_{gj}}\rangle}{\langle\alpha\rangle\langle \mathrm{j}\rangle}, \tag{9}$$

where v_{gj} is the drift velocity of the gas relative to the mean fluid velocity and j is the volumetric flux, that is $\mathrm{j} = V_{sg} + V_{sl}$, and $\langle\rangle$ represents the average value of the term within the brackets. The constant $\mathrm{C_o}$ is known as the "distribution coefficient" (more on this constant will be presented later). The second term on the right-hand side of Eq. (9) is the weighted mean drift velocity which takes into account the local relative velocity. Expressions for the weighted drift velocity of the gas with respect to the mean fluid were also given by Zuber and Findlay (1965) as:

$$\frac{\langle \alpha v_{gj} \rangle}{\langle \alpha \rangle} = 1.53 \left[\frac{\sigma g \Delta \rho}{\rho_l{}^2} \right]^{\frac{1}{4}} \tag{10}$$

for bubbly flow; and

$$\frac{\langle \alpha v_{gj} \rangle}{\langle \alpha \rangle} = 0.35 \left[\frac{g \Delta \rho D}{\rho_l} \right]^{\frac{1}{2}} \tag{11}$$

for slug flow. Eqs. (10) and (11) were developed based on analysis where the drift velocity, v_{gj}, was equated to the terminal velocity of a particle rising in an infinite medium.

Equation (9) is in a general form that could be applied to any flow regime to predict the void fraction. Since the phase velocities and the void fraction profiles (as well as the relative velocities between the phases) are taken into account, this equation is of a general use. The distribution coefficient, C_o, accounts for the non-uniform distribution of the void fraction over the cross section of the tube, and is obtained from:

$$C_o = \frac{\langle \alpha j \rangle}{\langle \alpha \rangle \langle j \rangle}. \tag{12}$$

The distribution coefficient, C_o, depends on both the flow regime and the void fraction profile. Depending on the radial void fraction distribution, C_o can be greater, equal to, or less than one. These conditions are summarized in Table 2, where α_c is the void fraction at the center of the tube, and α_w is the void fraction at the wall.

4.1 Comparisons of the Distribution Coefficient

It should be noted that the results in this section have been obtained, and are presented for two capacitance sensors. The capacitance sensor which was flown in February 1994 was a helical wound electrode sensor (see Elkow and Rezkallah, 1996, for details on the design and calibration of the sensor). Ground data, on the other hand, was collected with both the helical and a concave plate sensor that was later designed and calibrated, Elkow (1996). The flow settings were matched to those obtained during flight tests.

Table 2. The effect of α_w and α_c on C_o

Case 1	Case 2	Case 3
$\alpha_c < \alpha_w$ $C_o < 1$	$\alpha_w = \alpha_c = \alpha$ $C_o = 1$	$\alpha_c > \alpha_w$ $C_o > 1$

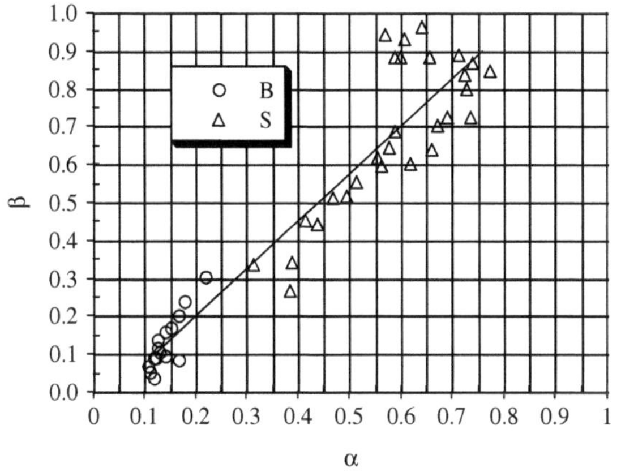

Figure 35. β as a function of α for bubble and slug flow at μ-g, $C_o = 1.25$

Eq. (9) was applied to both the ground and flight data. It has been already stated that due to the low gravity levels during the periods of μ-g, buoyancy effects would be minimized and thus, the second term on the right-hand side of Eq. (9) can be neglected. Therefore, by plotting β as a function of α, as shown in Fig. 35, C_o was found by linear regression to be 1.25 for the μ-g data in the bubble and slug flow regimes.

The results for slug flow at 1-g are shown in Fig. 36. From linear regression, C_o was found to be 1.17 and $\frac{\langle \alpha v_{gj} \rangle}{\langle \alpha \rangle} = 0.03$. Using Eq. (11) for slug flow, where the properties of water were evaluated at the average temperature of 30 °C, a value of 0.107 was obtained for $\frac{\langle \alpha v_{gj} \rangle}{\langle \alpha \rangle}$. Thus, the distribution parameter, C_o, for μ-g bubble and slug flow, and 1-g slug flow are 1.25 and 1.17, respectively. For water–air

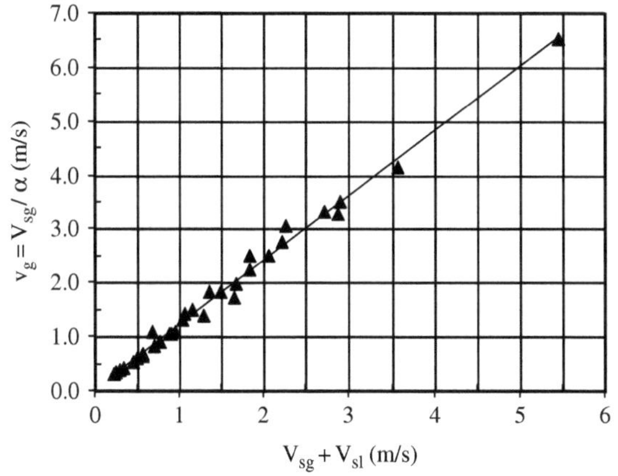

Figure 36. Results for slug flow at 1-g, $C_o = 1.17$

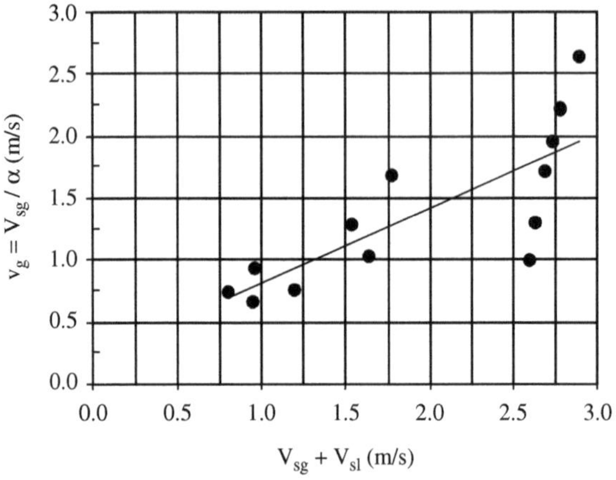

Figure 37. Bubble flow at 1-g, $C_o = 0.61$

flow at μ-g, Bousman (1995) obtained a C_o value of 1.27, which is in very good agreement with the present data considering that independent measurements were made using different sensor's geometry.

The bubbly flow results plotted in terms of the actual gas velocity, v_g, as a function of the total volumetric flux, $V_{sg} + V_{sl}$, can be seen in Fig. 37. From linear regression, C_o was found to be 0.61 and $\frac{\langle \alpha v_{gj} \rangle}{\langle \alpha \rangle} = 0.19$. As seen earlier, $C_o < 1.0$ occurs when $\alpha_c < \alpha_w$. The typical "saddle" shape profile is seen where the void fraction is higher at the wall of the tube and drops off toward the center. This agrees well with experimental results, where the radial void fraction profile for bubble flow was measured using local probes (e.g., Kamp et al., 1993).

At 1-g, Kamp et al. (1993) found that the peak in the void fraction profile occurred at a distance from the wall equal to approximately one bubble diameter. They also found that the peak void fraction near the wall was much higher, relative to the void fraction at the centerline, when the liquid velocity was significantly larger than the gas velocity.

An investigation into the development of the void fraction profile in a 38 mmi.d. vertical tube for bubble flow was done and reported recently by Grossetête (1995). Water–air mixtures were used with local measurements taken at L/D locations of 8, 55, and 155 from the mixer. Grossetête (1995) found that there was a dominant "saddle" shape profile at $L/D = 8$. This profile diminished as the L/D ratio increased, mainly due to gas expansion and coalescence. At $L/D = 55$, the "saddle" shape was still evident, however the void fraction at the center had also increased. At $L/D = 155$, the void fraction was highest at the center of the tube. Corresponding to the change in the void fraction profile, Grossetête (1995) found that the fluid velocity profile also changed. As the L/D ratio increased, the fluid velocity profile became more of a parabolic shape. The increase in the radial velocity gradient, combined with the expanding gas, a higher coalescence rate and added turbulence, cause the void fraction to increase at the centerline. In the present study the void fraction sensor was located at L/D = 65.1. Therefore a value of $C_o = 0.61$, which was obtained during ground tests, is in good agreement with the results of Grossetête (1995).

By comparing C_o for the flight and ground data, it can be seen that the distribution coefficient is higher for μ-g than for 1-g. This suggests that the void fraction distribution at μ-g tends to be maximum at the centerline, as opposed to near the wall (conditions which are very common at 1-g). The other interesting observation is that under μ-g conditions, C_o is the same for bubble and slug flows. This could be explained primarily in terms of the slip velocity. Since at μ-g the slip velocity is so small (almost zero), the difference between the two regimes should also be very small. For the 1-g case, on the other hand, a change in the flow regime results in a change in C_o since the void fraction profile and the velocity profile are different for different flow regimes. Drift velocity should also change for each flow regime since it depends on the momentum transfer between the two phases, and the shear stress due to interfacial interaction between the phases.

5. SIGNAL ANALYSIS AND PROBABILITY DENSITY FUNCTIONS

5.1 Introduction

For the flight data, the average time duration used for analysis consisted of approximately 400 samples (which corresponds to approximately 5.7 sec of void fraction measurements). The selection of the "window" length depends on the stability of the g-level and both liquid and gas flow rates. For the ground data, a "window" length of 625 samples was used, which represents approximately 8.8 sec of data gathering. Table 3 lists the data, in terms of the liquid and gas superficial velocities, the average void fraction, and the flow regime associated with each parabola.

Table 3. Data obtained from the flight and ground tests showing flow rates, gravity level, average void fraction, and flow regimes

Flight/Ground Parabola	V_{sl} (m/s)	V_{sg} (m/s)	Normalized Gravity g_z	Ave. α	Flow Regime
94F3P22	0.86	0.09	–0.042	0.139	B
94F3P13	0.24	0.19	–0.003	0.437	S
94F5P16	0.39	6.97	0.007	0.682	S-A
94F5P9	0.07	13.9	0.022	0.875	A
94G3P22	0.84	0.09	1.0	0.141	B
94G3P13	0.24	0.19	1.0	0.363	S
94G5P16	0.40	6.92	1.0	0.721	C
94G5P9	0.07	13.9	1.0	0.887	A

The data points listed in Table 3 represent typical flow regimes ranging from bubble (B), slug (S), slug-annular (S-A), churn (C), to annular (A) flow. They were used to compare the PDF plots obtained with the flight and ground data. The sample coefficient of variation, *cv*, was used to accept or reject regions of data (Barnes, 1994), and it was calculated from:

$$cv = \left(\frac{s}{\overline{x}}\right) 100\%, \tag{13}$$

where s is the sample standard deviation, and $\overline{x}$ is the sample mean. Data was retained if the sample coefficient of variation was 7% for V_{sl} and V_{sg}, and when the gravity level, g_z, was within 0.04. It was found that most of the data was approximately within 2–3 standard deviations, when the 7% range for V_{sl} and V_{sg} was used. To obtain a PDF for a discrete random variable, the ordinate is divided into equal class widths. If over the total time trace the number of samples is N, and n_i is the number of occurrences within $_i$, then the probability density function can be written as:

$$p(\alpha) = \frac{n_i}{N\Delta\alpha_i}. \tag{14}$$

Time trace signals, PDF plots, and images of the flow regimes are shown for the parabolas listed in Table 3. The PDF plots were obtained using Eq. (14), where a value of 0.01 was used for $\Delta\alpha$. Since a flow regime can be encountered at various flow rates it is impossible to show all of the differences between the 1-g and μ-g data for each case. Therefore, a discussion of the differences for each flow regime will be based on the collective results from video recordings, and the analysis of the void fraction signal. The discussion will also be based on the particular flow conditions listed in Table 3 and the PDF results shown.

5.2 Bubble Flow

In a comparison of the overall average void fraction values for bubble flow, just discussed, it was found that small differences exist between the 1-g and μ-g cases. However, from video images and from plots of the temporal change in the void fraction values, it was found that significant differences existed. At low flow rates and at μ-g conditions, bubbles are spherical in shape, have a uniform size, and travel through the tube at regular intervals. However, at the same flow rates at 1-g the bubbles have a rounded leading cap and a flat tail, vary in size, and flow through the tube at irregular intervals. At higher flow rates there is some added turbulence at both 1-g and μ-g conditions. At μ-g the bubbles are slightly distorted and somewhat elongated. There is some variation in bubble size but the majority of the bubbles are of similar length and still travel at regular intervals. At 1-g the bubbles can be longer than those seen at μ-g and even more distorted. They tend to travel irregularly in clusters with larger bubbles leading a wide range of bubble sizes in near proximity. This could be explained perhaps in light of the "physics" of the flow, both at 1-g and μ-g conditions. The predominant forces for this flow regime include buoyancy, surface tension, and turbulent stresses. At 1-g, forces due to gravitational acceleration tend to accelerate the bubbles and hence increase the probability of their interactions and coalescence. Thus, the void distribution varies as the bubbles agglomerate, collide, and coalesce.

Figure 38 shows the time trace and PDF plots for bubble flow at $V_{sl} = 0.85\,\text{m/s}$ and $V_{sg} = 0.09\,\text{m/s}$. The PDF plots show that the fluctuation in void fraction is much greater at 1-g compared to μ-g. This is clearly shown in Fig. 38 (d) where the fluctuations of α range from approximately zero to 0.40. On the other hand, at μ-g conditions, the slip velocity in bubbly flow is almost zero, and the bubble movement and liquid turbulence are highly suppressed. Therefore, the bubbles remain mostly intact, moving with a uniform velocity; that is, more or less equal to the liquid velocity. This is also evident from the video images that were analyzed for this flow. A consecutive set of images (covering approximately 0.3 sec), illustrating bubble

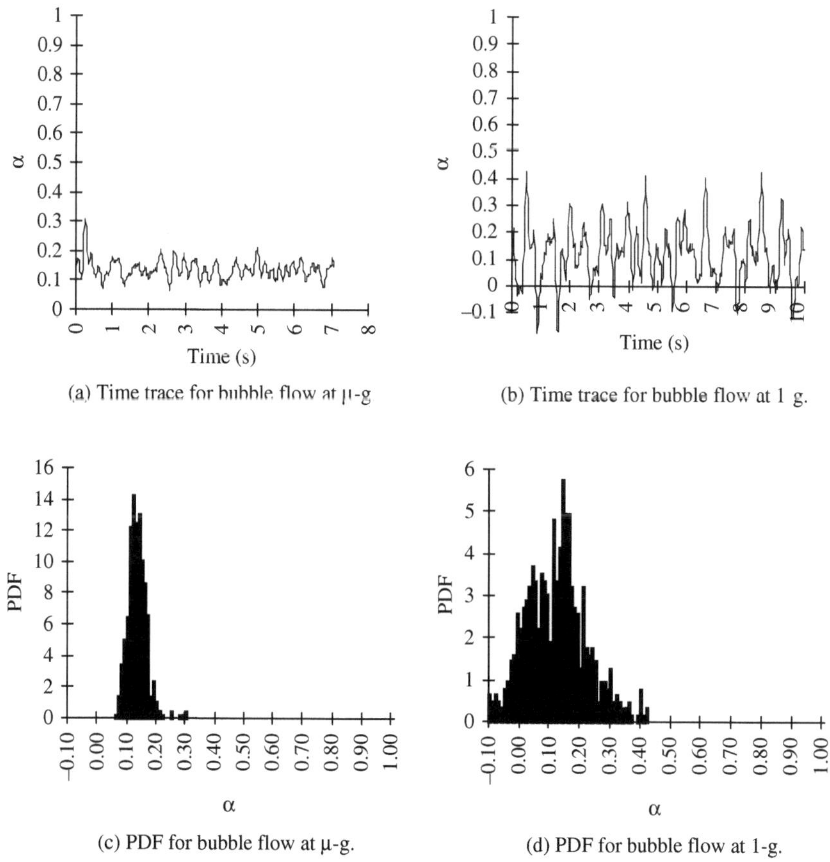

(a) Time trace for bubble flow at μ-g

(b) Time trace for bubble flow at 1 g.

(c) PDF for bubble flow at μ-g.

(d) PDF for bubble flow at 1-g.

Figure 38. Bubble flow regime time traces and PDF plots for data collected during Flight 3, P22, ((a) & (c)), and Ground 3, P22, ((b) & (d))

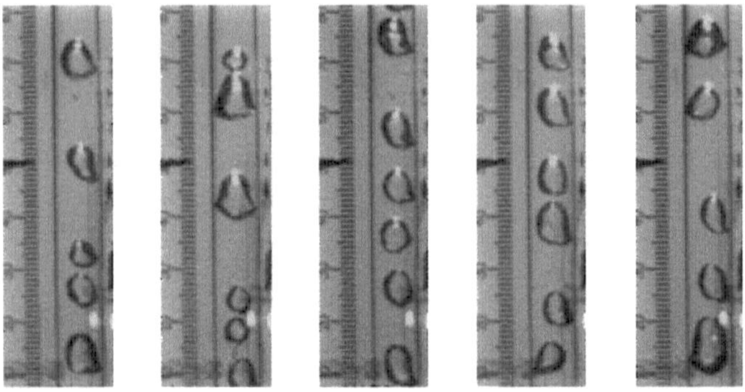

Figure 39. Sequential images for bubble flow at μ-g (Flight 3, Parabola 22)

flow can be seen in Figs. 39 and 40 for μ-g and 1-*g*, respectively. Although there is a limited number of frames due to space limitations, the important concepts discussed above are evident from those frames. At μ-g the bubbles are seen to be uniform in shape (approximately 1/2 tube diameter) and more equally spaced than those at 1-*g*. This uniformity at μ-*g* results in a narrow PDF. At 1-*g*, on the other hand, there is a wider variation in the bubble size and shape as well as the liquid separating them, causing a greater fluctuation in the PDF. Bubble diameters vary from a few millimeters to approximately 1 tube diameter. Liquid slugs separating the bubbles range from a few bubble diameters to spacing of about 6–7 tube diameters. In terms of the PDF distribution, as shown in Fig. 38 (c), the spectrum of variation of α at μ-*g* is limited in the range of $0.08 < \alpha < 0.20$. Within this void fraction range, it can be seen that the mean value that occurs at μ-*g* has a much higher probability of occurrence (approximately 2 times that of the 1-*g* counterparts).

These are important findings that can perhaps explain the lower heat transfer coefficients associated with bubbly flow at μ-*g* conditions. Rite (1995) reported that in the range of $0.10 < \alpha < 0.18$, the heat transfer coefficients for bubbly flow

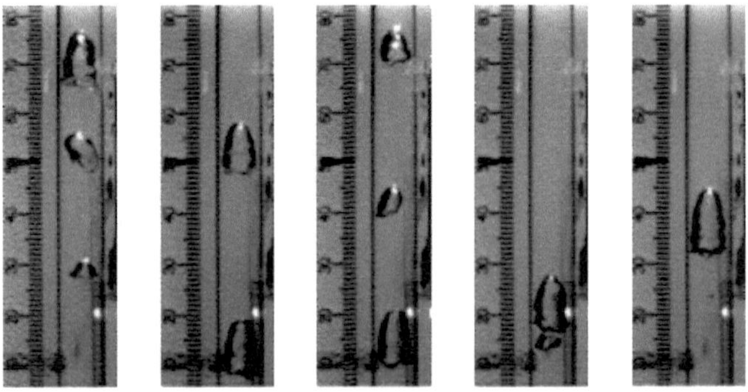

Figure 40. Sequential images for bubble flow at 1-g (Ground 3, Parabola 22)

at 1-g were 3–12% higher than those measured at μ-g. The higher heat transfer rates at 1-g are attributed to the larger bubble fluctuations and the higher mixture turbulence. Hence, better heat transfer rates are achieved with higher mixing of the flow.

5.3 Slug Flow

Typical time traces and PDF plots for slug flow are presented in Fig. 41 with images of the flow at μ-g and 1-g conditions shown in Figs. 42 and 43, respectively. The flow conditions are at relatively low liquid and gas flow rates ($V_{sl} = 0.24\,\mathrm{m/s}$ and $V_{sg} = 0.19\,\mathrm{m/s}$). The elapsed time covered in Figs. 42 and 43 was approximately 0.5 sec. A comparison of the average void fraction values for slug flow at 1-g and μ-g shows that the overall average void fraction values at μ-g conditions were

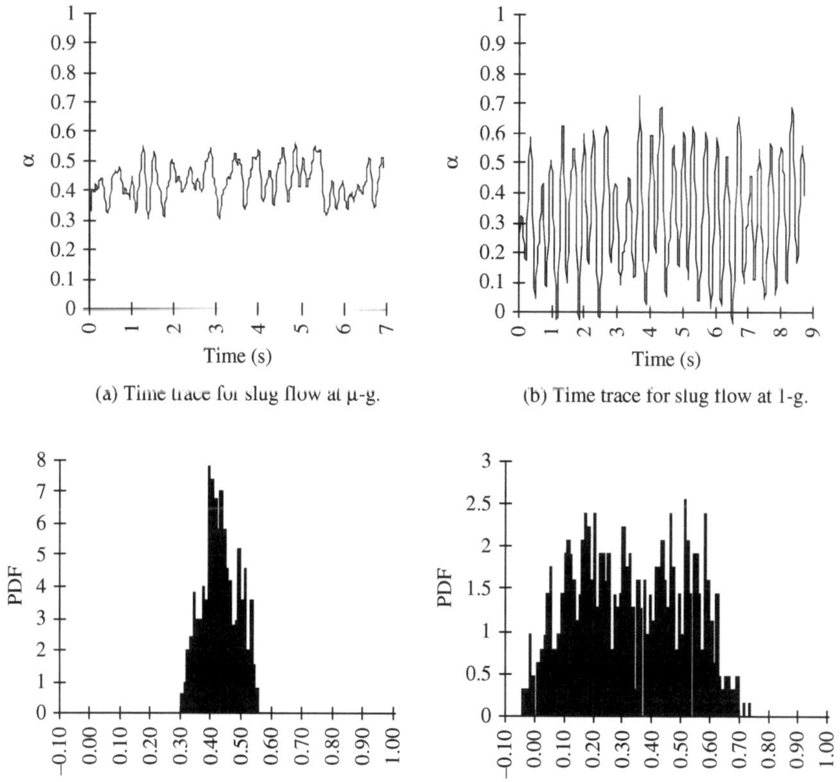

(a) Time trace for slug flow at μ-g.

(b) Time trace for slug flow at 1-g.

(c) PDF for slug flow at μ-g.

(d) PDF for slug flow at 1-g.

Figure 41. Slug flow regime time traces and PDF plots fow data collected during Flight 3, P13, ((a) & (c)), and Ground 3, P13, ((b) & (d))

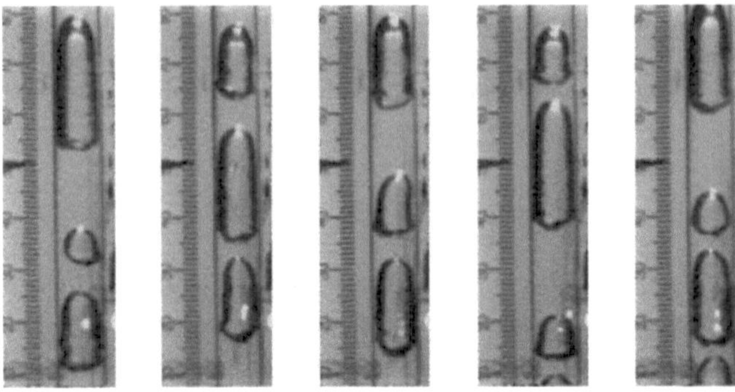

Figure 42. Sequential images for slug flow at μ-g (Flight 3, Parabola 13)

approximately 10% higher (shown above). The transition to slug flow occurs as the gas flow rate increases (α increases) to the point when significant coalescence of the smaller bubbles will form larger "bullet-shaped" bubbles, commonly known as "Taylor bubbles" for 1-g flows.

The signal output for slug flow is therefore characterized by high and low α values, depending on which phase is flowing in the conduit at that time. This is obvious from the signal trace shown in Figs. 41 (a) and (b). Due to buoyancy at 1-g, coalescence occurs rapidly under those conditions, and the time trace shows large fluctuations around a mean value. The coalescence observed at such low flow rates is primarily due to Taylor bubbles coming into contact with each other rather than coalescence, due to small bubbles merging into them (conditions that are very common at higher gas content at 1-g). Since buoyancy is negligible under μ-g conditions, the Taylor bubbles are seen to move with equal velocity resulting in little to no interaction, hence uniform bubble length and considerably smaller fluctuation in the void fraction is observed (Fig. 41 (a)). Hence, the spread of the void fraction signal is far more significant under 1-g compared to μ-g; see Fig. 41 (b).

The lesser fluctuations at μ-g are also reflected in the PDF plots (Figs. 41 (c) and (d)). The PDF from the μ-g data for slug flow covers a much narrower spectrum

Figure 43. Sequential images for slug flow at 1-g (Ground 3, Parabola 13)

when compared to 1-*g*, indicating more uniform bubbles with less variation in the void fraction in the liquid slug. Under μ-*g* conditions, α changes only between approximately 0.3–0.55, while at 1-*g* the range is between almost zero up to 0.75.

Ideally, slug flow should show two peaks on a PDF histogram: one associated with the liquid slug, and the other with the Taylor bubble. With the ground data, the development of the two peaks can be clearly seen in Fig. 41 (d) with one at approximately $\alpha = 0.20$, and the other at approximately $\alpha = 0.60$.

The distinctiveness of the peaks depends mainly on the geometry of the sensor, and to a lesser extent on the flow itself; that is, the length of the Taylor bubble and the length of the liquid slug. Since the sensor measures a volumetric void fraction, the signal is "filtered" for certain bubble sizes. This can be explained by tracing the progress of slug flow. As a bubble enters the sensor, and assuming it is longer than the sensor, a high void fraction reading is registered. This will account for the second peak in the PDF at a high void fraction. As the bubble exits, the liquid slug enters the sensor, and the void fraction begins to decrease. The decrease in the void fraction depends on the length of the slug and if any bubbles are entrained in the wake of the main Taylor bubble. This measurement will account for the first peak of the PDF at a low void fraction. For flow conditions (usually low flow rates) where there is no air entrained in the liquid slugs separating consecutive bubbles, the low void fraction values are due to a liquid slug slightly shorter than the void fraction sensor length. This arises from the trailing end of a Taylor bubble just leaving the sensor and a new Taylor bubble immediately entering it.

If the sensor was alternately filled with the gas phase followed by only liquid, two distinct peaks will register on the Data Acquisition System, with low void fraction values between the peaks (depending on the degree of entrained bubbles). However, if either the bubble or the liquid slug is shorter than the sensor, the sensor does not see the two extreme void fraction values but an "overall" average reading between the two. This was the case at μ-*g* conditions where only a single peak was registered. This is primarily due to the short liquid slugs separating the bubbles. In general, as seen in the recorded images, at no time was the sensor filled with liquid only. Thus, an average void fraction representing the bubble and slug was always read. Observations of the numerous video images at μ-*g* indicate that this type of flow is quite predictable in that the slugs were essentially stable and of equal length (seen in Fig. 42). This is also evident in the PDF plot, with a high probability at the average void fraction value and very little variation about the average.

Further discussion and interpretation of the PDF results can be made based on the flow images shown in Figs. 42 and 43. Taylor bubbles like flows at μ-*g* are more uniform in length. The Taylor bubbles at μ-*g* vary from only 1 to 3 tube diameters, whereas at 1-*g* the Taylor bubble for the same flow rates ranges from 1 to 6 tube diameters. Further evidence of the effects of buoyancy can be seen in the flatter trailing end of the Taylor bubble at 1-*g* (Fig. 43), whereas the trailing end of the Taylor bubble at μ-*g* (Fig. 42) where buoyancy is negligible is seen to be more rounded. Buoyancy causes a relative velocity difference between the liquid and gas phases. With the gas phase having the higher velocity, a low pressure region at the tail of the Taylor bubble would occur. From the video recordings in this region, it has been observed that coalescence is enhanced when another Taylor bubble approaches this low-pressure region. However, at μ-g there is essentially no slip, thus a low-pressure region at the edge of a Taylor bubble does not exist. It has been observed from these video recordings that Taylor bubbles within close proximity to each other rarely coalesce at μ-g.

Temperature measurements in slug flow, as reported by Rite (1995), indicate that the heat transfer coefficients at 1-g can be 30–40% higher than their counterparts at μ-g (corresponding to a void fraction from approximately 0.3 to 0.65). As the void fraction increased, the difference in the heat transfer coefficients was minimum. This could be perhaps explained in terms of the results shown in Figs. 41 (c) and (d). As already stated, the PDF scatter at 1-g is larger than that at μ-g, indicating that a wider range of bubbles are present in the 1-g flow. This could result in enhanced mixing in the slug region of the flow; thereby resulting in a higher heat transfer coefficient.

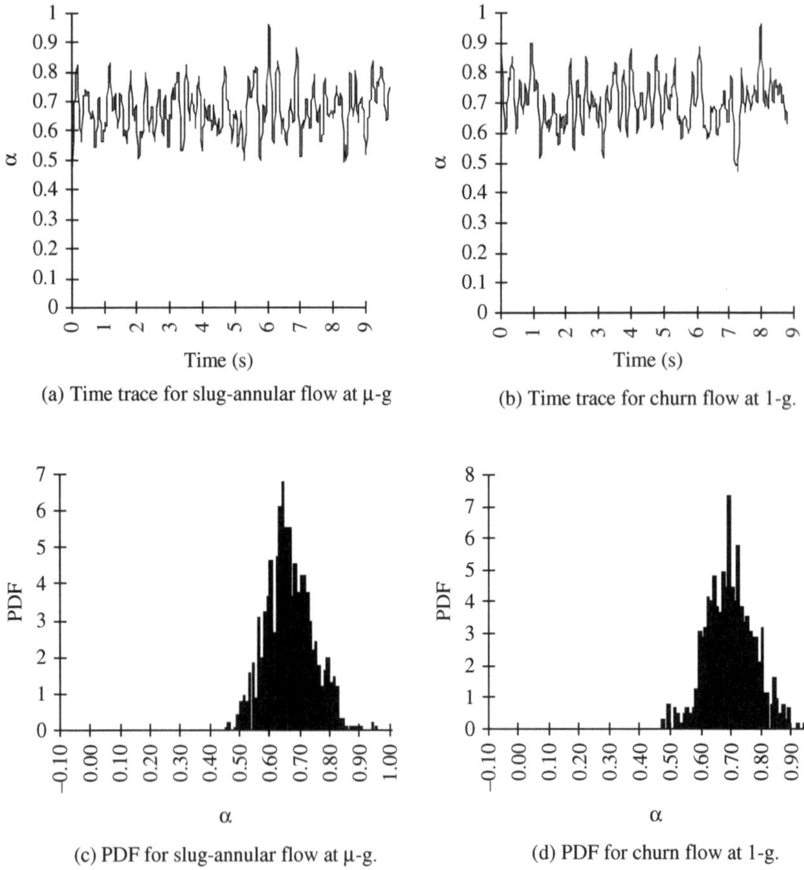

Figure 44. Slug-annular and churn flow regime time traces and PDF plots fow data collected during Flight 5, P16, ((a) & (c)), and Ground 5, P16, ((b) & (d))

5.4 Transitional Flow

The next series of plots, shown in Fig. 44, are those for the slug-to-annular transitional flow (churn flow regime at 1-*g*). Images of the flow at V_{sl}0.39 m/s and V_{sg}6.97 m/s can be seen in Figs. 45 and 46. This transition is the result of increasing the gas flow rate to the point where the gas phase becomes a continuous phase (annular flow) with little or no liquid bridging at the center. At 1-*g*, churn flow is observed in this region. Due to the pull of gravity on the liquid phase, churn flow is characterized by large oscillations and continuous liquid bridging. The average void fraction in both cases was found to be similar as shown earlier in this chapter. It can also be seen from Fig. 44 that the time trace and the PDF plots are also very similar. This is mainly due to the fact that inertia forces are starting to dominate, and hence gravity plays a much lesser role. The interface between the liquid and gas becomes wavy as the bubble length grows. As the waves are sheared by the incoming gas, liquid is entrained into the gas core and a "frothy" mixture is formed in the liquid regions. The similarities in this flow regime can be seen in the flow images given in Figs. 45 and 46 for μ-*g* and 1-*g* flows, respectively.

The heat transfer results showed that the slug-annular and annular heat transfer coefficients approached similar values for both 1-*g* and μ-*g* data (Rite, 1995). Some of the data showed that the μ-*g* heat transfer coefficients were slightly higher in the slug-annular regime (approximately 10% higher). However, this was approaching the measurement uncertainty of the heat transfer data. Although the PDF plots and

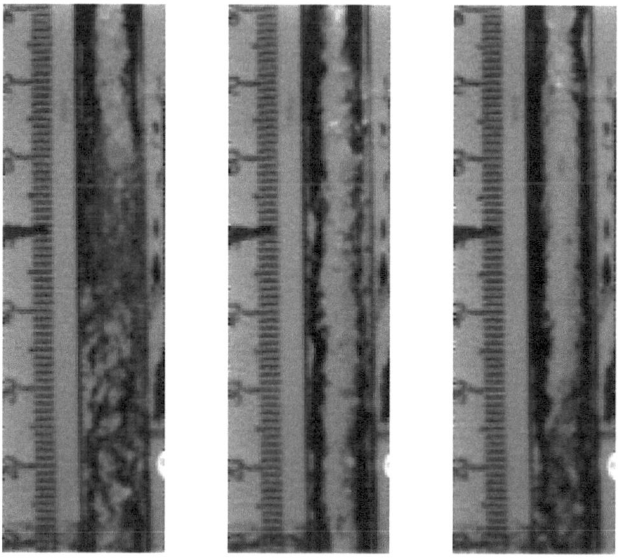

Figure 45. Images of slug-annular flow at μ-g (Flight 5, Parabola 16)

void fraction signals appear similar, the heat transfer for this regime would be determined by the froth present in the slug-annular flow at μ-g and the degree of churning at 1-g (since both are becoming inertia dominant flows).

5.5 Annular Flow

Finally, the annular flow results are shown in Fig. 47 for $V_{sl} = 0.07\,m/s$ and $V_{sg} = 13.9\,m/s$. Corresponding images of the flow are shown in Figs. 48 and 49 for μ-g and 1-g, respectively. In that flow, inertia forces are very significant and dominate by far all other body forces acting on the flow (e.g., surface tension force). Thus, it is expected that the effect of gravity is negligible in this regime. In both cases, a liquid film is retained at the tube wall, with the gas phase occupying the center of the tube. Thus, no great difference should be expected with respect to the volumetric void fraction results. This is seen in Fig. 47 with similar time traces, width, and maximum values of the PDF distribution. In general, the average void fraction values for this flow are also similar (see Table 3).

In addition to examining the individual PDF plots for the flight and ground data, Fig. 50 is used to show a complete presentation of the void fraction since the same flow regime can occur at various flow rates and void fraction. Fig. 50 is a plot of the maximum value of the first mode of the PDF as a function of void fraction for both sets of data. A mode of a PDF occurs whenever there is a local maximum in $f(\alpha)$. This means that a maximum value will occur in $f(\alpha)$ at some void fraction, α_1, and all values near α_1 will be less than $f(\alpha_1)$. More

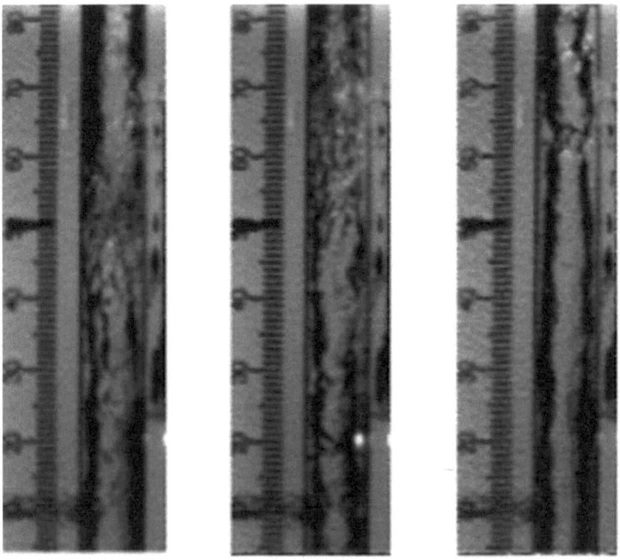

Figure 46. Images of churn flow at 1-g (Ground 5, Parabola 16)

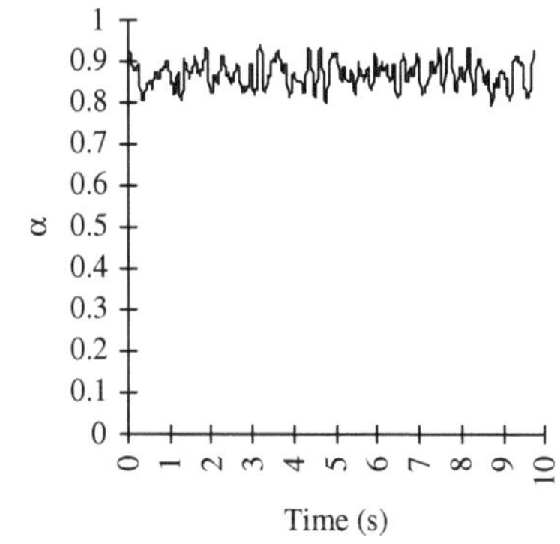

(a) Time trace for annular flow at μ-g.

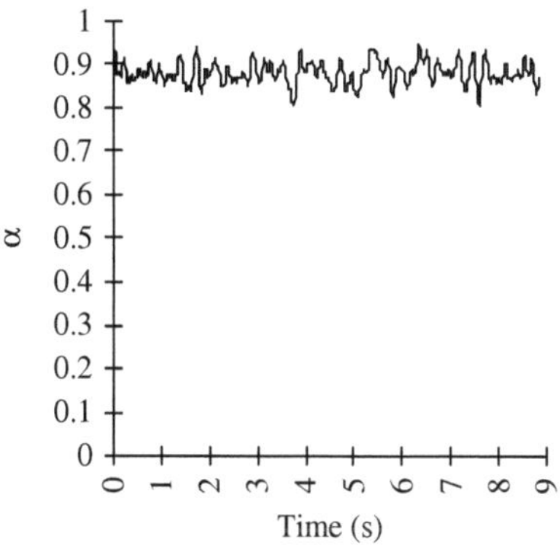

(b) Time trace for annular flow at 1-g.

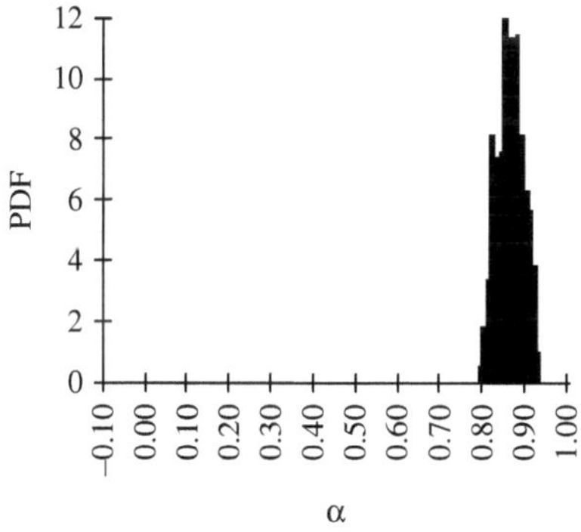

(c) PDF for annular flow at μ-g.

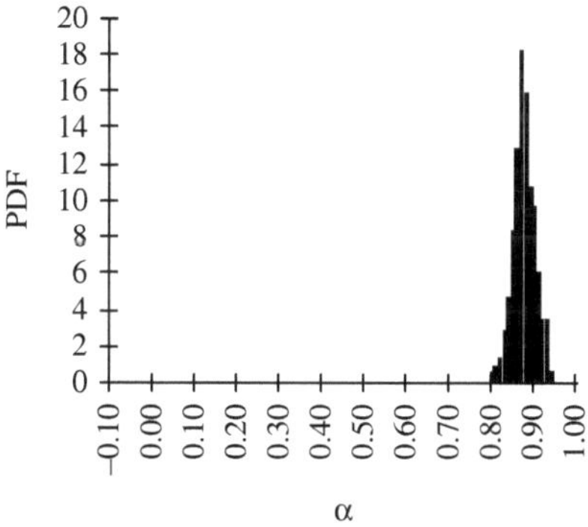

(d) PDF for annular flow at 1-g.

Figure 47. Annular flow regime time traces and PDF plots for data collected during Flight 5, P9, ((a) & (c)), and Ground 5, P9, ((b) & (d))

than one mode may exist, corresponding to each local maximum. The first mode is considered to be the greatest local maximum, the second mode is the next greatest local maximum, and so on. Fig. 50 shows only the first mode as a function of the measured void fraction. The area under a PDF is equal to one. Therefore, for bubble and annular flow regimes (where only one peak in the PDF occurs), a high value for the first mode will imply a narrower fluctuation in void fraction. Conversely, a lower value for the first mode will imply a wider void fraction fluctuation.

For the flight data, it can be seen from Fig. 50 that the first mode of the PDF is higher than that for the ground data (up to a void fraction of approximately 0.70). At a void fraction of 0.70 (which would represent the region where forces due to inertia begin to dominate), the two data sets merge into each other. For void fraction, values greater than 0.70, the 1-*g* and the μ-*g* results are almost identical.

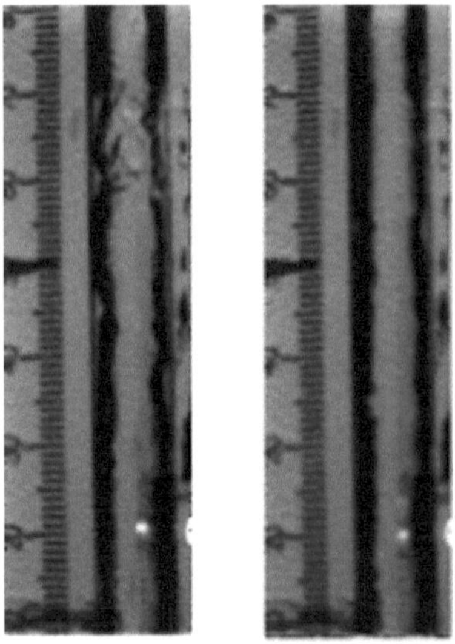

Figure 48. Images of annular flow at μ-g (Flight 5, Parabola 9)

6. CONCLUSIONS

Using a helical wound capacitance void fraction sensor, volumetric void fraction measurements were obtained for adiabatic two-phase water–air data at 1-*g* and μ-*g*. The data was collected for flow in a 9.53 mm i.d. tube over a wide range of liquid and gas flow rates, covering a range of void fraction from approximately 0.10 to 0.90. The test section was oriented vertically on-ground and with respect to the aircraft floor. Comparisons were made between the void fraction results collected at 1-*g* and μ-*g* and the Zuber and Findlay (1965) model. In addition, comparisons of time traces and PDF histograms for the void fraction signals at 1-*g* and μ-*g* were made. In summary, the following conclusions could be made:

- In comparing the average void fraction obtained for the bubble flow regime at 1-*g* and μ-*g*, it was found that for $\alpha < 0.16$ and $V_{sg}/V_{sl} < 0.1$, the μ-*g* void fraction was approximately 8–25% higher. This could be attributed to a significant reduction in the slip velocity between the phases, as well as the greater influence of surface tension under reduced gravity. For $0.16 < \alpha < 0.2$, the difference decreases until the trend is reversed with the 1-*g* data slightly higher, approximately 7–16% higher.
- The overall average void fraction values for slug flow were found to be 3–35% higher at μ-*g* compared to 1-*g* (the average was approximately 10%).
- The average values of void fraction for the transitional flow, slug-annular and churn flows at μ-*g* and 1-*g* were found to be similar. The percentage difference in the void fractions was within the uncertainty of the measurements. The similarities in the average void fraction values could be attributed to the fact that inertia forces are becoming more dominant in this region.

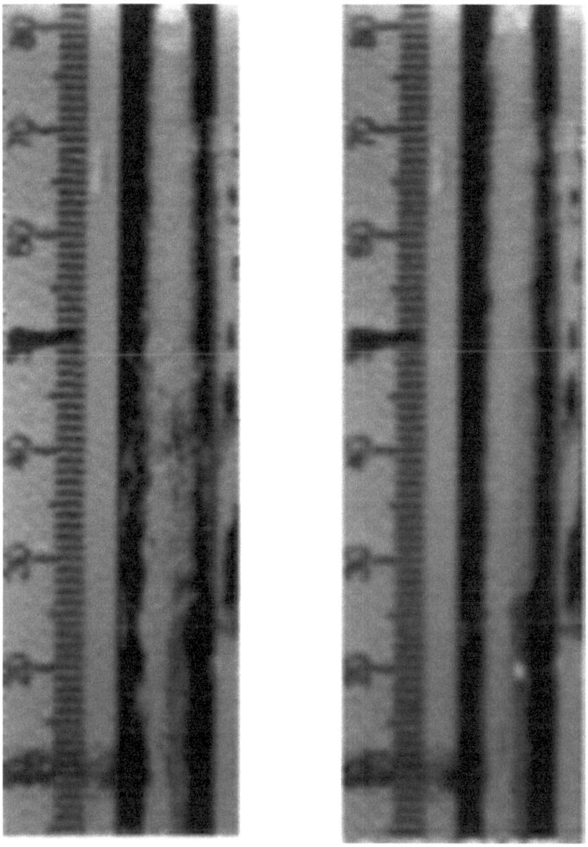

Figure 49. Images of annular flow at 1-g (Ground 5, Parabola 9)

- As with transitional flow, the average values of void fraction for annular flow, both at 1-*g* and μ-*g*, were also similar. The difference in the void fraction was within 5%, which is within the uncertainty of the measurement. Similar void fraction values would be expected since the flow is primarily dominated by inertia forces.
- The distribution coefficient, C_o, was determined for the 1-*g* and μ-*g* data. It was found to be higher than unity for the bubble and slug flows at μ-*g*($C_o = 1.25$). This was in excellent agreement with other results obtained at μ-*g*. For the bubble flow regime at 1-*g*, C_o was less than 1.0, indicating a higher concentration near the wall. This agrees well with the typical "saddle" shape void fraction profile present in vertical upward bubble flow at 1-*g*. For slug flow at 1-*g*, C_o was found to be 1.17.
- The PDF histograms for bubble flow showed a much greater fluctuation in void fraction at 1-*g* (in the range, $0 < \alpha < 0.4$); whereas the void fraction at μ-*g* showed

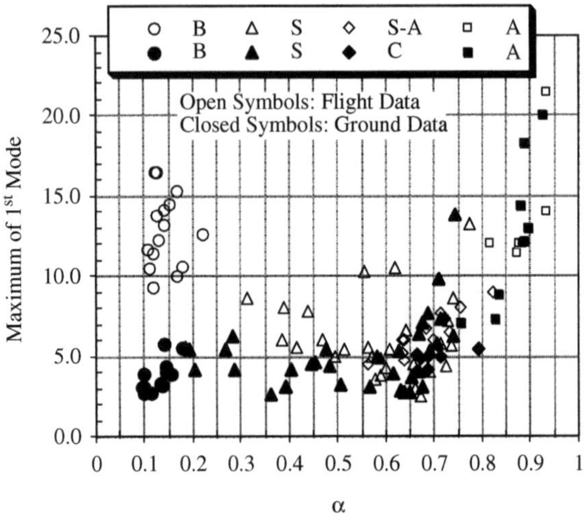

Figure 50. Maximum value of the first mode as a function of void fraction for the 1-g and μ-g data

a narrower range, ($0.1 < \alpha < 0.2$). At μ-*g*, the slip velocity between the phases is essentially zero, resulting in suppressed turbulence and lesser interactions between bubbles.

- The PDF histograms for slug flow also indicated a greater fluctuation in void fraction for the data collected at 1-*g*. Fluctuations were in the range $0 < \alpha < 0.75$ for 1-*g*, and restricted only to the range $0.3 < \alpha < 0.55$ for the μ-*g* data. In addition, the overall average of the void fraction values in this regime was approximately 10% higher in μ-*g*.
- As with the average void fraction values for slug-annular and churn flows, the PDF plots were also found to be similar. Void fluctuations from approximately 0.5 to 0.9 were observed for both 1-*g* and μ-*g* flows. This could be attributed to the fact that inertia is becoming dominant, thus reducing any differences that were previously seen.
- The PDF histograms for annular flow were found to be similar for both 1-*g* and μ-*g* flows. A very narrow fluctuation from approximately $0.8 < \alpha < 0.9$ was observed. This is expected since the flow in this regime is highly inertia-dominated.
- Analysis of the first mode of the PDF plots shows that the slug-to-annular transitional flow occurs at a void fraction of approximately 0.7. Beyond this region, differences between 1-*g* and μ-*g* data are extremely insignificant.

CHAPTER 6

GAS–LIQUID FLOW HEAT TRANSFER

Two-phase thermal management systems have been utilized and researched in the nuclear and petrochemical industries for many decades. However, the knowledge base for their use in space hardware has only recently been accumulating. Some experimental studies of two-phase convective heat transfer were performed under microgravity conditions, mostly for single-component flows (i.e., boiling and condensation). Among these are the studies by Papell (1962) and Feldmanis (1966). In his study, Papell reported a 15% increase in the heat transfer rate for a subcooled water system during reduced-gravity durations of his flight experiment. Feldmanis did not explicitly determine heat transfer coefficients, but based on temperature measurements that were made during his experiment he predicted that for forced convective condensation, gravity would have little influence on heat transfer.

Reinharts et al. (1992) worked with a boiling and condensing Refrigerant-12 test loop on-board NASA's KC-135. They reported that the condensation heat transfer coefficients were 26% lower for μ-g conditions as opposed to 1-g conditions. Boiler temperatures remained constant throughout the KC-135 flights. Thus, no conclusions could be drawn on the effect of gravitational acceleration on the boiling heat transfer coefficients. Ohta et al. (1994) completed a study of Refrigerant-113 in a convective boiling system with vertical, upward flow through an 8-mm i.d. circular tube. Parabolic flight experiments were used to generate the μ-g conditions. They reported that while boiling was occurring (vigorous bubble nucleation at the heated wall) in the bubbly and annular flow regimes, there was no difference in the heat transfer coefficients between 1-g and μ-g conditions. However, when boiling was suppressed due to the high flow rates and low heat flux, the heat transfer coefficients were lower at μ-g as compared with 1-g. They attributed this decrease to a reduction in the turbulence intensity at the liquid-vapor interface and an increase in film thickness. Based on video images that were recorded, it appeared that these changes were precipitated by an absence of disturbance waves at the surface.

From the studies cited above with their conflicting results, it is clear that further experimental investigation into the heat transfer behavior of two-phase flows under μ-g is warranted.

1. EXPERIMENTAL FACILITY AND PROCEDURE

A two-phase, two-component test apparatus was designed and built for the performance of microgravity experiments on NASA's KC-135 zero-gravity aircraft. The apparatus is instrumented such that simultaneous measurements of pressure drop and heat transfer data can be made, as well as continuous observation and recording of the two-phase flow patterns. In order to cover a wide range of test conditions, the facility allows for the independent control of three separate parameters during testing: air flow rate, water flow rate, and temperatures of the two-phase mixture in the flow loop.

A schematic of the test apparatus is shown in Fig. 51. On-board the plane the apparatus was situated such that the heated test section was oriented vertically with respect to the floor of the aircraft. A complete discussion of the flow loop may be found in Rite and Rezkallah (1993). In addition, more details concerning the heated test section will be provided here.

Air is injected radially into the mixer while water flows axially. After mixing, the two-phase mixture then proceeds through a 74-cm long flow developing section with an inside diameter of 9.53 mm ($L/D \approx 77$). A 16.2-cm long ($L/D \approx 17$) observation section follows as shown in Fig. 51. The observation section is constructed of a 9.53 mm i.d. acrylic tube. Thus, from the exit of the mixer to the heated test section, the total calming length is 110D. This provides for a fully developed velocity profile before heat transfer measurements are taken along the heated test section.

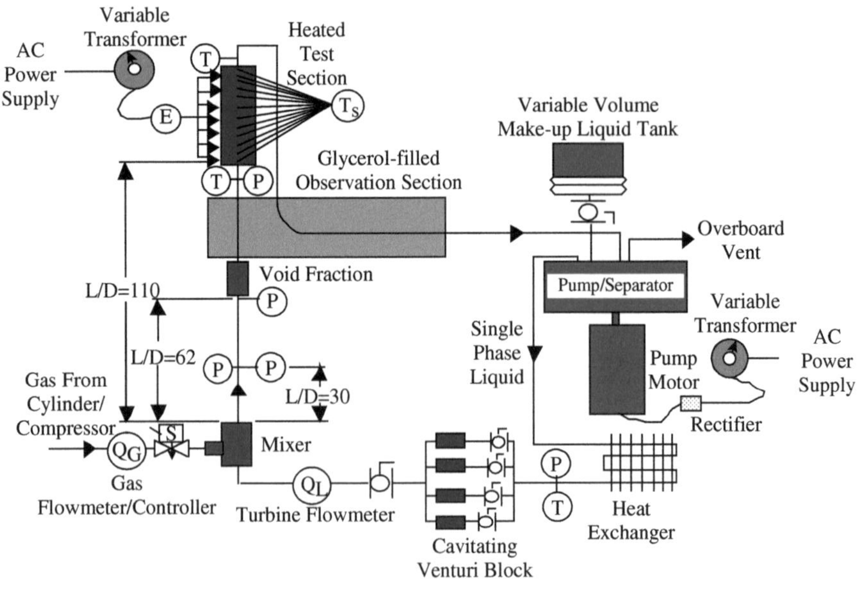

Figure 51. Schematic of a two-phase flow flight facility

The inlet temperature RTD (resistance temperature detector) fixture is followed by the heated test section. It consists of a 9.53 mm i.d. copper tube having a wall thickness of 1.59 mm. The copper tube has a total length of 39.4 cm, of which 35.6 cm are heated. Twelve copper-constantan thermocouples were soldered to the copper tube and used for the surface temperature measurements. The heater wire is a 28-gauge varnished copper wire. The wrapping was done in two separate sections wired in parallel in order to allow for the necessary power requirements without exceeding the amount of available amperage on-board the plane. This permits up to 1000 W of total heat addition to the two-phase mixture for a maximum power flux of 94 kW/m^2. In order to minimize heat losses to the ambient from the heater during operation, 50 mm of a high temperature, ceramic fiber insulation was packed around the heater. In addition, a 20-mm thick blanket of fiberglass insulation was wrapped around the entire structure. The latter was covered with an aluminum sheet metal casing.

At the outlet of the heated test section, another 3.18 mm RTD is used to measure the outlet temperature. This temperature probe is inserted axially into the outlet flow, allowing for complete immersion in the gas–liquid mixture. Interference with the flow pattern was not a consideration here.

All data acquisition functions were handled by a 80486-based PC equipped with 12-bit analog-to-digital conversion using a successive approximation algorithm. A C^{++} computer program was used to acquire the heat transfer data once in every 0.03 sec and to control the air flow rate.

The independent measurements for the heat transfer study were the flow rates of the liquid and gas phases and the heat flux at the surface of the heated test section, while the dependent measurements were the temperatures along the heated test surface and the bulk fluid temperature. During the flights, measurements were collected continuously during the low-gravity portions of each parabola. The data collected were then averaged over time for each parabola. Before the averaging process was performed, three conditions were imposed on the independent variables collected for each data point. These conditions were as follows: (1) the liquid and gas flow rates (or velocities) and heater power will have minimum fluctuations for the duration of data collection, (2) the gravity level was between $\pm$ 0.04 of standard Earth gravity (g_0), and (3) quasi-steady-state conditions for heat transfer measurements existed.

The quasi-steady-state condition addresses the question of whether the time response of the heated test section and the flow is such that steady-state thermal conditions could be possibly achieved given the short duration of reduced gravity on-board the flight aircraft. This was investigated experimentally, first on-ground and later during the course of the first flight campaign in 1991. Details on the above mentioned tests are presented in Section 2.

Once the averaged data were obtained, the 12 surface temperatures $T_{w,i}$ were used to calculate the local convective heat transfer coefficients (h_i) from the equation:

$$(1) \qquad h_i = \frac{E}{\pi \, LD \, (T_{w,i} - T_{b,i})},$$

where E is the total heat input into the test section (W), L is the length of the heated test section (m), and $T_{b,i}$ is the local, average bulk fluid temperature, (°C).

The measured electrical power to the heater and the heated test section surface area were used to calculate the heat flux in Eq. 1. Ground and flight tests showed that the test section outer insulation provided enough thermal resistance such that there was insignificant heat loss to the ambient. This conclusion was reached by a comparison of the heater power output (E) to the heat input to the fluid mixture (E') calculated by:

$$E' = \left(w_L C_{p,L} + w_G C_{p,G}\right)\left(T_{b,out} - T_{b,in}\right), \tag{2}$$

where w_L = the mass flow rate of the liquid, $C_{p,L}$ = the specific heat of the liquid, w_G = the mass flow rate of the gas, $C_{p,G}$ = the specific heat of the gas, $T_{b,out}$ = the inlet bulk temperature of the fluid, and $T_{b,in}$ = the outlet bulk temperature of the fluid.

It should be mentioned that the surface temperature of the heater was maintained well below the boiling point of the liquid, and mass transfer between the liquid and gas was found to be negligible over the short length of the test section.

Returning to Eq. 1, the inner tube wall surface temperature was calculated from the measured outer surface temperature by assuming one-dimensional conduction through a cylindrical tube wall. The gas–liquid mixture temperature could only be measured at the inlet and outlet of the heated test section, so local bulk fluid temperatures were interpolated from the inlet and outlet fluid temperatures using a linear temperature profile.

After determining the local heat transfer coefficients with Eq. 1, an average heat transfer coefficient was calculated based on integrating the local values along the length of the test section. For this integration, the two local heat transfer coefficients near the inlet and the outlet of the heated test section were discarded on account of axial conduction effects. Therefore, the reported average two-phase heat transfer coefficients are based on the remaining 10 local heat transfer measurements which were integrated as follows:

$$h_{TP} = \frac{1}{L - dx} \int_{\frac{dx}{2}}^{L-\frac{dx}{2}} h_i \, dx, \tag{3}$$

where dx = a differential length of the heated test section.

The Sieder-Tate correlations for single-phase laminar and turbulent flows were used to normalize the two-phase heat transfer coefficient h_{TP}, Kays and Perkins (1985) and Kakaç (1987). These correlations may be expressed in terms of the single-phase Nusselt number (Nu_L) as:

$$\text{Laminar}: Nu_L = \frac{h_L D}{k_L} = 1.86(Re_{SL} Pr_L D/L)^{0.33} \left(\frac{\mu_{L,b}}{\mu_{L,w}}\right)^{0.14}. \tag{4}$$

$$\text{Turbulent}: Nu_L = \frac{h_L D}{k_L} = 0.023 Re_{SL}^{0.8} Pr_L^{0.33} \left(\frac{\mu_{L,b}}{\mu_{L,w}}\right)^{0.14}. \tag{5}$$

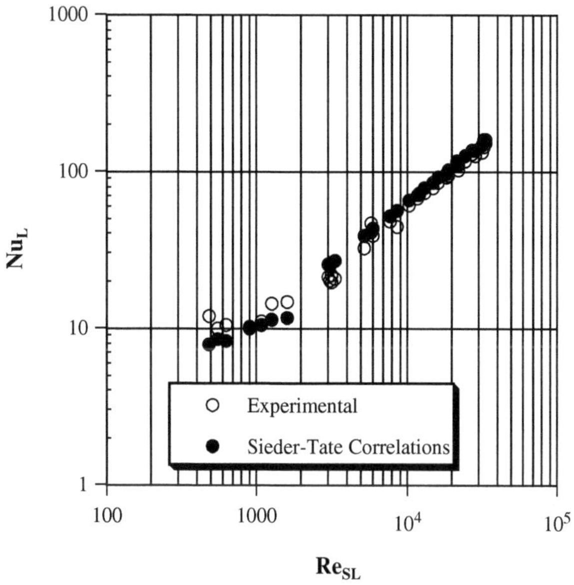

Figure 52. Comparison of single-phase experimental data with Sieder-Tate correlations: laminar and turbulent flows

where k_L is the liquid thermal conductivity, Pr_L is the liquid Prandtl number, and $\mu_{L,b}$ and $\mu_{L,w}$ are the dynamic viscosities of the liquid at bulk fluid and tube wall conditions, respectively. These correlations were used to normalize the data based on a comparison of the correlations with single-phase data collected with the flight apparatus. Results of this comparison are shown in Fig. 52. In this figure, experimental ground data and Sieder-Tate generated Nusselt numbers are plotted as functions of Re_{SL}. It can be seen that the agreement was quite good with a root-mean-square deviation of 9.9% for turbulent flow and 14.9% for laminar flow.

2. TRANSIENT EFFECTS

2.1 Nature of the Transients

Due to the short duration of reduced gravity on-board the KC-135 (20–25 seconds), it is important to consider whether "quasi-steady" heat transfer coefficients for two-phase, convective flow can be measured. Dukler et al. (1988) have concluded from flight experiments that the flow patterns reach a steady-state condition in 1.0–1.2 sec. Thus, as far as flow regimes are concerned a steady condition can be attained. But, what about thermal steady-state conditions? Does the system respond quickly enough to allow for such measurements?

There are two sources of transients during any flight test: one is due to the variation of the gravity field, and the other is due to the changing gas and liquid flow rates and the heater power. Changes in these parameters will affect the five basic measured quantities which are of interest in this study: the flow rates of the liquid and gas phases, the heat flux at the surface of the heated test section, the temperatures along this surface, and the bulk fluid temperature.

Since the heat transfer coefficient is based on the difference between the surface temperature and the fluid temperature, the "quasi-steady-state" condition will be determined based on when this difference approaches a constant value. Although the absolute values of both of these quantities tended to slowly increase throughout the duration of testing, it was found that the difference between them did reach an approximately constant value. Hence, this difference will be used to define "quasi-steady-state" in thermal terms.

An estimate of the transient time of the heated test section was undertaken using empirical two-phase, 1-g, heat transfer correlations with the transient heat exchanger analysis of Kays and London (1984) for flow through an insulated duct with a step-change in inlet fluid temperature. This study indicated a time constant on the order of 20 seconds for the most extreme flow conditions that would be tested on-board the KC-135. A mathematical model of the test section based on the one-dimensional energy equations developed by Zhao and Rezkallah (1993b) indicated a transient response of 10 sec for the flight facility test section.

In order to more thoroughly quantify the transient response and address the concerns raised by the RTD and thermocouple specifications, an experimental investigation to assess the transient was pursued. This investigation involved two processes: gathering ground data to specifically examine transient thermal response of the test section and an assessment of flight data.

2.2 Assessment of the Transients — Ground Data

As stated in the previous section, two factors create a transient situation on-board the experimental flight aircraft: set points and gravity. The first is by far the easiest to contend with because it involves independent variables which are controlled by the experimenter. The gravity level, however, is dictated by the experimental platform.

To minimize the impact of the set point transient, changes to the set points were always made at the completion of a parabola. This allowed a settling time of approximately 60 sec before the next data collection window was opened. Therefore, the time constant that the test section must have in order to reach "quasi-steady-state" conditions after a change in set point is on the order of a minute. The transient response of the heated test section to changes in the set point was determined in the laboratory to see if this was indeed feasible. This study was carried out using two methods. First, the liquid and gas flow rates were set at constant values and the heater power was instantaneously increased. In Fig. 53, a sample of the results for the case of air–water, bubble-slug transition flow with $V_{SL} = 1.2\,\text{m/s}$ and

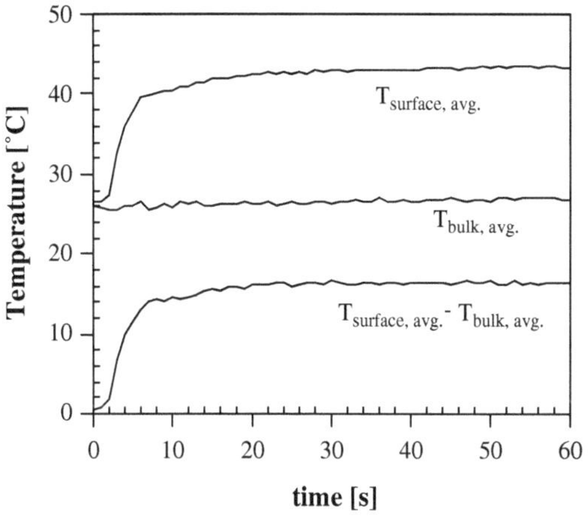

Figure 53. Transient response to a step-change in heat input

$V_{SG} = 0.5\,\text{m/s}$ and an increase in heat input from 0 to 520 W at time = 0 sec are presented. Three temperatures are plotted in the figure: the average surface temperature of the heated test section, the average fluid temperature in the test section, and the difference between the two. After 10 sec the difference between the surface and fluid temperature is within approximately 90% of its "quasi-steady-state" value.

Another approach that was used to assess transient response involved imposing a constant liquid flow rate and heater power on the system, and then increasing the gas flow instantaneously. Air and water were the two fluids tested. The results at $V_{SL} = 1.2, 0.5$, and 0.1 m/s are shown in Figs. 54–56, respectively. In the figures, the difference between the average surface and fluid temperatures, as well as the superficial gas and liquid velocities are plotted for 60 sec after an increase in the gas velocity at time = 0 sec from 0.2 m/s to 10 m/s. The three quantities have been normalized as follows:

$$\Theta = \frac{(T_{s,avg.} - T_{b,avg.}) - (T_{s,avg.} - T_{b,avg.})_{initial}}{(T_{s,avg.} - T_{b,avg.})_{final} - (T_{s,avg.} - T_{b,avg.})_{initial}} \tag{6}$$

$$V^*_{SL} = \frac{V_{SL}}{V_{SL,avg.}} \tag{7}$$

$$V^*_{SG} = \frac{V_{SG} - V_{SG,initial}}{V_{SG,final} - V_{SG,initial}}, \tag{8}$$

where $T_{s,avg.}$ = the average surface temperature along the length of the heated test section, $T_{b,avg.}$ = the mean bulk fluid temperature in the test section, $V_{SL,avg.}$ = the mean liquid superficial velocity, $V_{SG,initial}$ = the gas superficial velocity at the start of a test, and $V_{SG,final}$ = the gas superficial velocity at the end of a test.

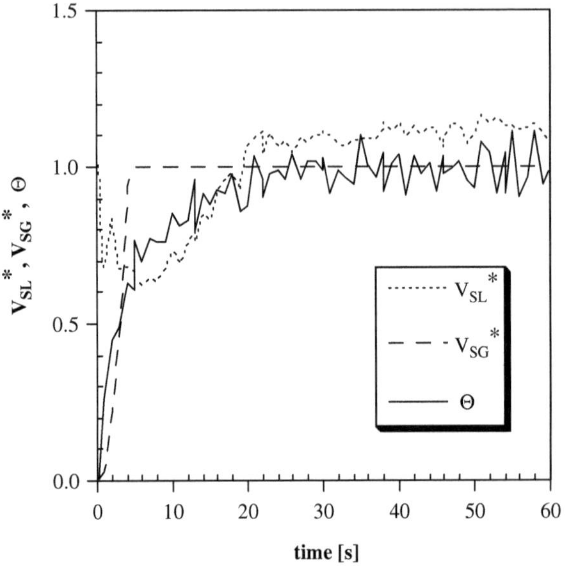

Figure 54. Transient response at $V_{SL} = 1.2\,m/s$, $V_{SG} = 0.2$–$10\,m/s$ – air–water

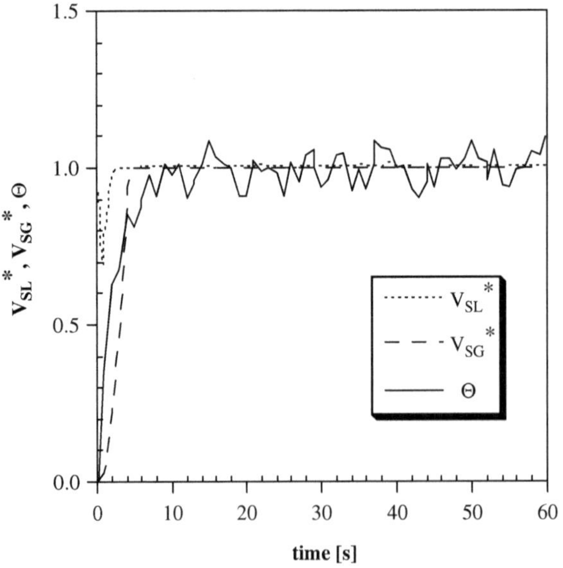

Figure 55. Transient response at $V_{SL} = 0.5\,m/s$, $V_{SG} = 0.2$–$10\,m/s$ – air–water

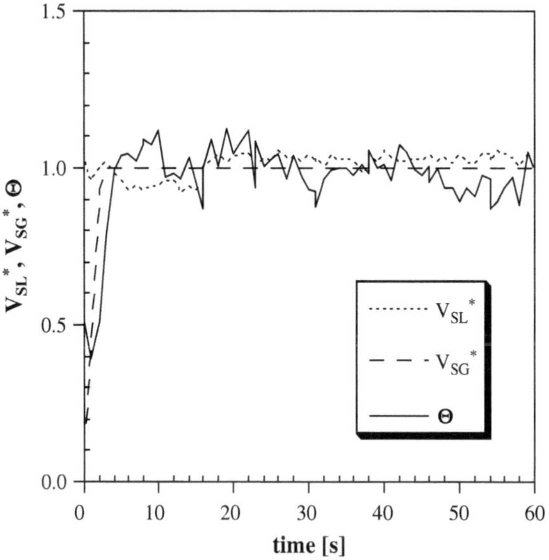

Figure 56. Transient response at $V_{SL} = 0.1\,m/s$, $V_{SG} = 0.2–10\,m/s$ — air–water

In each of the cases in Figs. 54–56, a gas velocity change induces a change in the flow pattern as follows: slug to annular flow for $V_{SL} = 1.2\,m/s$, bubble-slug transition to churn-annular transition flow for $V_{SL} = 0.5\,m/s$, and slug-churn transition to annular flow for $V_{SL} = 0.1\,m/s$. It can be seen that as the liquid velocity increases the time required for steady-state temperatures to be achieved increases. This follows from the fact that the total heat capacity (*W*) becomes larger as the liquid mass flow rate increases. The total heat capacity of the two-phase mixture is defined as:

$$(9) \qquad W = \left(wC_p\right)_L + \left(wC_p\right)_G,$$

where $w =$ the mass flow rate of the liquid or gas and $C_p =$ the specific heat capacity of the liquid or gas.

The unsteady, one-dimensional conservation of energy equation for the fluid in the test section may be written in differential form as:

$$(10) \qquad \mathrm{WL}\frac{\partial T}{\partial z} + \mathrm{AL}\left[\alpha\rho_G C_{p,G} + (1-\alpha)\,\rho_L C_{p,L}\right]\frac{\partial T}{\partial t} = h\pi\,\mathrm{DL}\,(T_S - T)$$

where $\partial T/\partial z =$ the partial derivative of the fluid temperature with respect to the axial direction, z, and $\partial T/\partial t =$ the partial derivative of the fluid temperature with respect to time, t.

The energy equation shows that the thermal response of the heated test section will be slower as the heat capacity of the two-phase mixture is increased. This is provided that the convective heat transfer coefficient (h) does not increase at a faster rate as the flow velocity increases. Based on these experimental results, this does not appear to be the case. The important conclusion from this analysis, however, is the fact that even in the worst case ($V_{SL} = 1.2\,\mathrm{m/s}$), the temperature difference between the surface and the fluid reaches 90% of its "quasi-steady-state" value after 10 sec.

In two-phase flow, showing that the thermal inertia of the heated test section and the fluids inside of the tube is such that severe transient conditions are not affecting measurements is not enough. The residence times of the fluids inside of the test section must also be determined and must be less than the time constant of the heat exchanger. This requirement permits any change in void fraction and flow regime caused by a change in gravity or set point to reach equilibrium.

The residence time or dwell time (τ_d) is defined as:

$$(11) \qquad \tau_d = \frac{L}{V},$$

where V = the superficial velocity of either the liquid or the gas.

The dwell times for the three transient test cases presented above for both the liquid and the gas flows are given in Table 4. The dwell times for both the liquid and the gas are less than the transient time that was observed. Therefore, the liquid inventory inside of the tube should be rearranged in accordance to the new equilibrium void fraction during the relatively short data collection window.

In addition to results for air–water, two-phase, two-component flows, the heat transfer characteristics of air and glycerin/water mixtures with glycerin weight percentages of approximately 50, 59, and 65 will also be reported. Since these mixtures have different thermal and transport properties than pure water, an analysis of the transients for these cases is also required. A summary of the properties of these mixtures for a typical testing temperature of 35 °C is given in Table 5. The properties of the mixtures were determined by direct measurement of the kinematic viscosities of the mixtures in order to verify the concentration of glycerin. Property tables from the Glycerine Producers Association (1973) were used to determine mixture density, thermal conductivity, surface tension, and specific heat. The Prandtl number (Pr) was calculated from its definition:

$$(12) \qquad Pr = \frac{C_p \mu}{k},$$

where μ = the dynamic viscosity, and k = the thermal conductivity.

Table 4. Dwell times for the liquid and gas flows

V_{SL} (m/s)	V_{SG} (m/s)	$\tau_{d,L}$ (s)	$\tau_{d,G}$ (s)
0.10	10.0	3.56	0.036
0.50	10.0	0.71	0.036
1.20	10.0	0.30	0.036

Table 5. Water and glycerol/water properties at 35 °C

Glycerol Wt. Pct. (%)	Density (kg/m^3)	Dynamic Viscosity (Ns/m^2)	Thermal Conductivity (W/m K)	Surface Tension (dynes/cm)	Specific Heat (J/kg K)	Prandtl Number
0	994	0.00072	0.63	72.1	4177.9	4.84
50	1120	0.0036	0.41	67.1	3459.4	30.9
59	1147	0.0056	0.38	66.4	3330.3	50.0
65	1161	0.0080	0.36	65.9	3250.0	72.8

It should be mentioned that all of the thermal and transport properties for the other fluids used in this study were evaluated at bulk fluid conditions based on tabulated properties measured at 1-g conditions, Incropera and DeWitt (1981) and ASHRAE (1981). The effect of a microgravity environment on fluid properties such as surface tension, density, viscosity, and thermal conductivity is not completely known at this time, although some research has been published (e.g., Greger and Rath (1995)) suggesting that it may be significant.

An examination of the conservation of energy in Eq. 10 indicates that in order to determine the change in transient response of the test section with these new mixtures of fluids, the important quantity to compare with pure water is the product of the density and the specific heat. Comparing ρC_p for water to a 65% glycerin/water mixture, it can be seen that the glycerin mixture has a ρC_p approximately 9% lower than pure water. This indicates that the thermal response of the test section with a glycerin mixture compared to pure water should be quicker, depending on the range of flow rates that will be used during testing and the resulting effect on the convective coefficient. Clearly, it is prudent to experimentally examine this response for the range of testing conditions.

The results of a transient study for glycerin/water mixtures are given in Figs. 57–60 for glycerin percentages of 50 and 59 at the lowest and the highest liquid velocities that were encountered during testing. In the figures, the same non-dimensionalization scheme has been implemented for velocity and temperature as presented in Eqs. 6–8. A 30-sec window is shown in all of the figures.

In Fig. 57, a liquid velocity of 0.40 m/s was set as the steady-state value for the data shown. The gas velocity was then increased at time = 0 sec from 0.23 to 10.0 m/s. This changed the flow regime from bubble-slug transition to slug-annular transition flow. Within 10 sec the non-dimensional temperature approximately reaches its "quasi-steady-state" value. It is also apparent that there was fluctuation in the liquid flow as the gas flow was increased before a final value was achieved. This transient obviously also affected the normalized temperature Θ (defined in Eq. 6) and the actual transient response is probably less than 10 sec.

Fig. 58 shows the transient response as the gas velocity was increased from 0.23 to 10.0 m/s at $V_{SL} = 1.2\,\text{m/s}$. This corresponded to a flow regime transition of bubble to churn-annular transition. In this figure, the transient time appears to be on the order of ten seconds again, although it is somewhat difficult to judge based

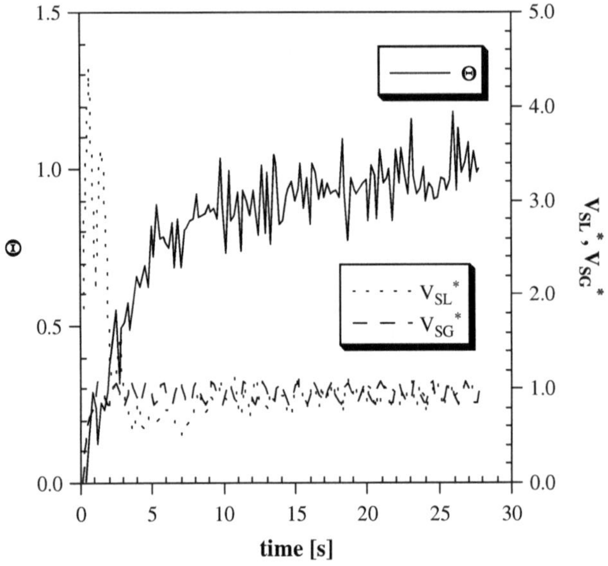

Figure 57. Transient response at $V_{SL} = 0.40\,m/s$, $V_{SG} = 0.23–10.0\,m/s$ — air–50% glycerin/water

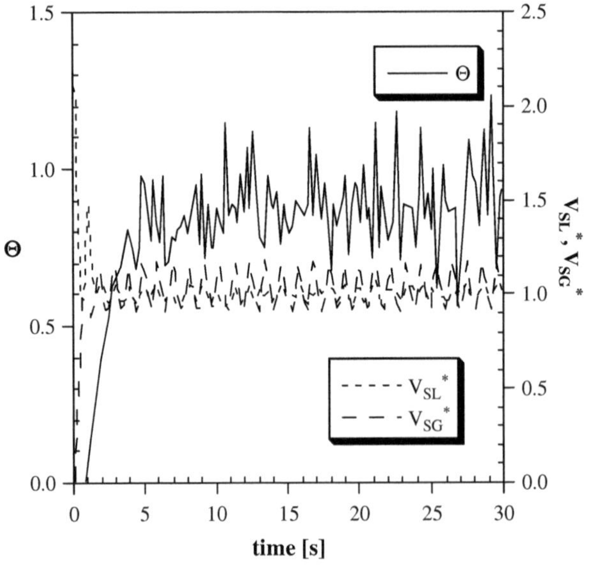

Figure 58. Transient response at $V_{SL} = 1.2\,m/s$, $V_{SG} = 0.23–10.0\,m/s$ — air–50% glycerin/water

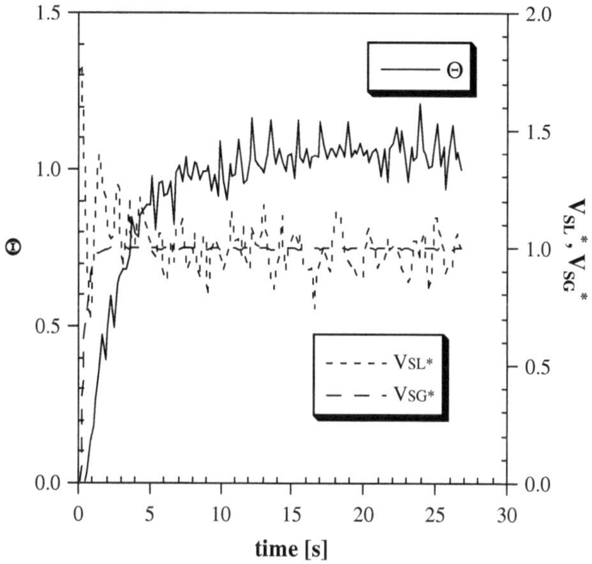

Figure 59. Transient response at $V_{SL} = 0.55\,m/s$, $V_{SG} = 0.20$–$9.9\,m/s$ — air–59% glycerin/water

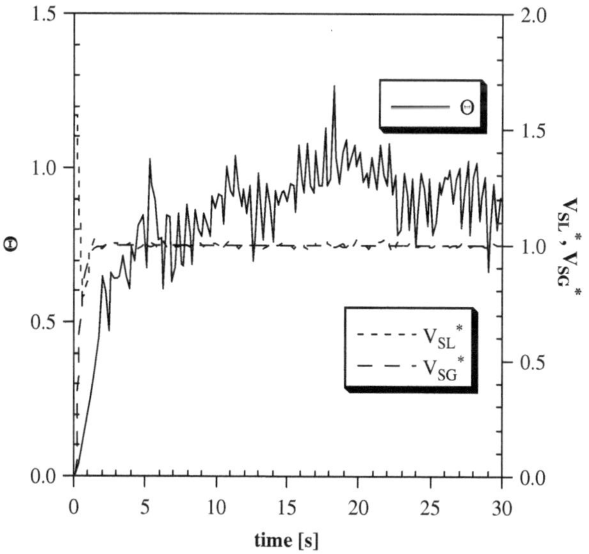

Figure 60. Transient response at $V_{SL} = 1.6\,m/s$, $V_{SG} = 0.24$–$10.0\,m/s$ — air–59% glycerin/water

on the large fluctuation in Θ at this flow rate. The settling time of the liquid flow rate is again fairly long as it was back in Fig. 58.

The weight percentage of glycerin is approximately 59% for the results presented in Figs. 59 and 60. For the results in Fig. 59, the liquid superficial velocity was set at 0.55 m/s and the gas velocity was increased to 9.9 m/s from 0.20 m/s. A flow pattern transition of slug to slug-annular transition was observed to occur. In the figure, the liquid flow rate appears to be much more stable than was evident in Figs. 57 and 58. The transient time is also less. Approximately 7 sec pass before Θ reaches its "quasi-steady" value of unity.

Finally, in Fig. 60 data for a liquid superficial velocity of 1.6 m/s with a gas velocity increase from 0.24 to 10.0 m/s at time = 0 sec and a flow regime change of bubble-slug to annular flow are shown. The response time is slightly longer than for the lower liquid flow rate of Fig. 59. Approximately 10 sec are required in this case.

The ground experiments described above for air–water and air–glycerin/water mixtures showed that a set point change (heat input/gas flow rate change) would take up to approximately 10 sec to reach the thermal "quasi-steady-state" condition for the range of liquid flow rates that were covered in the experimental test matrix. In addition, the tests in which the gas flow rate was instantaneously stepped-up also give an indication of the transient response due to a change in gravity. Varying the gas flow rate was shown to change the flow regime. This along with the increase in actual liquid velocity will change the convective heat transfer coefficient in the tube. This sort of change thus simulates what gravity can be expected to do to the flow, i.e., change the flow pattern and the heat transfer. So, the ground tests discussed above give a good indication as to the transient response of the system. However, actual flight data still needs to be considered.

2.3 Assessment of the Transients — Flight Data

During the flights of October 1991, data were gathered continuously throughout the duration of the flight for both the microgravity and hypergravity ($g/g_0 \approx 1.8$) portions. Thus, the temperature response of the heated test section as the gravity varied in the KC-135 can be examined and a clear indication of its response to gravity obtained. However, it should be mentioned that heat transfer coefficients for the hypergravity periods were not calculated. The temperatures considered were the surface temperatures (at the inlet, the outlet, and approximately midway between the two), and the gas–liquid mixture outlet temperature. Fig. 61 shows these temperatures and the ratio of on-board gravity to normal Earth gravity as a function of time over three parabolas (October 3, morning flight, Parabolas 32–34). In the plot, $T_{s,out}$ is the outlet surface temperature, $T_{s,in}$ is the surface temperature at the inlet, $T_{s,mid}$ is the surface temperature at approximately the middle of the test section, and T_{out} is the outlet temperature of the two-phase mixture. The water velocity during all three parabolas was approximately 3.5 m/s and the air velocity was increased, step-wise, from 0.3 to 0.4 m/s and then to 0.6 m/s. The heat flux in

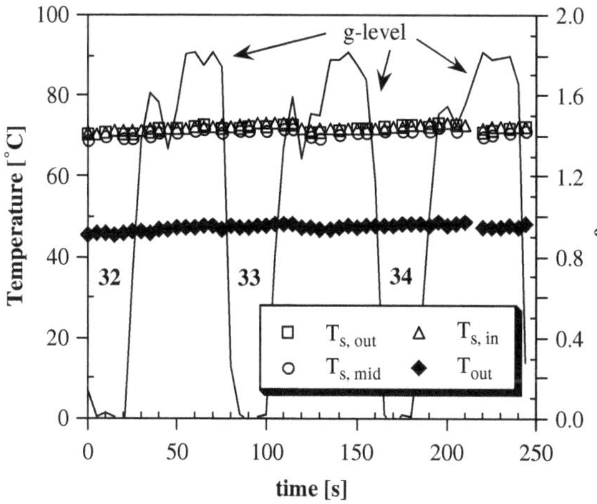

Figure 61. Transient response to gravity for air–water at $V_{SL} = 3.5\,m/s$ — bubble flow

the heated test section was fixed at $93\,kW/m^2$ for all three parabolas. The figure shows essentially no variation in any of the surface temperatures as the gravity field changes from 1.8-g_0 to approximately 0.01-g_0. Also, the outlet fluid temperature does not seem to be affected by the gravity field. This indicates that for these particular flow conditions the convective heat transfer coefficient is not significantly affected by gravity and little information about transient response is available.

The flow conditions described above coincide with a flow regime of bubble flow for all three parabolas. This can be seen in Fig. 62. In this figure, digitized images from the Sony 8 mm video camera recording of the flow through the visual, vertical test section are presented for the microgravity and hypergravity portions of Parabola 32. In the figure, three images taken approximately 3 to 5 sec apart in the microgravity portion of the parabola are shown in chronological order from left to right. To the immediate right of these figures, three images from the preceding hypergravity portion are shown. Bubble flow exists during both hypergravity and microgravity; thus, no flow regime transition occurred.

In Fig. 63, four microgravity and three hypergravity time periods are shown. These periods correspond with Parabolas 23–26 in the morning flight of October 3. For these parabolas, the microgravity periods have flow regimes of bubble, bubble-slug transition, and slug, respectively. The bubble-slug transition and slug flows are shown in Figs. 64 and 65, where the images for the microgravity portions of Parabolas 24 and 25 are presented as well as images from the hypergravity conditions immediately preceding. In Fig. 64, the bubble-slug flow present at μ-g in Parabola 24 is more slug flow-like at the higher gravity levels. However, it is still classified as bubble-slug transition flow. In Fig. 65, slug flow is present at both gravity levels. In either situation, the slug flow consists of very poorly defined elongated gas bubbles with a large amount of turbulence at the gas–liquid interface.

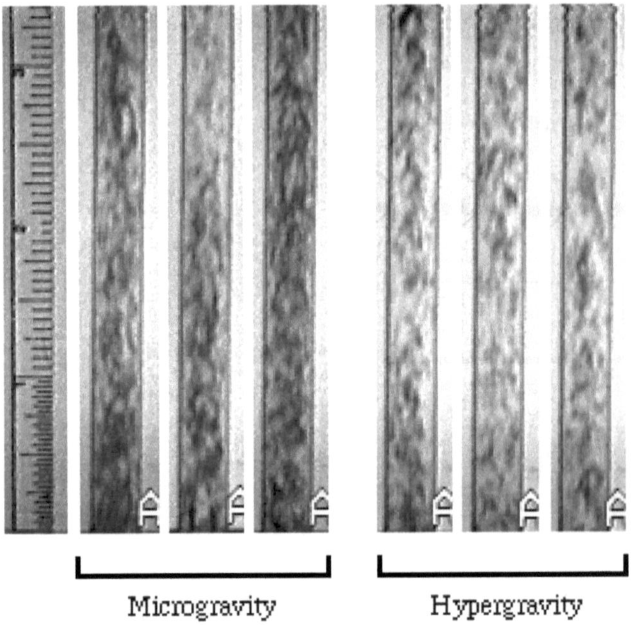

Figure 62. Video images of the bubble flow regime — micro- and hypergravity portions of Parabola 32

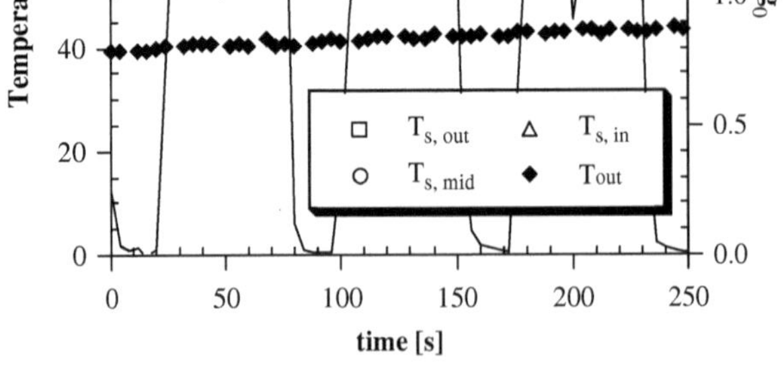

Figure 63. Transient response to gravity for air–water at $V_{SL} = 2.3\,m/s$ – bubble, bubble-slug transition, and slug flow

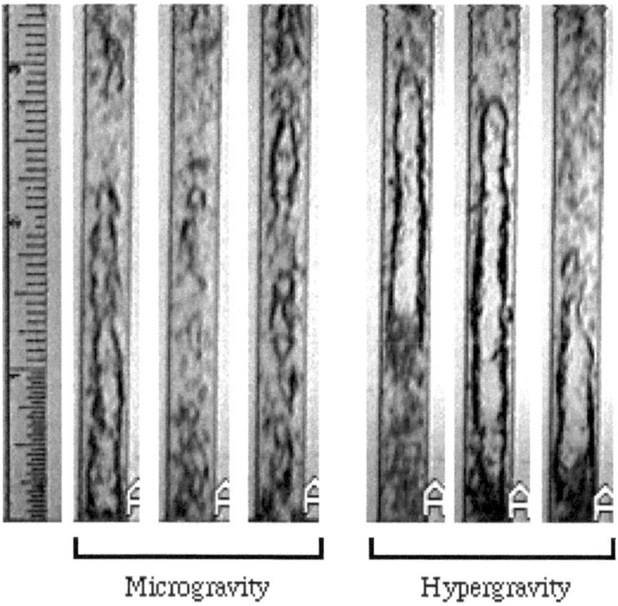

Figure 64. Video images of bubble-slug transition flow regime — micro- and hypergravity portions of Parabola 24

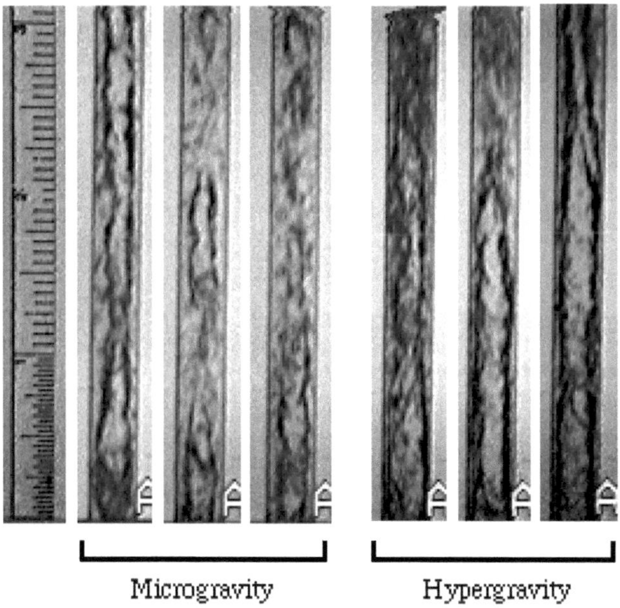

Figure 65. Video images of slug flow regime — micro- and hypergravity portions of Parabola 25

Referring to Fig. 63, the superficial water velocity is approximately 2.3 m/s for all four microgravity periods and the superficial air velocity starts at 0.5 m/s for Parabola 23 and then increases to 0.6 m/s, 0.8 m/s, and finally to 2.0 m/s during Parabola 26. Heat flux for all four parabolas is approximately $93\,kW/m^2$. None of the temperatures are responding to the change in the gravity field and, therefore, the heat transfer coefficients are also not varying measurably.

Fig. 66 shows the temperature response of the heated test section for the October 3 afternoon flight over Parabolas 30–32. The V_{SL} for these cases is 0.2 m/s for Parabola 30 and 0.3 m/s for Parabolas 31 and 32. V_{SG} starts at 10.0 m/s for Parabola 30 then decreases to 0.2 m/s for Parabola 31 and ends up at 0.26 m/s during Parabola 32. Heat flux is $37.5\,kW/m^2$ for all three parabolas. The flow regime proceeds from annular during the first microgravity period to slug for the next two periods.

The annular flow regime illustrated in Fig. 67 consists of a very wavy interface between the gas and liquid with a significant amount of liquid entrainment in the gas core. The hypergravity flow is in a slug-churn transition flow pattern. Churn flow is characterized by its oscillatory nature with an up-and-down pulsating motion caused by a "falling-film" of liquid at the inner tube surface as described in Chapter 2. In the slug-churn transition region, distinct elongated gas bubbles are still evident in addition to the distinct oscillating frequency of churn flow. No change in flow regime occurred during the elevated gravity periods before Parabolas 31 and 32. Slug flow was present during both of these hypergravity periods.

In Fig. 66, it is difficult to determine if the temperatures are responding to gravity changes over the first two parabolas because of the large reduction in the air flow between these two as well as the change in liquid flow. However, it does appear that there is a slight change in temperature with gravity for the two microgravity periods in slug flow. Small increases in the surface temperatures, $T_{s,out}$ and $T_{s,mid}$,

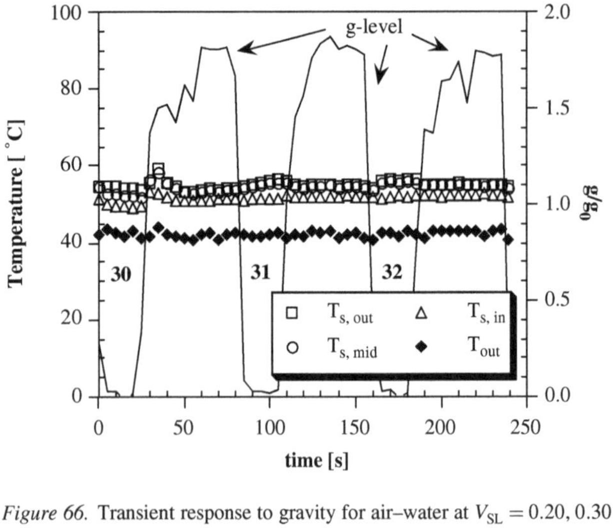

Figure 66. Transient response to gravity for air–water at $V_{SL} = 0.20, 0.30$ m/s — annular and slug flow

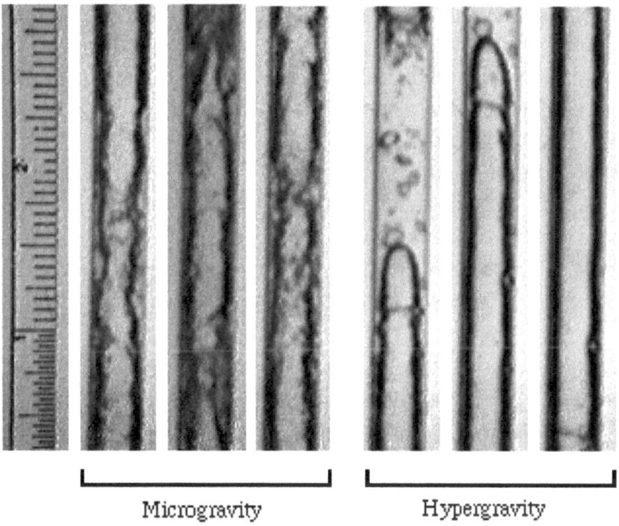

Figure 67. Video images of annular and slug-churn transition flow regimes — micro- and hypergravity portions of Parabola 30

are evident. These suggest that the heat transfer coefficients are affected by gravity in slug flow. Some information about the thermal response of the test section may also be gleaned from Parabola 30.

Fig. 68 shows a close-up of Parabola 30. In this close-up, it can be seen that there is a change of temperature profile with gravity at the midway and outlet

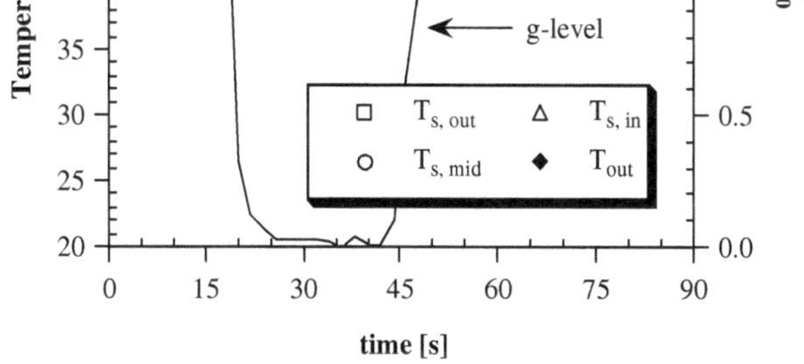

Figure 68. Close-up of Parabola 30, October 3, 1991 — air–water

points on the surface of the heated test section. In the plot, it does appear that the surface temperatures approach a crest within the low-gravity period aboard the aircraft. The relatively small μ-g window aboard the KC-135 will allow useful heat transfer coefficients to be determined, but several more flow conditions need to be examined before a conclusion may be drawn.

The next parabolas to be examined are Parabolas 14–16 for the afternoon flight of October 3. A plot of heater surface temperature, fluid temperature, and gravity field versus time is shown for these three parabolas in Fig. 69. Superficial liquid velocity for all three parabolas is 0.13 m/s. Superficial gas velocity increases from 0.8 to 1.0 m/s and then, finally, to 2.0 m/s. The heat flux over the heated test section is $28\,\mathrm{kW/m^2}$.

The flow regime progresses from slug during Parabola 14 to slug-annular transition flow during the microgravity portions of Parabolas 15 and 16. In Fig. 70, for the slug-annular transition flow of Parabola 15, there is a relatively smooth air–water interface over most of the tube length in each image. However, in the liquid between gas pockets, severe turbulence exists in the flow with high mixing between the two components. The hypergravity video images in Fig. 70 correspond to the slug-churn transition flow mentioned earlier. Decreasing gravity has produced a flow regime transition in this case.

In Fig. 69, it is very clear that the surface temperatures are responding to the short duration of μ-g. There is a marked increase in the outlet and midway surface temperature and a decrease in the inlet surface temperature. A trend in the outlet fluid temperature is much more difficult to discern. These results indicate that the slug and slug-annular flow regimes are sensitive to the gravity field at these relatively low flow rates. In addition, this gravitational effect can be measured

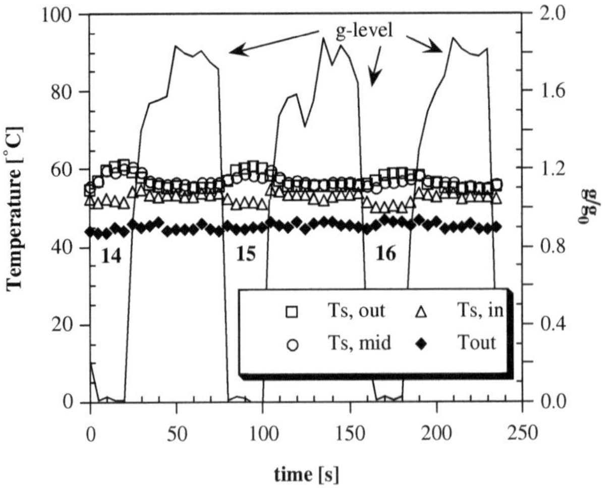

Figure 69. Transient response to gravity for air–water at $V_{SL} = 0.13\,\mathrm{m/s}$ — slug and slug-annular transition flow

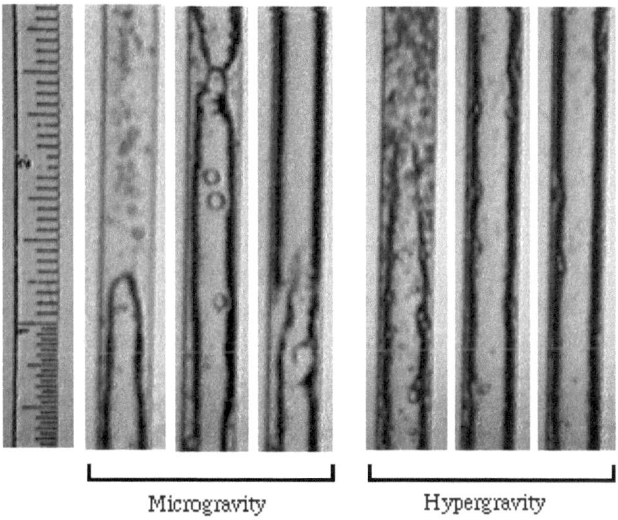

Figure 70. Video images of slug-annular and slug-churn transition flow regimes — micro- and hyper-gravity portions of Parabola 15

during the short duration of time dealt with here. In the figure, "quasi-steady-state" is approached in approximately half the μ-g period.

Fig. 71 shows the temperature response over Parabolas 36–38 of the afternoon flight of October 3. For these parabolas, V_{SL} is 0.3 m/s and V_{SG} increases from

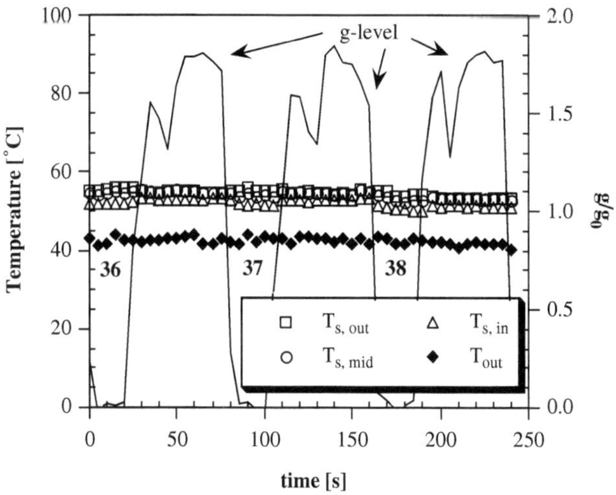

Figure 71. Transient response to gravity for air–water at $V_{SL} = 0.30\,\mathrm{m/s}$ — slug-annular transition and annular flow

2.0 to 4.0 m/s and then to 6.0 m/s. Heat flux is approximately 37.2 kW/m^2. Slug-annular transition flow is occurring during the microgravity portion of Parabola 36. The flow regime then changes to annular flow for Parabolas 37 and 38. The annular flow for these parabolas involves a relatively smooth interfacial region between gas and liquid with occasional "necking" of the gas core accompanied by substantial turbulence as illustrated by the microgravity video image of the flow during Parabola 38 presented in Fig. 72. In the hypergravity time period immediately preceding Parabolas 36 and 37, the flow regime is slug-churn transition. Before Parabola 38, churn-annular transition flow exists as shown in the three video images on the right-hand side of Fig. 72. As with slug-churn transition flow, the flow configuration is also characterized by oscillatory motion. However, instead of the distinct elongated bubbles associated with slug flow, in this situation an annular flow arrangement exists with a liquid film at the inner wall of the tube.

Referring to the temperature profiles in Fig. 71, there is a surface temperature response to the gravity field change. It is not as pronounced as the change that was apparent in the slug flow regime parabolas shown in Fig. 69, however, a change is present and indicates that gravitational acceleration has an influence on heat transfer during the annular flow regime. Although it is somewhat difficult to tell, given the small change in the surface temperatures, it does appear that the change has been more or less completed during the μ-g period.

The flight data from the February 1993 campaign in which the three glycerin/water mixtures were tested were also scrutinized for response time with a change in gravity.

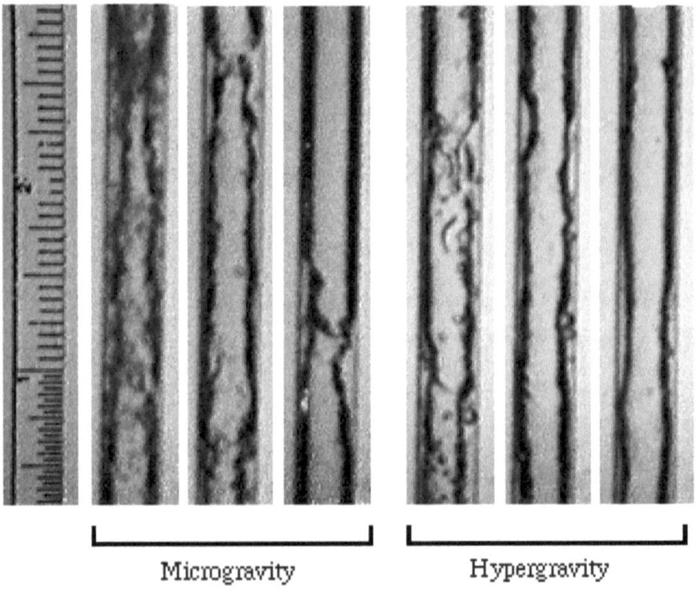

Figure 72. Video images of annular and churn-annular flow regimes — micro- and hypergravity portions of Parabola 38

Examples of the results for a low liquid velocity and a high liquid velocity set point are presented in Figs. 73–75 with the highest percentage of glycerin tested, 65%. In Fig. 73, temperature and gravity profiles of Parabolas 4 through 6 for the flight of February 27, 1993, are plotted as functions of time of day (h is the hour, m is the minute, and s is the second). As with the air–water data, three surface temperatures and the outlet fluid temperature are given along with the gravity level normalized

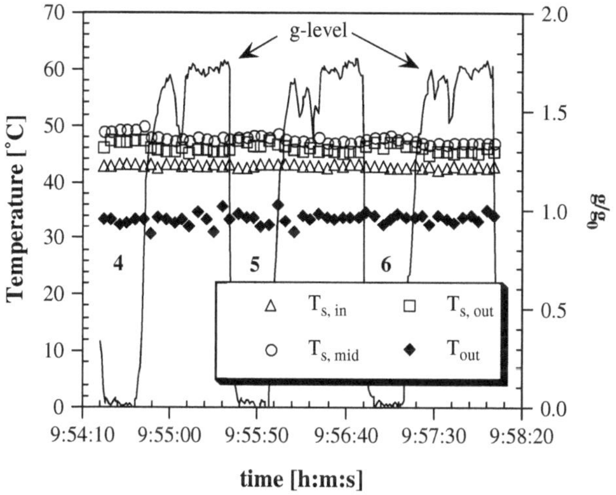

Figure 73. Transient response to gravity for air–65% glycerin/water at $V_{SL} = 0.8\,m/s$ — slug and slug-annular transition flow

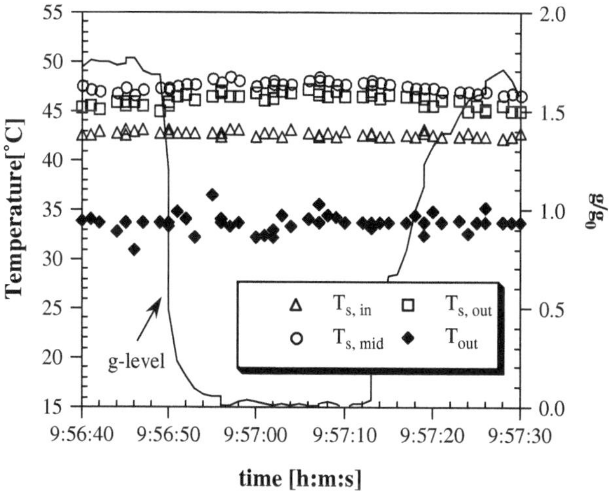

Figure 74. Close-up of Parabola 6, February 27, 1993 — air–65% glycerin/water

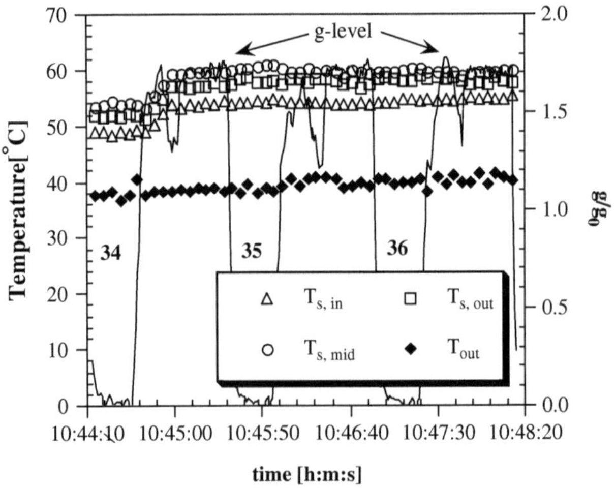

Figure 75. Transient response to gravity for air–65% glycerin/water at $V_{SL} = 1.9\,\mathrm{m/s}$ — slug flow

with respect to standard Earth gravity. The three surface temperatures are at the inlet ($T_{s,in}$), near the midpoint of the test section ($T_{s,mid}$), and at the outlet ($T_{s,out}$). The liquid superficial velocity for all three parabolas was set at 0.8 m/s, as the gas velocity was increased from 2.3 m/s for Parabola 4 to 3.4 m/s during Parabola 5, and then finally to 5.8 m/s for Parabola 6. Slug flow was evident in Parabolas 4 and 5, slug-annular in Parabola 6. In the figure, there is a response in the surface temperature to the gravity change at the midpoint and at the outlet. It is small but is still apparent. The time for this change to reach a crest is also within the 20-sec window as shown in the close-up of Parabola 6 in Fig. 74. Once again, 10 sec is the apparent time duration for the transient.

Finally, in Fig. 75 a higher liquid flow rate is examined for its transient response. This time the liquid superficial velocity is approximately 1.9 m/s. Data at this liquid velocity are shown since it is the highest that was tested with the glycerin/water solutions. Parabolas 34 through 36 of the same flight on February 27, 1993, are presented. For these plots, the gas velocity was stepped-up from 1.6 to 2.4 m/s and then, finally, to 3.6 m/s for Parabola 36. Slug flow was present in each parabola. A large adjustment in the temperature occurs just after Parabola 34 due to the change in the gas flow and the heater power. This adjustment reaches a crest after approximately 20 sec while gravity is at its peak. The sharp reduction of gravity then results in a small increase in the surface temperatures. This increase seems to crest near the end of the μ-g window. This same pattern is repeated in Parabola 36. As was expected based on the ground transient results, there is not a major difference between the air–water and air–glycerin/water thermal response.

Based on the findings from the numerical models, the ground tests, and the flight tests, it was decided to use only the data collected during the last 8 to 10 sec of the microgravity portions of each parabola. This provides a large enough pool of data from each parabola to calculate a valid average with respect to time and also minimizes the impact of the thermal transient on the reported heat transfer measurements. Although the actual transient response will, of course, vary with the liquid and gas flow rates, the flow regime, void fraction, fluid properties, and test section properties, this approach was adopted for all data points collected because it appears to cover the "worst-case" scenarios and permits a uniform treatment of all the data to be presented.

3. EXPERIMENTAL RESULTS

3.1 Average Heat Transfer Coefficients – Water–Air

The results of a comparison of two-phase heat transfer coefficients gathered in-flight and on-ground for 1994 air–water data are shown in Figs. 76–79. In the figures, the two-phase Nusselt number (Nu_{TP}) measured on-ground and in-flight at approximately the same liquid velocities using the same flight apparatus are plotted as functions of pseudo-void fraction (α'). The pseudo-void fraction is calculated based on the liquid and gas superficial velocities:

$$\alpha' = \frac{V_{SG}}{V_{SL}}. \qquad (13)$$

For the results to be presented here, the Nusselt number is defined as:

$$Nu_{TP} = \frac{h_{TP}D}{k_L}, \qquad (14)$$

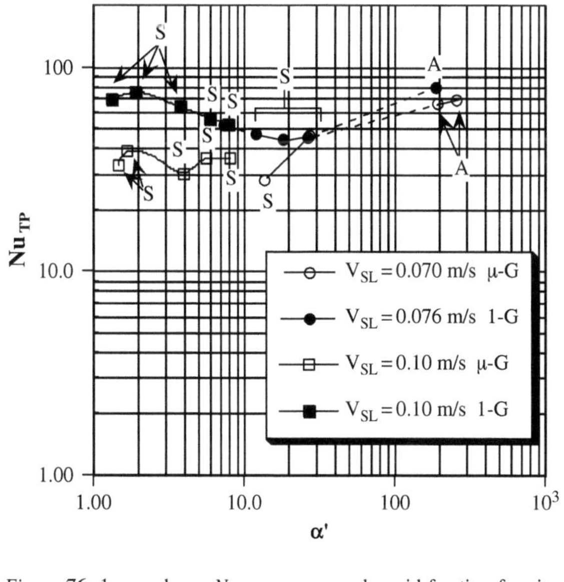

Figure 76. 1-*g* and μ-*g* Nu_{TP} versus pseudo-void fraction for air–water — $V_{SL} \approx 0.07$ and 0.10 m/s

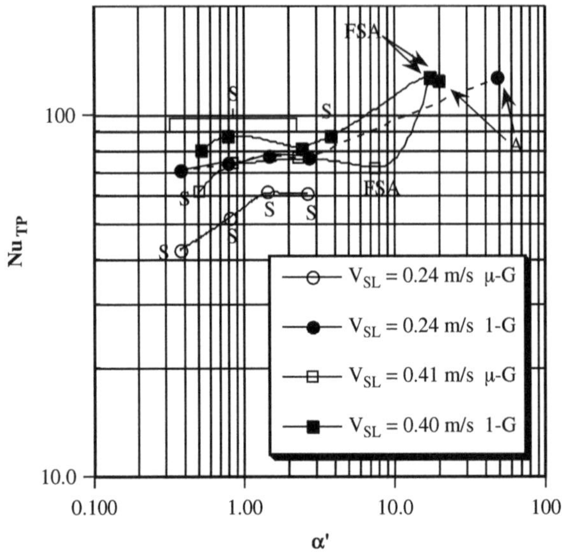

Figure 77. 1-g and μ-g Nu_{TP} versus pseudo-void fraction for air–water – $V_{SL} \approx 0.24$ and 0.4 m/s

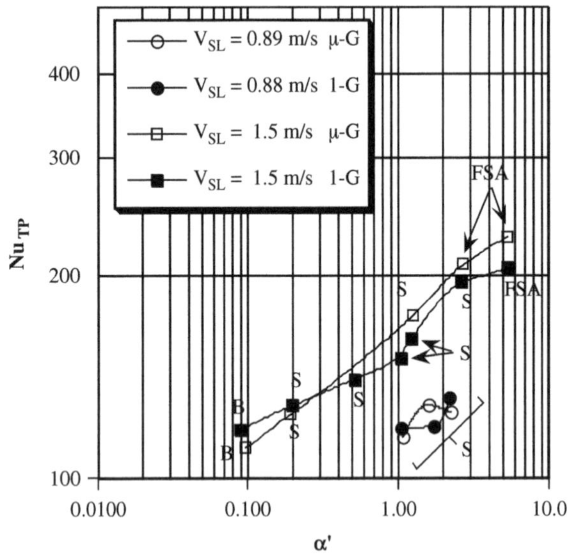

Figure 78. 1-g and μ-g Nu_{TP} versus pseudo-void fraction for air–water – $V_{SL} \approx 0.9$ and 1.5 m/s

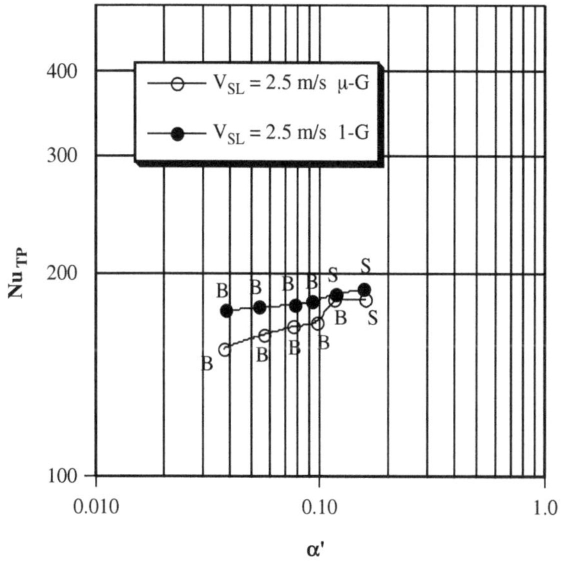

Figure 79. 1-g and μ-g Nu_{TP} versus pseudo-void fraction for air–water — $V_{SL} \approx 2.5\,\mathrm{m/s}$

where h_{TP} = the average, two-phase convective heat transfer coefficient and k_L = the thermal conductivity of the liquid-phase. The superficial liquid Reynolds number is

$$(15) \qquad Re_{SL} = \frac{\rho_L V_{SL} D}{\mu_L}$$

Data at superficial liquid velocities of approximately 0.07, 0.10, 0.24, 0.41, 0.89, 1.5, and 2.5 m/s with V_{SG} varying from 0.1 to 26 m/s are presented. The flow regime for each data point is also marked. The abbreviations for the various flow regimes are as follows: B – bubble, S – slug, FSA – frothy slug-annular, and A – annular.

In Fig. 76, comparisons of the Nusselt numbers under 1-g and μ-g conditions for liquid superficial velocities of approximately 0.07 and 0.10 m/s ($Re_{SL} \approx 840$ and 1200, respectively) are shown, as the gas velocity is increased from 0.6 to 18 m/s at $V_{SL} = 0.07\,\mathrm{m/s}$ and from 0.14 to 0.8 m/s at $V_{SL} = 0.10\,\mathrm{m/s}$. For the $V_{SL} = 0.07\,\mathrm{m/s}$ data, the velocity was maintained to within 3.9% of this value at both μ-g and 1-g conditions. For the $V_{SL} = 0.10\,\mathrm{m/s}$ data, the velocity was maintained to within 7.6% of the average value given. It can be seen that for α' less than approximately 20.0 the 1-g Nusselt numbers are almost twice their values under μ-g conditions. The two groups of data then converge, and above $\alpha' = 20.0$ the 1-g and μ-g Nu_{TP} are nearly the same. It can also be seen in the figure that as the gas velocity increases in the slug flow regime the Nusselt numbers actually become smaller until annular flow is reached. This is especially apparent for the 1-g data as has been found by other researchers, e.g., Collier (1972) and Rezkallah and Sims (1989). This trend can also be seen in the μ-g data at $V_{SL} = 0.10\,\mathrm{m/s}$, but is not apparent for the $V_{SL} = 0.07\,m/s$ data. This may be due to the scarcity of data points above $\alpha' = 20$. Dashed-lines are used in the figure for both the 1-*g* and μ-*g* data points at $V_{SL} = 0.07\,m/s$ to emphasize the lack of sufficient information in this area of the plot.

Moving to higher liquid velocities in Fig. 77, with $V_{SL} = 0.24\,m/s$ ($Re_{SL} \approx 2900$) and a superficial gas velocity ranging from 0.1 to 12.0 m/s the 1-*g* coefficients are once again higher than the μ-*g* for $\alpha' < 3.0$. For these data, the liquid velocity was maintained within 3.3% of the average. The difference between the 1-g and μ-g heat transfer coefficients declines from 40% to approximately 28% as the pseudo-void fraction increases from 0.4 to 3.0. Above $\alpha' = 3.0$, only ground data were obtained, so a direct comparison is not possible. As the liquid superficial velocity is increased to 0.40 m/s ($Re_{SL} \approx 4900$), the differences between 1-*g* and μ-*g* Nusselt numbers are not as pronounced. The fluctuation in velocity was also slightly more severe, up to 7.4% for the μ-*g* data and 5.4% under 1-*g*. For $\alpha' = 0.5$ the 1-*g* results are 23% greater than the μ-*g* results. This difference decreases to practically zero at $\alpha' = 2.5$. Above 2.5, the μ-*g* Nusselt number decreases to approximately 70 at $\alpha' = 7.5$, and then the final data point surpasses 100 at a pseudo-void fraction of 17.0 as shown in the figure. The 1-*g* values, however, continue to increase in a steady fashion rising above 100 for slug-annular flow as α' increases.

Fig. 78 shows the Nusselt numbers for much higher liquid flow rates. In this figure, the liquid superficial velocities are approximately 0.9 and 1.5 m/s ($Re_{SL} \approx$ 11,000 and 19,000, respectively) and the gas velocities range from 1.0 to 2.0, and 0.1 to 8.2 m/s, respectively. Superficial liquid velocity was maintained to within 6.2% of the average at $V_{SL} \approx 0.9\,m/s$ and 3.9% at $V_{SL} = 1.5\,m/s$. At these higher liquid velocities, the difference between 1-*g* and μ-*g* heat transfer is considerably less pronounced. For $V_{SL} \approx 0.90\,m/s$, the Nusselt numbers for 1-*g* and μ-*g* are within 7%. Given the uncertainty in the measurements (which will be addressed in Section (4), this may not be a significant difference. At the lowest void fractions for $V_{SL} = 1.5\,m/s$ the Nu_{TP} at 1-*g* is still larger than its corresponding μ-*g* value, but the difference is again within 5%. As α' is raised to 2.0 and the slug flow regime changes to frothy slug-annular, the μ-*g* coefficients actually become larger than the 1-*g* ones by approximately 10%. This is approaching the measurement uncertainty, but the reduction in 1-*g* heat transfer may be significant in terms of the influence of flow reversal in slug flow which was very evident in the 1-*g* results presented in Fig. 76.

Finally, in Fig. 79 the results for a liquid superficial velocity of 2.5 m/s ($Re_{SL} \approx$ 34,200) are presented as V_{SG} increases from 0.09 to 0.4 m/s. The deviation from the average velocity for both sets of data is less than 1%. For these data points, which are all in the bubble flow regime, the 1-*g* Nusselt numbers are approximately 12% higher than the μ-*g* ones at the lowest pseudo-void fraction examined. As the gas flow increases, this difference becomes smaller until at the highest α' point the Nu_{TP} for 1-*g* and μ-*g* are within approximately 3%. Given the very minimal fluctuation in the liquid velocity at this flow rate, the effect of gravity on the heat transfer coefficients in the bubble flow regime for $\alpha' < 0.1$ is real. It can be mechanistically explained by the fact that at these low gas flow rates the effect of buoyancy forces on the gas bubbles at 1-g will have a profound effect on turbulence intensity in the liquid. However, as the gas velocity is increased buoyancy forces will become less important. Again, the trend of μ-g heat transfer coefficients increasing at a greater rate than 1-g values as void fraction becomes larger which was seen for the lower liquid flow rates is also apparent.

It can be seen in Figs. 76 through 79 that at low liquid velocities and void fractions, 1-g heat transfer data are greater than μ-g data. As the liquid velocity is increased this was also found to be the case, but only at very low void fractions in the bubble flow regime. As the void fraction is increased (increasing gas velocity), the 1-g and μ-g data points approach one another until the two are nearly identical (μ-g marginally greater than 1-g by 10–15%). This suggests that the difference between 1-g and μ-g behavior is flow regime dependent (with all the hydrodynamic and thermal variation that a change in flow regime entails) as was shown in Section 2 when the surface and fluid temperatures were plotted as a function of time as the gravitational acceleration on-board the aircraft varied between 0.01-g_0 and 1.8-g_0. In Figs. 53, 66, 69 and 71, large temperature changes as the gravity varied were only seen for slug, slug-annular, and annular flows at low liquid flow rate conditions. This behavior is also evident for the low liquid velocity cases in Figs. 76 and 77 where slug flow registers the largest differences between 1-g and μ-g with the 1-g Nusselt numbers being higher. For the moderate liquid flow cases in Fig. 78, the bubble flow points are very similar, while as the flow pattern proceeds to slug and slug-annular and finally annular, the μ-g heat transfer data approach and become nearly the same. Bubble flow in Fig. 79 again shows a difference of, at most, 12%.

Preliminary results from the void fraction measurements that were taken during the 1994 flight campaign for air–water mixtures, Elkow (1996), may offer a clue as to the reasons for the heat transfer findings discussed above. These results indicate that the void fractions for low gas velocity slug flows ($\alpha' < 5.0$) are higher under μ-g conditions as compared with the 1-g data by, on average, 17.1%. This difference in actual void fraction would indicate larger elongated gas bubbles at μ-g (shorter liquid-slugs) and, hence, lower heat transfer coefficients. However, the uncertainty in the void fraction measurements, on the order of 10%, makes these findings less than clear-cut. Also, differences in void fraction for bubble, slug-annular, and annular flow regimes were even less pronounced, so these preliminary measurements in these regions are of minimal value.

Vijay et al. (1982) related the enhancement of two-phase heat transfer compared with single-phase heat transfer directly to the increase in two-phase frictional pressure drop. The heat transfer results presented in Figs. 76 to 79 agree with this conjecture to a certain extent. The pressure gradient results for the new data reported by Zhao and Rezkallah (1995b) and summarized in Chapter 4 indicated that the frictional pressure drops at μ-g and 1-g conditions were comparable at high liquid velocities similar to the heat transfer coefficients. Unfortunately, pressure drop data at lower velocities suffer from a large scatter due to the limits of the instrumentation. Therefore, a comparison cannot be made over the range of flow rates in which large differences in heat transfer were obtained.

3.2 Average Heat Transfer Coefficients – Glycerin/Water–Air

The purpose of looking at two-phase mixtures besides air–water is twofold. First, by varying the properties of the mixtures, different forces (such as surface tension and viscous forces) that can affect pressure gradient and heat transfer characteristics of flows through conduits can be evaluated through the choice of appropriate dimensionless numbers. Secondly, empirical correlations can be developed by allowing one to hold some non-dimensional quantities constant while varying others in a systematic parametric study. As was illustrated in Table 5 with the three glycerin/water mixtures, the largest change in the properties is in relation to the viscosity and, consequently, the Prandtl number. The glycerin mixtures used to generate the flight and ground heat transfer data included combinations of 50% glycerin (25 flight data points), 59% glycerin (33 flight points), and 65% glycerin (38 flight points). These mixtures have dynamic viscosities greater than pure water by up to an order of magnitude, ranging from 400 to 1011% higher. Based on these property changes, the liquid Prandtl number increases by 538% with 50% glycerin as compared with pure water, 933% with 59% glycerin, and 1404% with 65% glycerin. This enables one to assess the influence of the viscous forces and their relations to gravity, inertia, and surface tension forces and to manipulate the various dimensionless groups that are important to convective heat transfer. By using a two-component system, the flexibility in varying properties is much greater than in a single-phase system where the difference in the properties of the gas phase to the liquid phase cannot be controlled to as wide an extent.

The heat transfer results for the three air–glycerin/water mixtures of 50%, 59%, and 65% that were tested are presented in Figs. 80 through 83. For these results, instead of non-dimensionalizing the two-phase convective heat transfer coefficients by using the Nusselt number, the values are normalized with the single-phase heat transfer coefficient (h_L) at the same Reynolds number. This normalization allows one to isolate the effect of the variation in physical properties on the two-phase enhancement of the heat transfer without a change in the heat transfer coefficients due to different liquid Prandtl numbers and thermal conductivities interfering with the interpretation of the results.

The single-phase coefficient used in the normalization was obtained from a curve-fit of experimental single-phase data gathered with the same test apparatus. The normalized function is thus expressed as:

$$\psi^2 = \frac{h_{TP}}{h_L}. \tag{16}$$

The data gathering procedure was the same as for the air–water data in Figs. 76–79; a constant liquid superficial velocity was set and then the gas superficial velocity was increased incrementally.

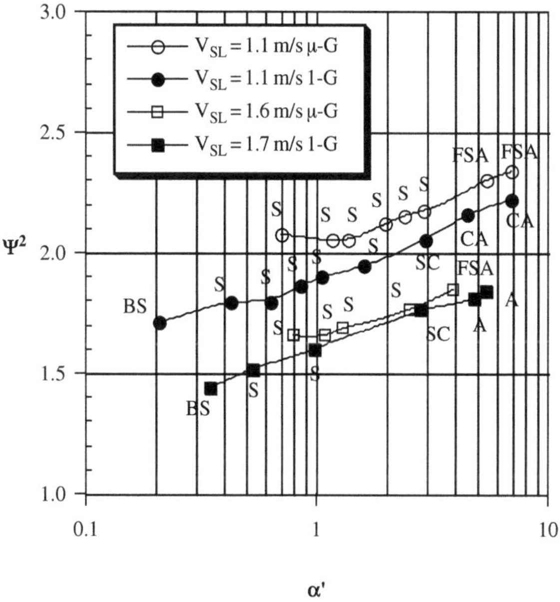

Figure 80. 1-g and μ-g ψ^2 versus pseudo-void fraction for air–50% glycerin/water — $V_{SL} \approx 1.1$ and 1.6 m/s.

In Fig. 80, the normalized heat transfer coefficients for superficial liquid velocities of approximately 1.1 and 1.6 m/s ($Re_{SL} \sim 3120$ and 4890) with a 50% glycerin/water mixture are shown plotted as functions of the pseudo-void fraction. For all of the data points shown in the figure, the liquid velocity was maintained to within 3% of the given value. It can be seen in the figure that normalized heat transfer coefficients during μ-g are approximately 3–14% higher than their 1-g counterparts for $V_{SL} = 1.1\,\mathrm{m/s}$ and less than 5% higher at $V_{SL} = 1.6\,\mathrm{m/s}$. These results are in contrast to the air–water results at approximately the same Reynolds number in Fig. 77 where the 1-g Nusselt numbers were consistently higher than the μ-g values by nearly 50% in the slug flow region. It would be expected that increasing the viscosity markedly would lessen some of the effects of buoyancy forces on the movement of the gas bubbles and on the liquid phase itself due to increased viscous damping. Therefore, the μ-g and 1-g heat transfer coefficients would approach one another as viscosity is increased. However, a mechanism by which high viscosity would tend to depress the 1-g coefficients to a level below the μ-g ones (or conversely enhance the μ-g above the 1-g) is not readily apparent. Indeed, Vijay (1977) found that for 1-g flows with air and pure glycerin mixtures having a liquid viscosity of almost 500 times that of pure water, the heat transfer behavior matched that of single-phase flows until the gas velocities were raised to extremely high levels. The high liquid viscosity damped any perturbations caused by the gas bubbles. This behavior is similar to the effect of reduced gravity: two-phase enhancement of heat transfer is minimized. Based on this as well as the measurement uncertainty, the significance of the μ-g heat transfer coefficients being higher than those at 1-g is minimal.

In Figs. 81 and 82, the percentage of glycerin in the liquid mixture has been increased to 59%. Heat transfer data for liquid superficial velocities of 1.3 and 1.9 m/s ($Re_{SL} = 2450$ and 4070) are given in Fig. 81, and for $V_{SL} = 0.9$ and 1.5 m/s ($Re_{SL} = 1540$ and 3120) in Fig. 82. For these results, V_{SL} was maintained within 2% of its given average value. In the figures, the μ-g normalized heat transfer coefficients are again marginally higher than the 1-g; by approximately 5% for $V_{SL} = 1.3$ m/s and 8% for the $V_{SL} = 1.5$ m/s case. For $V_{SL} = 1.9$ m/s, the μ-g normalized heat transfer coefficients are up to 7% higher than those under 1-g. Again, these differences are small, but the trends are consistent – increasing the viscosity of the liquid reduces the enhancement of heat transfer by the gas phase under 1-g conditions. However, the level of this reduction is not constant for all Reynolds number values. The effect of gravity on the heat transfer rate changes as the flow conditions change. This will be illustrated by the 65% glycerin data in Fig. 83.

Fig. 83 shows two sets of data for a glycerin/water mixture of 65% glycerin. In the figure, the superficial liquid velocities were maintained at 0.81 and 1.9 m/s ($Re_{SL} = 1030$ and 3060) within 4%. For $V_{SL} = 0.81$ m/s, the 1-g heat transfer data are higher than the μ-g data by as much as 20% at an α' of 1.0 and, as α' is increased, the heat transfer coefficients converge until at $\alpha' \approx 10.0$ they are the

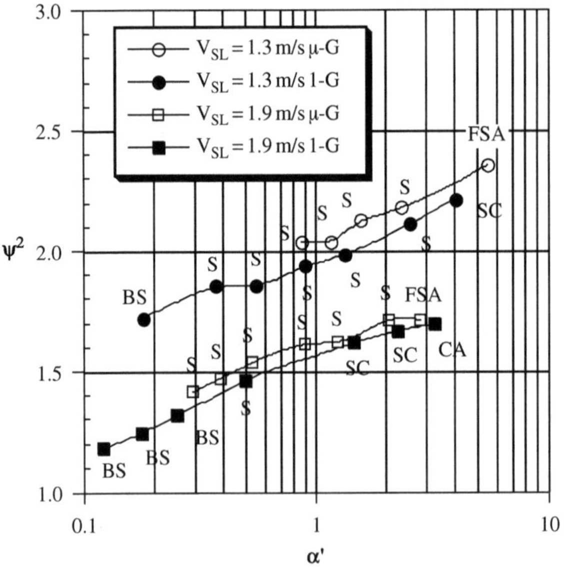

Figure 81. 1-g and μ-g ψ^2 versus pseudo-void fraction for air–59% glycerol/water — $V_{SL} = 1.3$ and 1.9 m/s

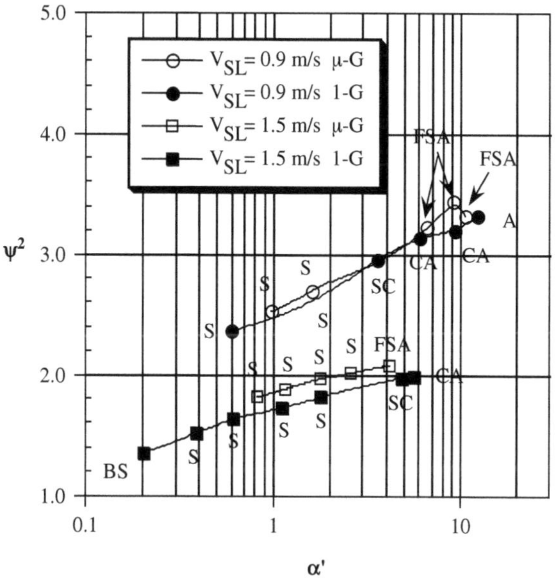

Figure 82. 1-*g* and μ-*g* ψ^2 versus pseudo-void fraction for air–59% glycerin/water — $V_{SL} = 0.9$ and 1.5 m/s

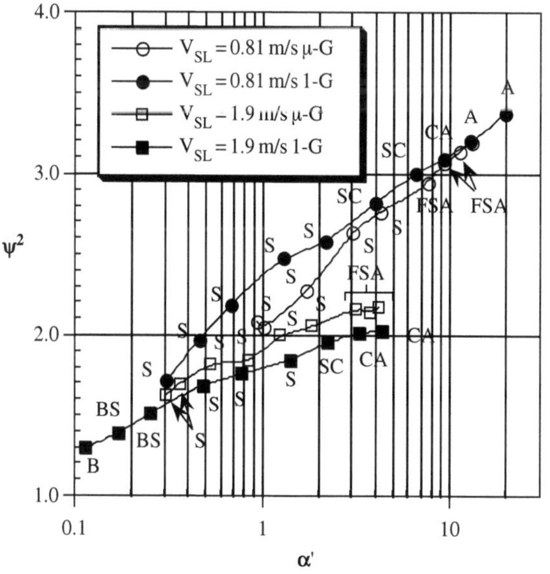

Figure 83. 1-*g* and μ-*g* ψ^2 versus pseudo-void fraction for air–65% glycerin/water — $V_{SL} = 0.81$ and 1.9 m/s

same. This is similar to the behavior found with the pure water and air mixtures at $V_{SL} = 0.24\,m/s$ in Fig. 77. Thus, the 1-g heat transfer enhancement effect is evident with the higher viscosity flows, but only at much lower Reynolds numbers. Once again, however, at higher superficial liquid Reynolds numbers the 1-g and μ-g heat transfer coefficients are very close in magnitude. In the figure, at a liquid velocity of 1.9 m/s, ψ^2 under μ-g conditions is higher than its 1-g counterpart by only 6%. In Fig. 83, it can be seen that the flow is mainly in the slug flow regime for both 1-g and μ-g. No major differences in flow classification are evident besides the appearance of the "falling-film"-driven oscillations of churn-type flows for the 1-g data at approximately the same α' as the appearance of slug-annular flow in μ-g.

Comparing the $V_{SL} = 1.1\,m/s$ results in Fig. 80 and the pure water results in Fig. 77 at $V_{SL} = 0.24\,m/s$ (with nearly the same Reynolds numbers of 2950 and 2900, respectively), it can be seen that the 1-g normalized heat transfer coefficients have been calculated to be marginally lower than the μ-g data for α' between 0.7 and 7.0 in Fig. 80, while for the air–water data over a similar range of α' (0.4–2.0) the 1-g heat transfer data points are significantly higher than the μ-g points. Nearly identical μ-g and 1-g heat transfer coefficients are also evident for the 59% glycerin/water data in Fig. 82 at V_{SL} of 1.5 m/s ($Re_{SL} = 3120$) and for 65% glycerin at $V_{SL} = 1.9\,m/s$ ($Re_{SL} = 3060$) in Fig. 83. Based on these findings, it would appear that the higher viscosities of the glycerin/water mixtures are suppressing the 1-g heat transfer coefficients since viscosity is the most significantly varying parameter due to damping effects discussed earlier. This viscosity effect is further supported by the results given in Fig. 83 for 65% glycerin/water at $V_{SL} = 0.81\,m/s$ or $Re_{SL} = 1030$. In the figure, ψ^2 at 1-g is higher than at μ-g by almost 20% at $\alpha' \approx 1$, and then the heat transfer coefficients tend to converge and become essentially equal at $\alpha' \approx 10.0$. This trend of convergence was also found for the $Re_{SL} = 4900$ case in Fig. 77.

To summarize, the trends in 1-g versus μ-g heat transfer that occur at a certain Reynolds number with air–water mixtures happen in highly viscous air–glycerin/water mixtures at much lower Reynolds numbers. The question that needs to be addressed is: What is the mechanism responsible for this effect? The traditional two-phase heat transfer dimensionless groups of Reynolds number, Prandtl number, and superficial velocity ratio do not answer this question completely. Other effects due to gravity and surface tension must be included in the analysis, although their effects may be of secondary importance. This approach will be further explored in Sections 5 and 6, but first the uncertainties in the results that have been presented so far need to be defined.

4. MEASUREMENT ERROR AND UNCERTAINTY

All the flight and ground heat transfer data that were collected over the course of this research project were utilized in an assessment of the total uncertainty of the normalized heat transfer coefficient ($\psi^2 = h_{TP}/h_L$). In this analysis, the bias uncertainties of the thermocouples used were 0.5 °C and those of the RTD measurements were 0.26 °C. The heater power measurement had an uncertainty of approximately 1.1%. The liquid and gas flow measurements had biases of 1.0 and 1.8%, respectively.

It was found that the total uncertainty in the measurements, including both the bias of the instrumentation and repeatability of ψ^2 over 3 years worth of flight and ground experiments was 14.3% for turbulent flow cases and 6.4% for laminar flow cases. Details on this uncertainty analysis may be found in Rite and Rezkallah (1994a), and Rite (1995).

5. LOCAL HEAT TRANSFER COEFFICIENTS

The first step taken in the analysis of the heat transfer results was an examination of local heat transfer coefficients. The results of this investigation are shown graphically in Figs. 84 through 91 for air–water at both 1-*g* and μ-*g* conditions. In these figures, the local heat transfer coefficients (h_{local}) calculated from each of the 12 surface temperature measurements on the heated test section are plotted as functions of location from inlet to outlet (stations 1–12). The shown data are those obtained during the 1994 flight campaign.

Fig. 84 and 85 show the local convective coefficient for $V_{SL} \approx 0.07\,\text{m/s}(Re_{SL} \approx 840)$ and gas velocities ranging from 0.96 (slug flow) to 14.0 m/s (annular flow) for 1-*g* and μ-*g* conditions, respectively. The "two-phase" Reynolds number is estimated from:

$$(17) \qquad Re_{TP} = \frac{Re_{SL}}{(1-\alpha)}.$$

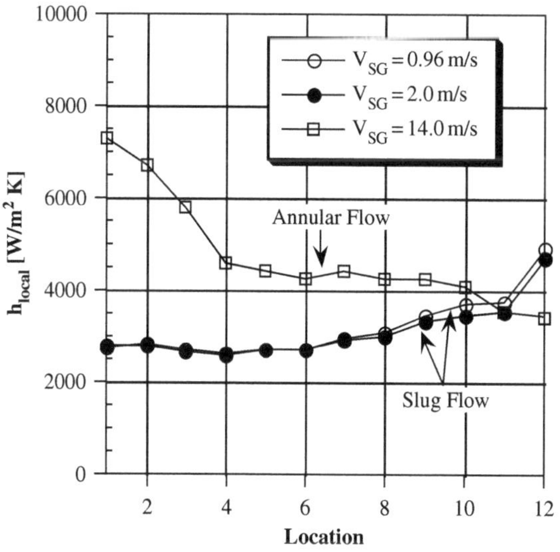

Figure 84. Local, two-phase heat transfer coefficients: $V_{SL} = 0.076\,\text{m/s}$, 1-*g*

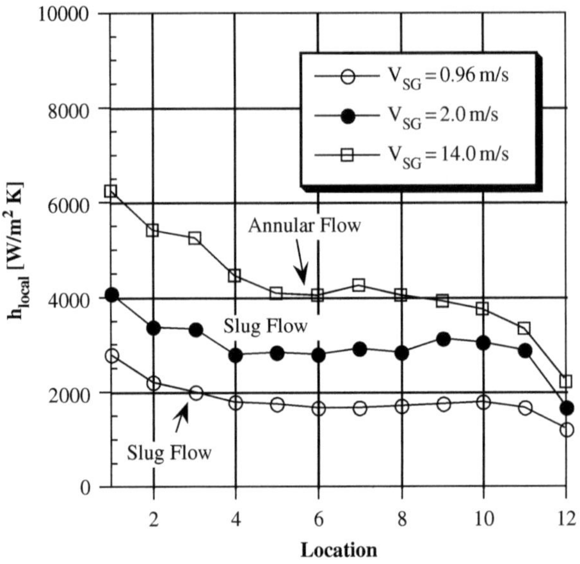

Figure 85. Local, two-phase heat transfer coefficients: $V_{SL} = 0.070\,\text{m/s}$, μ-$g$

From the experimental measurement of $\boldsymbol{\alpha}$ during flight, the value of Re_{TP} at this low liquid superficial velocity is approximately 3231 (for $\boldsymbol{\alpha} \approx 0.74$). This is the region marked "transitional" flow in single-phase systems. It can be seen that for 1-g flows the local heat transfer coefficient slightly increases from inlet to outlet for the slug flow points ($V_{SG} \leq 2.0\,\text{m/s}$). With annular flow and a two-phase Reynolds number near the turbulent region, a longer distance is required for the flow to reach fully developed conditions, and thus h_{local} decreases from inlet to outlet. At μ-g conditions (Fig. 85), the heat transfer coefficients are still developing along the length of the heated test section. This is true for all gas flows and flow regimes. There is also a significant increase in the local heat transfer coefficient as the gas velocity is increased from 0.96 to 2.0 m/s when compared to the 1-g results. These results indicate that for all the μ-g cases and for the annular flow case at 1-g there is a relatively long thermal entry length. This is common for developing laminar and turbulent flow conditions in single-phase flows, Kays and Crawford (1980). But, for low gas flows under 1-g the bubble's movement and interaction with the fluid layer near the wall cause an added turbulence to the flow and thus minimizes the thermal entry length; i.e., it resembles turbulent flow behavior.

The disturbance of the temperature profiles shown in Figs. 84 and 85 is also evident in Figs. 86 and 87 for a constant V_{SL} of 0.10 m/s ($Re_{SL} \approx 1200$ and $Re_{TP} \approx$ 3183). In these figures, slug flow is present in the tube at all times for a superficial gas velocity range of 0.14–0.79 m/s. In general, the 1-g data shows an increase in the local heat transfer coefficient from inlet to outlet while the μ-g results show a

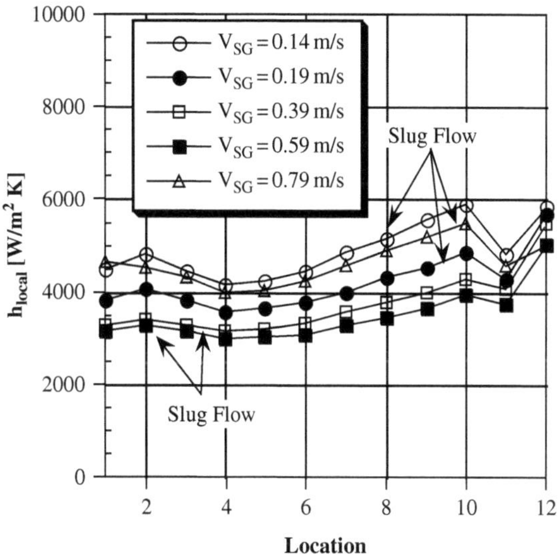

Figure 86. Local, two-phase heat transfer coefficients: $V_{SL} = 0.10\,\text{m/s}$, 1-$g$

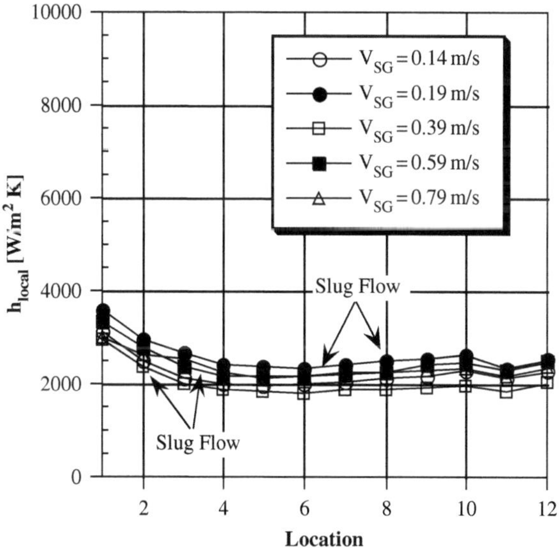

Figure 87. Local, two-phase heat transfer coefficients: $V_{SL} = 0.10\,\text{m/s}$, μ-g

slight decrease in h_{local}. Once again, the behavior of h_{local} at μ-g resembles that at 1-g, single-phase flow in the viscous region. It is noticed from these two figures that the two-phase heat transfer coefficients actually decrease in slug flow as the gas flow rate becomes larger. This was also shown in the Nusselt number plots

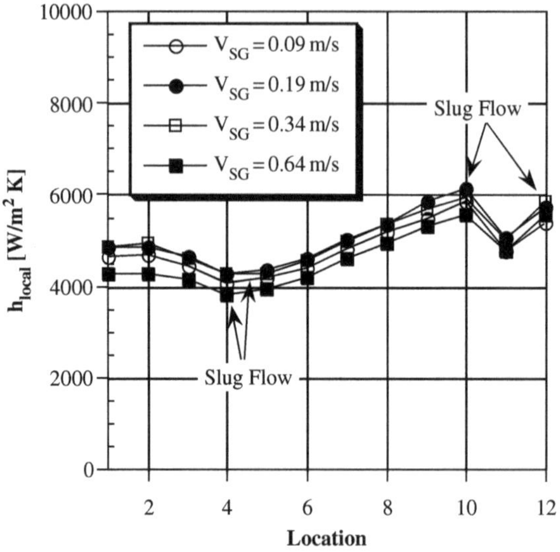

Figure 88. Local, two-phase heat transfer coefficients: $V_{SL} = 0.24\,\mathrm{m/s}$, 1-$g$

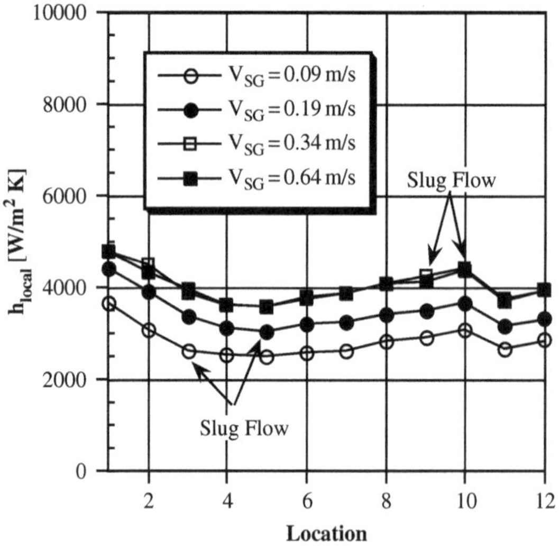

Figure 89. Local, two-phase heat transfer coefficients: $V_{SL} = 0.24\,\mathrm{m/s}$, μ-g

(Fig. 76). It should be noted that the local maxima and minima at stations 2 and 11 are not related to a certain phenomena. It is simply an artifact of the heater design, where at those two locations an overlap in the wrapped electrical wire occurred.

A further increase in the superficial liquid velocity to 0.24 m/s ($Re_{SL} \approx 2900$ and $Re_{TP} \approx 5577$) in Figs. 88 and 89 still gives the same trend with a "viscous flow behavior" at μ-g conditions, and a "turbulent-like" flow at 1-g conditions. However, the heat transfer coefficient profiles are beginning to more closely resemble one another as the liquid deviates from a "laminar-type" flow to a "transitional/turbulent-type" flow.

Lastly, at $V_{SL} \approx 0.4\,\text{m/s}(Re_{SL} \approx 4900,\ Re_{TP} \approx 10,278)$, as the flow develops to "fully turbulent-type" flow the difference in the heat transfer coefficient profiles between the 1-g and μ-g data is significantly minimized. The results of Fig. 90 and 91 show, in addition to a similar trend, very similar heat transfer values. In these figures, the flow pattern starts as slug flow at $V_{SG} \approx 0.19\,\text{m/s}$ and then changes to slug-annular at $V_{SG} \approx 7.0\,\text{m/s}$. As mentioned earlier, the local minima and maxima near stations 2 and 11 are artifacts of the heater design. Despite this and considering the scale of the results, it is apparent that a more uniform profile is developing as would be a characteristic of turbulent, single-phase flows. It should be remembered that superficial velocities are being used to determine Reynolds numbers, and the actual liquid Reynolds numbers especially at high gas flows will be significantly higher due to the reduced liquid flow area.

It appears that under 1-g conditions the gas is influencing the local heat transfer by disturbing the liquid flow at the wall of the tube. This agrees with the findings of Vijay et al. (1978), who hypothesized that the effect of buoyancy forces on the gas bubbles caused the gas to have a greater velocity than the surrounding liquid. This "slip" between the phases led to a breakdown in the laminar sub-layer near the tube wall, thus leading to a movement away from laminar flow-type behavior. However, under μ-g conditions this would not happen. The reduction of gravity greatly lessens the buoyancy forces and therefore the turbulence-generating ability

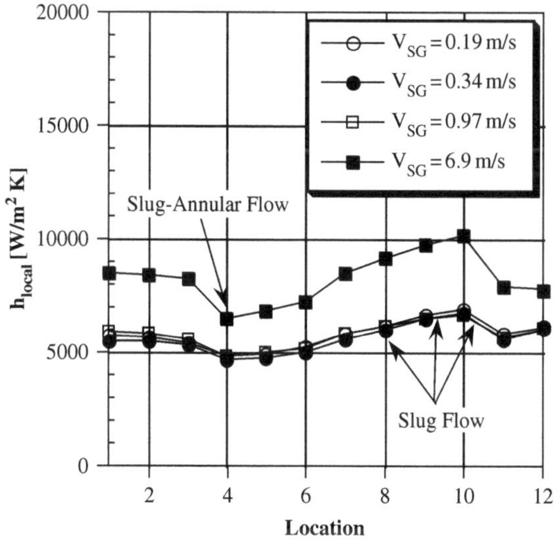

Figure 90. Local, two-phase heat transfer coefficients: $V_{SL} = 0.41\,\text{m/s}$, 1-$g$

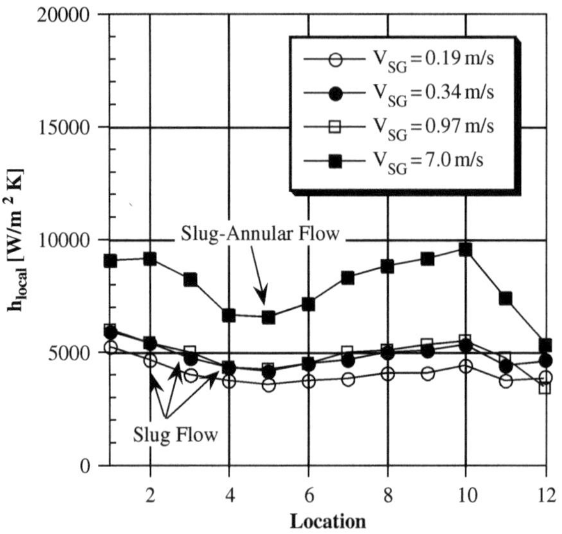

Figure 91. Local, two-phase heat transfer coefficients: $V_{SL} = 0.41\,\text{m/s}$, μ-g

of the gas bubbles. The bubble dynamics at μ-g do not seem to interfere with the maintenance of the laminar sub-layer.

The work of Serizawa et al. (1975b) further supports the hypothesis of Vijay. Serizawa reported for air–water flows through a 17.5 mm diameter vertical tube a "saddle-shaped" void fraction distribution. For bubbly flow, the bubbles were distributed such that the largest concentration was very near the tube wall. As the flow transitioned to slug flow, the void fraction distribution moved toward a more "bullet-shaped" profile with a maximum at the tube centerline. This concentration of bubbles near the tube wall would account for the disturbance of the liquid layer at the wall.

More recently, a difference in the void distribution at 1-g and μ-g has been found by Kamp et al. (1993). In that work, void distributions for air–water upflow through a vertical tube with a diameter of 40 mm are reported. Superficial liquid velocities ranging from 0.27 to 0.99 m/s with superficial gas velocities of 0.023 and 0.044 m/s were tested. Comparing results obtained in the laboratory at 1-g conditions with those for μ-g conditions aboard a low-gravity aircraft, they also found the "saddle-shaped" distribution of Serizawa at 1-g, but did not at μ-g.

The effect of gravity on the turbulence associated with the gas phase at the near-wall layer partially explains the higher Nusselt numbers that were found in Figs. 76–79 for 1-g as compared with μ-g flows at low liquid flow rates and gas qualities. However, the influence of natural convection on the liquid must also be considered at the lowest liquid flow rates.

6. MIXED CONVECTION

Following the criterion for mixed convection given by Petukhov (1988) for vertical, upward flow in a tube heated with a uniform heat flux, it was determined that free convective forces could be significant for some of the data points collected on-ground and even a few collected on-board the flight aircraft. The criterion that was utilized is calculated as follows:

$$\text{(18)} \qquad \frac{Gr_L^*}{Re_{SL}^{2.75} Pr_L^{1.15}} > 9 \times 10^{-5},$$

where Gr_L^* is the Grashof number based on heat flux:

$$\text{(19)} \qquad Gr_L^* = \frac{g\beta D^4 q''}{\nu_L^2 k_L},$$

where g is the local gravitational acceleration, β is the volumetric thermal expansion coefficient which is approximated as the reciprocal of the film temperature, q'' is the heat flux at the surface of the heated test section, and ν_L is the liquid kinematic viscosity. Using this criterion, it was found that all the ground data in the range of $Re_{SL} < 2300$ for air–water were influenced by the presence of buoyancy forces. It was also found that even with a few of the low-gravity cases, buoyancy forces were present. For the data points with $Re_{SL} > 2300$, it was found that all air–water ground data with a superficial liquid Reynolds number less than approximately 8000 failed to pass the criterion outlined in Eq. 19 and, thus, mixed convection is occurring to some degree.

In the literature, it is shown that the influence of mixed convection on heat transfer can be a positive or a negative one. According to Aung (1987), for laminar flow the buoyancy forces for vertical up-flow with uniform heating clearly aid the transfer of heat. Hence, for the results presented in Fig. 76 at low pseudo-void fraction buoyancy forces in the liquid phase are augmenting 1-g heat transfer. However, for the turbulent flow cases, Aung (1987) reports that over a range of the $Gr/Re^{2.75}$ ratio of 10^{-5} to approximately 10^{-3} there is a depression of the heat transfer coefficient compared with pure forced convection. The mechanism for this effect is thought to be a reduction of turbulence production in the flow. This reduction is caused by a decrease in the shear stress between the flow at the wall and away from the wall. The buoyancy forces are accelerating the near-wall flow, thus minimizing the difference in fluid velocity. For the range of $Gr/Re^{2.75}$ in this study, it was found that the amount of this heat transfer reduction gradually increases as Re_{SL} increased from 3000 to 8000. This corresponds with the two-phase results discussed above. The difference between the 1-g and μ-gNu_{TP} decreased as gas velocity was increased for $V_{SL} = 0.24$, 0.41, and 0.89 m/s ($Re_{SL} \approx 2900$, 4900, and 11,000, respectively). The heat transfer augmentation that the gas phase provides is being offset by heat transfer reduction due to turbulent mixed convection at 1-g.

It is apparent that the influence of gravity in the results found here is a two-phase as well as a single-phase phenomena: two-phase when low-gravity inhibits the production of turbulence by the gas bubbles; and single-phase when reduced gravity lessens the influence of mixed convection.

7. EMPIRICAL CORRELATIONS

The influence of flow patterns on heat transfer in two-phase flows has been recognized for many years. Shortly after the behavior of two-phase flows was first explored in-depth for terrestrial applications such as in the nuclear, petroleum, and process industries, it was discovered that the geometry of the flow was important in determining useful engineering parameters such as pressure drop and heat transfer, Govier and Aziz (1972). This information has since been used by other researchers in the development of empirical and semi-empirical heat transfer correlations that incorporate flow regime. Among these researchers are Vijay et al. (1982) and Soliman (1986).

In Vijay et al.'s (1982) work, a large data set of two-phase, two-component air–water, air–glycerin, and air–glycerin/water mixtures flowing through vertical, circular tubes at 1-g was used in the development of their flow regime dependent heat transfer correlations. The two-phase heat transfer coefficients from these data were normalized with the corresponding single-phase heat transfer coefficient at the same flow rate. These normalized coefficients were then correlated in terms of the commonly used two-phase flow pressure drop multiplier raised to a certain power that varies with each flow regime.

Using a steam and Refrigerant-12 and -113 database, Soliman (1986) also considered flow regime in the development of an empirical correlation for the condensation process in the mist flow regime. This correlation includes a two-phase Reynolds number and a pseudo-Prandtl number incorporating latent heat and saturation temperature.

Likewise, for microgravity research, the need to account for the influence of the gravitational level as well as the flow regime on the two-phase heat transfer has been recognized. In his early microgravity heat transfer experiments, Papell (1962) reported that reduced gravity had an influence on the two-phase heat transfer which he attributed to a change in the flow regime that was present in the conduit. Since then, there have been some correlations for pool boiling heat transfer that account for gravitational acceleration, most notably: Usiskin and Siegel (1961), and Merte and Clark (1963). However, to the best of the author's knowledge, no correlations incorporating the effect of gravity and flow regime on forced convective two-phase flows are available in the literature. It is essential that such correlations be available to the designers of space-based two-phase thermal systems. Such correlations incorporating the influence of gravity and flow regimes were developed based on the large heat transfer data set collected at μ-g and are presented here.

It should be noted that although the correlations to be presented are for gas–liquid flows (not boiling or condensing flows), it has been shown by researchers such as

Kudirka (1964) and Wallis (1966) that a complete hydrodynamic analogy exists between the two systems. In addition to this and considering the small contribution of the boiling heat transfer compared to the forced convective term (Chen, 1963), it was also shown by the above-mentioned researchers that outside the nucleate boiling regime, the heat transfer values are very comparable in both systems. Obviously, the Critical Heat Flux (CHF) cannot be determined from the results presented in this study.

Heat transfer data from four flight campaigns as well as the corresponding ground data were used in the development of the heat transfer correlation presented here. These data consist of air–water flows and three glycerin/water mixtures with liquid viscosities of 36, 56, and 80 N-s/m^2 with air as the gas phase.

From the glycerin/water data, it was learned that the commonly used dimensionless groups for heat transfer at 1-g do not account for the effect of gravity, Rite and Rezkallah (1994b). A comparison of flight and ground heat transfer data with the glycerin mixtures indicated that heat transfer under 1-g conditions was higher than that for μ-g at low liquid and gas flow rates, but lower than μ-g at higher flow rates. These are similar results as were found for pure water and air flows. However, the superficial velocity ranges over which 1-g heat transfer was higher than μ-g heat transfer differed significantly between air–water and air–glycerin data and could not be explained by applying the usual dimensionless parameters of Reynolds and Prandtl numbers. Forces that are usually unimportant at 1-g need to be considered with μ-g data in order to explain the phenomena which were observed.

The two-phase heat transfer coefficient (h_{TP}) was normalized with the corresponding single-phase coefficient (h_L) at the same Re_{SL} and transport properties to calculate (ψ^2). The liquid-phase heat transfer coefficient (h_L) is the actual measured single-phase coefficient in all cases, except when the agreement with single-phase correlations (such as those by Sieder-Tate) was extremely good, the latter was used instead. Analysis of the data collected over the duration of the project, including repeatability data, indicated an uncertainty of less than 15% for the normalized coefficients as described in Section 4.

7.1 Pertinent Dimensionless Groups

In order to develop useful empirical correlations to model observed phenomena, the identification of dimensionless groups that will describe the phenomena is required. For low-gravity two-phase flow heat transfer, in addition to the most commonly used groups (e.g., Reynolds and Prandtl numbers) the effect of gravitational and surface tension forces on the arrangement of the gas and liquid phases must also be considered in the dimensional analysis.

The flow regime transitions proposed by Zhao and Rezkallah (1993b) and introduced in Chapter 3 were incorporated into the analysis in order to account for the influence of the change in the flow regime due to reduced gravity on the two-phase heat transfer. In their analysis, it was reported that the flow regime transitions were very much dependent on the force balance between inertia forces and surface tension forces. This balance was well represented as a function of the gas Weber number, such that annular flow was shown to be present for $We_{SG} > 20$ (inertia dominated region), slug-annular flow for $1 < We_{SG} < 20$, and slug and bubbly flows for $We_{SG} < 1$ (surface tension dominated region). In this last range of Weber numbers, another transition has been added at $We_{SG} = 0.1$ to identify the region where true bubbly flow can only exist. The associated heat transfer coefficients for such homogeneously dispersed flows would be different from that of typical Taylor-type slug flows.

In addition to segregating the flows in terms of flow regime using the dimensionless group We_{SG}, the data were also divided based on the liquid superficial Reynolds number Re_{SL}. For internal, single-phase flows, a Reynolds number of 2300 is commonly used to indicate a change from laminar to transitional/turbulent flow. This is not the case for two-phase flows, since the presence of the gas phase will accelerate the liquid velocity and also affect its turbulence intensity. These effects will tend to lower the value of superficial liquid Reynolds number at which the transition from laminar to non-laminar liquid flow occurs. However, because the actual liquid velocity is not known for all of the heat transfer data to be correlated, it was decided to still maintain a dividing point between low-velocity liquid flow rates and high-velocity liquid flow rates at $Re_{SL} = 2300$. This will enable the normalization process with single-phase correlations to be accomplished in a straightforward manner (i.e., the laminar single-phase correlation of Sieder-Tate will be used for $Re_{SL} < 2300$ and their turbulent single-phase correlation for $Re_{SL} > 2300$).

From a dimensional analysis, it was determined that for $Re_{SL} < 2300$ the following dimensionless groups are pertinent: Graetz number (Gz_{SL}) based on the liquid superficial velocity and properties, the ratio of the superficial velocities or the pseudo-void fraction (α'), and a property group known as the liquid Morton number (Mo_L). The Graetz number is defined as:

$$(20) \qquad Gz_{SL} = Re_{SL} Pr_L \frac{D}{L},$$

where Pr_L is the Prandtl number of the liquid phase, and L is the length of the heated test section. The Morton number is defined by Delil (1991) as:

$$(21) \qquad Mo_L = \frac{\rho_L \sigma^3}{g \mu_L^4},$$

where g is the gravitational acceleration and μ_L is the liquid dynamic viscosity. As seen in the above equation Morton number is strictly dependent on fluid properties and gravity. This incorporates the effect of gravitational forces and surface tension forces in the analysis of the heat transfer data.

With liquid superficial Reynolds numbers above 2300, the non-dimensional numbers that are important include the following: Re_{SL}, α', liquid Froude number

(Fr_{SL}), and Mo_L. Froude number, based on the liquid superficial velocity, is defined as:

$$Fr_{SL} = \frac{V_{SL}}{\sqrt{gD}}. \quad (22)$$

Before least-squares curve-fitting was done to the data to determine the relationship between the various dimensionless groups and the normalized heat transfer coefficient (ψ^2), two adjustment were made in the database. First, all the data points in which mixed convection was significant were omitted from the data set used to generate the correlations. The criterion that was used to distinguish these points was described in detail in Section 6. Secondly, because gravitational acceleration was negative for some flight data points, a value of g/g_0 of 0.01 was used in all the calculations of Fr_{SL} and Mo_L of the flight data.

7.2 Empirical Correlations

Comparisons of the empirical correlations for two-phase flow having $Re_{SL} < 2300$ with all of the experimental data (both μ-g and 1-g) are shown in Figs. 92–95. In the figures, ψ^2 is plotted as a function of the pertinent dimensionless groups raised to the appropriate powers. Microgravity data points have been separated from those collected at 1-g. Ranges at 25% higher and 25% lower than the curve-fit are shown in Fig. 92 as dashed lines. These ranges were chosen for this figure based on the

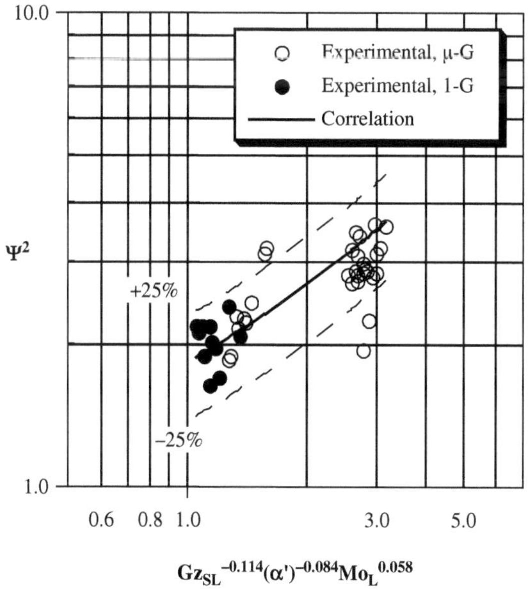

Figure 92. Two-phase empirical correlation: $We_{SG} < 0.1$, $Re_{SL} < 2300$

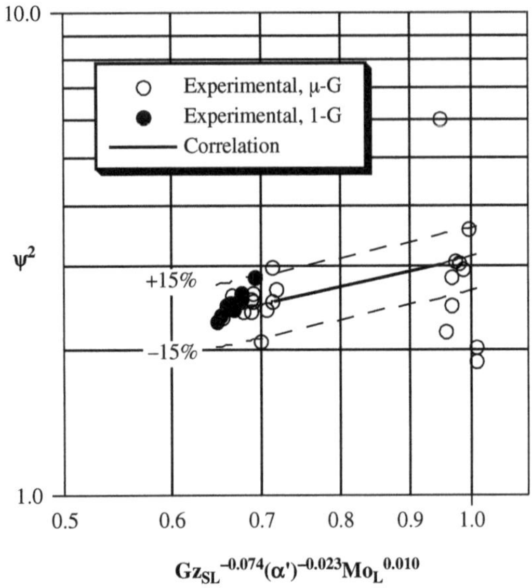

Figure 93. Two-phase empirical correlation: $0.1 < We_{SG} < 1, Re_{SL} < 2300$

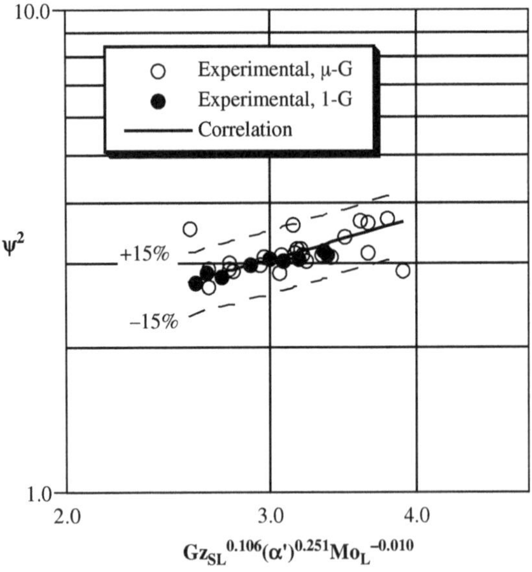

Figure 94. Two-phase empirical correlation: $1 < We_{SG} < 20$, $Re_{SL} < 2300$

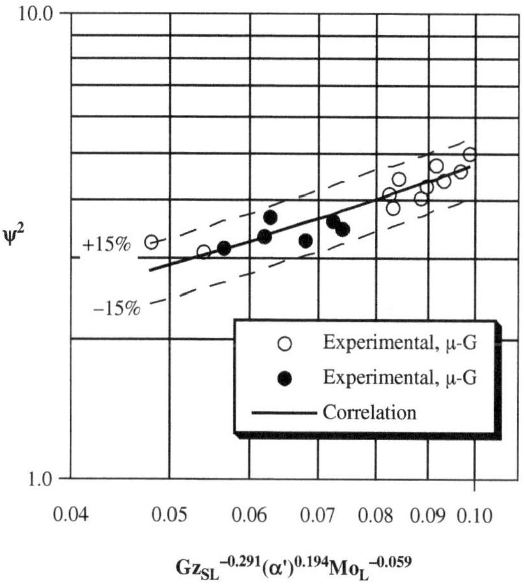

Figure 95. Two-phase empirical correlation: $We_{SG} > 20$, $Re_{SL} < 2300$

comprehensive evaluation of two-phase heat transfer correlations with a very large experimental data set by Rezkallah and Sims (1987), who showed that for all of the correlations tested, the best possible predictions were within approximately $\pm 25\%$.

Fig. 92 gives the results for $We_{SG} < 0.1$. The correlation developed for this case is given as:

$$(23) \qquad We_{SG} < 0.1 : \psi^2 = 1 + 0.8 Gz_{SL}^{-0.114} (\alpha')^{-0.084} Mo_L^{0.058}.$$

In Fig. 92, it can be seen that all but four of the data points are within the $\pm 25\%$ band. The root-mean-square error (e_{rms}) was calculated to be 17.5% and the mean error (e_{avg}) is -5.5%. The root-mean-square error was calculated from:

$$(24) \qquad e_{rms} = 100 \left[\frac{1}{N} \sum_{i=1}^{N} \left(\psi_{i,exp.}^2 - \psi_{i,corr.}^2 \right)^2 \right]^{1/2},$$

and the mean error is defined as:

$$(25) \qquad e_{avg} = \frac{100}{N} \sum_{i=1}^{N} \left(\psi_{i,exp.}^2 - \psi_{i,corr.}^2 \right),$$

where i is the summation index, N is the total number of data points, $\psi_{i,exp.}^2$ is the experimental heat transfer coefficient, and $\psi_{i,corr.}^2$ is the heat transfer coefficient given by the correlation.

For the next range of gas Weber numbers, the empirical correlation has the form:

$$(26) \qquad 0.1 < We_{SG} < 1 : \psi^2 = 1 + 2.1 Gz_{SL}^{-0.074} (\alpha')^{-0.023} Mo_L^{0.010}.$$

The comparison of the data to the curve-fit is presented in Fig. 93. For this correlation, e_{rms} is slightly higher than for the previous case (25.3%) owing largely to significant scatter in the data at the lowest Re_{SL} values (right-hand side of the graph.)

Moving on to the next range of data for which the flow regime is predominantly slug, the recommended correlation is given as:

$$(27) \qquad 1 < We_{SG} < 20 : \psi^2 = 1 + 0.7 Gz_{SL}^{0.106} (\alpha')^{0.251} Mo_L^{-0.010}$$

A plot of the correlation and the experimental data is given in Fig. 94. Only two data points are outside the 15% markers shown in the figure and the error in prediction is quite low: $e_{rms} = 7.7\%$ and $e_{avg} = -0.1\%$.

In Fig. 95, the results of the comparison between the curve-fit at the highest Weber numbers for the gas phase are given. The empirical correlation referred to in the figure is expressed as:

$$(28) \qquad \mathrm{We}_{SG} > 20 : \psi^2 = 1 + 37.2 Gz_{SL}^{-0.291} (\alpha')^{0.194} Mo_L^{-0.059}.$$

In the figure, the discrepancies between the data and the correlation are very small ($e_{rms} = 6.1\%$ and $e_{avg} = 0.8\%$). This is within the uncertainty in the measurements and is therefore an excellent prediction.

A summary of the results of the comparisons of the experimental data with the correlations for $Re_{SL} < 2300$ is presented in Table 6.

In the correlations presented in Eqs. 27 and 30 through 28, it can be seen that the functional dependency of ψ^2 on Graetz number varies over the range of gas Weber numbers (i.e., over the range of flow regimes). The changes in the exponents are clearly flow regime dependent.

Graetz number is commonly defined as the ratio of thermal capacity to convective heat transfer. It is directly related to the length of the thermal entry region: a large Graetz number indicates that the thermal entry length is relatively long. It follows that as Gz_{SL} increases, the heat transfer enhancement effect due to the presence

Table 6. Comparisons of the new empirical correlations to the experimental data — $Re_{SL} < 2300$

We_{SG}	α'	Re_{SL}	N	e_{rms}	e_{avg}
0.001–0.1	0.2–11.3	879–2299	42	17.5	−5.5
0.1–1.0	0.7–32.4	872–2274	31	25.3	0.8
1.1–19.8	2.8–76.1	966–2280	33	7.7	−0.1
22.1–81.4	9.2–183.8	523–2231	17	6.1	0.8

of gas bubbles will be less pronounced because any change in the convective heat transfer coefficients will tend to be overshadowed by the larger thermal capacitance. This explains the negative exponent on Gz_{SL} for $We_{SG} < 1$.

As was mentioned previously, for $1 < We_{SG} < 20$ the exponent on Gz_{SL} is positive. This is related to the fact that in this range, churn and slug-annular transition flows are present. Due to the chaotic nature of the flow in this range, the relationship between ψ^2 and Graetz number is somewhat different from the rest of the flow regimes. A homogenous-type flow is usually present in the "frothy" region between the elongated gas bubbles. However, separated flow exists (similar in nature and appearance to the annular flow regime) in the elongated gas bubbles themselves. It is evident that this dual-nature (or intermittence) of the flow prohibits the formation of a long thermal entry length (in both the 1-g and μ-g cases.) The enhancement of the heat transfer coefficients is increased with larger Gz_{SL} due to a significant increase in the gas–liquid mixing. This high mixing is associated with an increase in Re_{SL} effect, which would be primarily responsible for the increase in the liquid Graetz number.

In annular flow ($We_{SG} > 20$), the resistance to heat transfer is primarily concentrated in the liquid film at the wall. Thus, the relationship between ψ^2 and Gz_{SL} is based on the heat transfer coefficient through this film. This heat transfer coefficient is a function of the actual thickness and thermal conductivity of the film. At large gas velocities, the film thickness is much less gravity-dependent. This suggests that the correlation for this range (Eq. 28) should be similar to the 1-g correlation of Rezkallah and Sims (1987) for $Re_{SL} < 2000$. The latter may be expressed as:

$$(29) \qquad \psi^2 = 1 + 4(\alpha')^{0.25} Pr_L^{-0.33}.$$

In this correlation, instead of considering the effects of Re_{SL}, Pr_L, and D/L on the heat transfer through the liquid film, only the effect of Pr_L was examined. However, the heat transfer coefficient through this film is the essential quantity that is being assessed. The relationship between this heat transfer coefficient and ψ^2 is in fact very similar in the new correlation and in the correlation of Rezkallah and Sims (1987) based on the exponents for Gz_{SL} in the former, and Pr_L in the latter.

The effect of increasing the phase velocity ratio (or the pseudo-void fraction) on the enhancement of the heat transfer varies significantly with the transition from one flow regime to the next. For $We_{SG} < 1$, α' is related to ψ^2 by a small, negative exponent. Since the turbulence intensity is relatively small for these low superficial liquid Reynolds number flows, the addition of more gas into the liquid phase would tend to decrease the normalized heat transfer coefficient due to the lower thermal conductivity of gas compared to liquid. In addition, as the liquid velocity is increased, the gas bubbles may also decrease the turbulence intensity of the liquid flow rather than increasing it. This effect has been reported by Serizawa et al. (1975a). In their work, they observed an initial reduction in turbulence intensity with the addition of gas bubbles to a single-phase liquid flow under 1-g conditions. They attributed this phenomenon to energy dissipation through the work required to float and rotate the bubbles, and what they termed the "energy absorbing" character of the gas phase. For the μ-g data, the floatation energy is assumed to be minimal based on the very low gravitational accelerations recorded and flow pattern observation, but the others may not be.

With $We_{SG} > 1$, the flow is mainly controlled by inertia forces (annular flows) and thus it is expected that the heat transfer mechanism at μ-g will not be much different from that for 1-g annular flow. The new empirical correlations that is presented here are closely related to previous 1-g correlation that have been developed by, e.g., Rezkallah and Sims (1987), and Knott et al. (1959). The latter has the form:

$$\psi^2 = (1+\alpha')^{0.25}. \tag{30}$$

The difference that has been found in the functional relationship between ψ^2 and α' over the range of gas Weber numbers reinforces the need for heat transfer correlations that are dependent on flow regimes.

Since the flows that were examined are forced-convective, mainly inertia-driven, gravity has a minor effect on heat transfer as was shown in Figs. 92– 95. This is reflected in the relatively weak exponents to which the Morton number is raised in each correlation. Morton number essentially reflects the influence of gravity and the liquid properties (particularly the liquid viscosity) on surface tension effects; i.e., bubble-size in bubbly and slug flows, and film thicknesses in annular-type flows. It is also an indicator of turbulence generation in all flow regimes, since turbulence by necessity must be maintained by some force such as buoyancy or shear, Tennekes and Lumley (1972).

The influence of Morton number is relatively small ($Mo_L^{0.058}$ to -0.059), but there is a difference depending on the flow regime. For bubbly and slug flows, there is a positive relationship between ψ^2 and Mo_L. Increased Mo_L (lower viscosity) results in an increase in turbulence generation since less viscous damping occurs. In this case, the effect of increased viscosity is more dominant than the influence of gravity on the turbulence generation of the bubbles since the Morton number is a stronger function of viscosity (μ^{-4}) than gravity (g^{-1}).

When the gas void fraction is increased ($\boldsymbol{\alpha}'$ increases) and intermittent and separated flows are present, the influence of higher viscosity tends to increase turbulence generation at the interface between the gas and the liquid because of greater interfacial shear (the shear force being proportional to μ_L). It has also been reported by Bousman and Mcquillen (1994) that larger disturbance waves exist at the interface for higher viscosity fluids under μ-g conditions. Hence, one would expect negative exponents for Mo_L.

The comparisons of the new heat transfer correlations with the data for $Re_{SL} > 2300$ are presented in Figs. 96– 99. In these figures, the experimental data are again plotted in the form of $\boldsymbol{\psi}^2$ versus the pertinent non-dimensional numbers given in Section 7.1 along with the new curve-fits. For these data, Graetz number was not used in the correlations because thermal entry length is of less consideration in this relatively high Reynolds number range. It was also found that the degree of

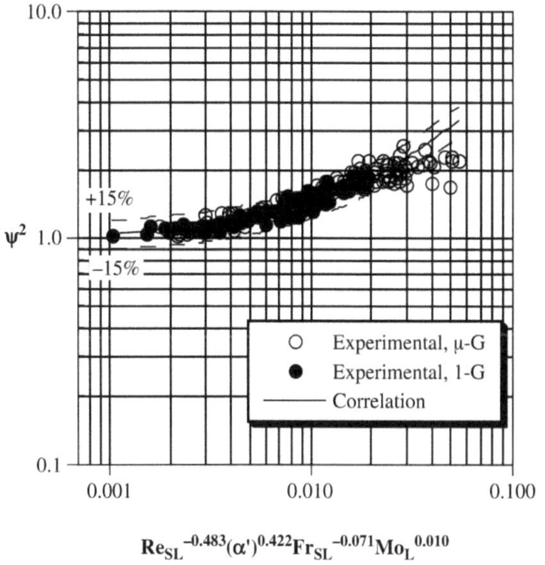

Figure 96. Two phase empirical correlation: $We_{SG} < 0.1$, $Re_{SL} > 2300$

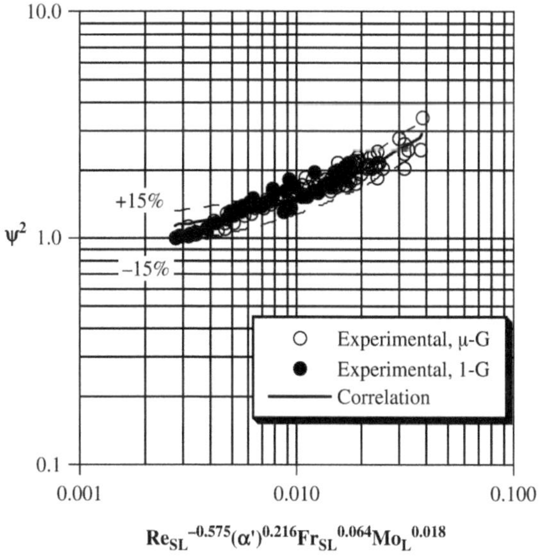

Figure 97. Two-phase empirical correlation: $0.1 < We_{SG} < 1$, $Re_{SL} > 2300$

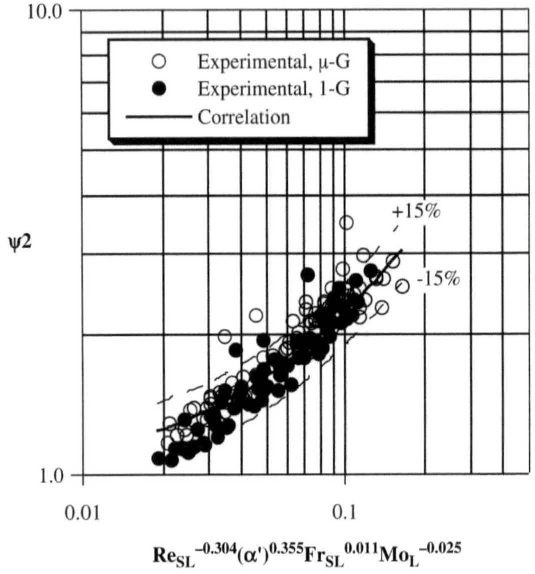

Figure 98. Two-phase empirical correlation: $1 < We_{SG} < 20$, $Re_{SL} > 2300$

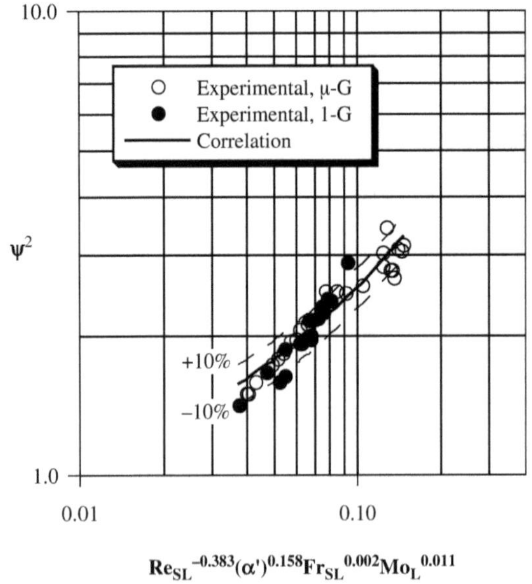

Figure 99. Two-phase empirical correlation: $We_{SG} > 20$, $Re_{SL} > 2300$

heat transfer enhancement by the addition of the second phase is not a function of the liquid Prandtl number. Therefore, in these cases, Re_{SL} is the important variable.

In Fig. 96, the comparison for $We_{SG} < 0.1$ is shown. In the figure, the correlation shown may be expressed as:

$$We_{SG} < 0.1: \quad \psi^2 = 1 + 42.0 Re_{SL}^{-0.483} (\alpha')^{0.422} Fr_{SL}^{-0.071} Mo_{L}^{0.010}. \tag{31}$$

It can be seen that the curve-fit is quite good for all points except at the far right on the graph. In this area, there is a noticeable drop-off in the normalized heat transfer coefficients compared with the correlation. The Re_{SL} for these points is very low, approaching 2300, which perhaps explains the discrepancy between the correlation and the data. The e_{rms} and e_{avg} for the correlation were found to be 15.1% and 0.7%, respectively. A complete listing of errors, Weber and Reynolds number ranges, pseudo-void fractions, and the number of data points (N) is given in Table 7 for this curve-fit and for the ones with $We_{SG} > 0.1$.

For We_{SG} between 0.1 and 1, the correlation that was developed is as follows:

$$0.1 < We_{SG} < 1: \psi^2 = 1 + 48.0 Re_{SL}^{-0.575} (\alpha')^{0.216} Fr_{SL}^{0.064} Mo_{L}^{0.018}. \tag{32}$$

The comparison of this correlation with the data is presented in Fig. 97. For this range of Weber numbers, the curve-fit is excellent with no data outside the 15% error range. This results in a root-mean-square deviation of only 9.7% and a mean deviation of 3.9%. It is also apparent that at these higher gas flow rates, the drop-off in ψ^2 that was evident in Fig. 96 is not occurring. The higher gas flow rates in the slug flow regime are increasing the actual liquid velocities such that they are well within the fully turbulent regime.

Fig. 98 shows the comparison of the correlation and the experimental data for the next range of Weber numbers which are associated with frothy slug-annular and churn-type flows. In this range, the empirical correlation is calculated by:

$$\begin{aligned} &1 < We_{SG} < 20: \\ &\psi^2 = 1 + 12.2 Re_{SL}^{-0.304} (\alpha')^{0.355} Fr_{SL}^{0.011} Mo_{L}^{-0.025}. \end{aligned} \tag{33}$$

It can be seen that few data points are outside the 15% limits, shown as dashed lines in the figure. The e_{rms} for this correlation is 11.5% and the e_{avg} was calculated to be 3.4%.

Table 7. Comparisons of the new empirical correlations to the experimental data – $Re_{SL} > 2300$

We_{SG}	α'	Re_{SL}	N	e_{rms}	e_{avg}
0.001–0.1	0.021–5.7	2332–62, 556	208	15.1	0.7
0.1–0.9	0.24–16.8	2331–60, 955	140	9.7	3.9
1.0–18.8	0.84–47.8	2354–49, 123	164	11.5	3.4
20.8–80.8	3.9–118.5	2400–23, 780	42	8.0	0.4

Finally, the last correlation for all data with gas Weber numbers greater than 20 and $Re_{SL} > 2300$ is

$$(34) \qquad We_{SG} > 20: \quad \psi^2 = 1 + 15.8 Re_{SL}^{-0.383} (\alpha')^{0.158} Fr_{SL}^{0.002} Mo_L^{0.011},$$

and Fig. 99 shows the quality of the curve-fit. In the figure, it can be seen that very nearly all of the data points are within ± 10% for an e_{rms} and e_{avg} of 8.0 and 0.4%, respectively. These deviations between the correlation and the data are quite small, but it can be seen in the figure that the correlation will tend to overpredict ψ^2 for data in the lower left-hand corner of the plot. This tendency is due to the fact that all of the correlations have been developed to approach a value of unity as the gas velocity approaches a value of zero, while some of the data were found to have a value of ψ^2 slightly below unity due to measurement uncertainty and single-phase correlation error.

Reviewing all of the correlations in Eqs. 30 through 33, the relationship between Re_{SL} and ψ^2 for all flow regimes is the most straightforward. In the correlations, it can be seen that the heat transfer augmentation by the addition of gas decreases as Re_{SL} increases. This is to be expected since as Re_{SL} becomes larger, the turbulence in the liquid phase increases (h_L becomes larger), and the influence of the gas phase becomes less.

The influence of the gas-to-liquid velocity ratio is positive throughout the range of gas Weber numbers; i.e., increasing the gas velocity increases the heat transfer augmentation (ψ^2 becomes larger). Since the liquid velocities for $Re_{SL} > 2300$ are much higher than they were for the results presented earlier in Figs. 92 and 93, the liquid flow area reduction and turbulence generation provided by the introduction of gas bubbles outweighs their effect on the properties of the mixture or their "energy absorbing" characteristics. The effect of this increased turbulence and mixing is most evident for the predominantly bubbly flow regime ($We_{SG} < 0.1$), where α' is raised to a large positive power (0.422). This was also noted by Serizawa et al. (1975b) when they measured the eddy diffusivity of heat (ε_H) and related it to quality and velocity. They found ε_H increased with liquid velocity and quality, which is directly related to α'.

Over the range of gas Weber numbers from 0.1 to 1, slug flow is predominant and the main effect of increasing the gas content is a further lengthening of the elongated Taylor gas bubbles. The presence of gas thus has a two-faceted effect on the heat transfer. Increased gas velocity will again increase the liquid velocity as well as mixing. But, in addition, since the Taylor bubbles are increasing in length, the length of the liquid slugs between the bubbles tends to decrease. It has been shown by studies cited in Collier (1972), that as the liquid slugs become shorter, heat transfer coefficients can actually decrease. The result of these opposing effects is α', having a weaker relationship with ψ^2.

Flows in the range of $1 < We_{SG} < 20$ are almost as strongly influenced by velocity ratio as the bubbly flows. This is due to the fact that this ratio determines whether there is slug flow (with very long gas bubbles) or, as the gas-to-liquid ratio is increased, frothy slug-annular or churn-type flows. A change of flow regime is occurring in this Weber number range and hence one would expect a strong functional relationship with heat transfer.

For annular flows, $We_{SG} > 20$, the relationship is only to an exponent of 0.158 because the resistance to heat transfer is mainly due to the liquid film. The liquid film will not be very sensitive to the void fraction, provided, of course, that the gas velocity is not raised so high as to lead to a complete removal of the film at the tube wall. This was not the case for the experimental results presented here. No "dry-out" at the tube wall occurred and the tube wall was always wetted by the liquid phase. This was observed in the video images collected during experiments.

The influence of the ratio of inertia forces and gravity forces, Fr_{SL}, is also of interest. Over all four We_{SG} ranges, it can be seen in the correlations that the dependence of ψ^2 on this group decreases as the gas inertia increases until $We_{SG} > 20$. At this point, the liquid Froude number has minimal influence on the heat transfer coefficient ratio ($Fr_{SL}{}^{0.002}$). This may be explained physically by the fact that at the higher gas velocities associated with annular flow, the thickness of the fluid film (which will control heat transfer in this regime) is controlled by shear forces, and gravity becomes less important.

Finally, the Morton number must be considered. It can be seen in the correlations that its influence is once again of a lower order than with the other two-phase flow dimensionless groups. However, it was possible to correlate the normalized heat transfer coefficients against Morton number, and it did change as the flow regime varied. In the correlations given in Eqs. 14–17, for all of the ranges of We_{SG} except 1–20, ψ^2 is a function of Morton number raised to approximately 0.01–0.02. For gas Weber numbers between 1 and 20, the heat transfer coefficient ratio is a slightly stronger function of Mo_L and is also negative, −0.025. As was found with Graetz number and the low Reynolds number flows, the fundamental nature of two-phase flow changes during the frothy slug-annular and churn flow regimes. A stronger inverse relationship between ψ^2 and Mo_L replaces the weaker direct relationship that exists in the other flow regimes. This implies that unlike the other flow regimes, slug-annular flow is a "quasi-equilibrium" state where surface conditions are easily perturbed by changes in gravity and physical properties.

CHAPTER 7

MODELING PERIODIC SLUG FLOWS USING A VOLUME OF FLUID METHOD

The mathematical model governing fluid motion is based on the Navier-Stokes equations and the incompressible continuity equation. For the purposes of this model, these equations are set in a cylindrical co-ordinate system, thus allowing a two-dimensional axisymmetric slug flow simulation to be performed. The model solves for the discrete primitive field variables u, v, P, and F at discrete points in the solution domain. For a two-component gas–liquid flow, inertial and viscous effects are dominated by the liquid phase. Thus, the governing equations are solved in the liquid phase only. A fractional volume-based interface-tracking method is used to specify the position of the gas–liquid interface. The interface-tracking algorithm employed is a variation of the standard VOF method pioneered by Hirt and Nicholas (1981). The model is used to simulate laminar slug flows in normal and zero-gravity situations.

1. ASSUMPTIONS

Since the flow of interest occurs in a pipe and is composed of the liquid and gas phases, a number of assumptions are made to simplify the model. These are as follows:

1. The bubbles are two-dimensional and axisymmetric. The bubbles captured on videotape appear highly axisymmetric and strongly resemble 1-g Taylor bubbles.
2. The flow is incompressible. The simulation employs periodic boundary conditions to model the repetitive nature of the flow. Since the same bubble continuously enters and exits the solution domain, the modeled bubble is assumed to be constant in volume.
3. The fluid is Newtonian. Air and water are Newtonian fluids.
4. The flow is laminar. To ease the computational load, a turbulence model is not implemented. This should not be a major problem as the liquid side Reynolds number based on the superficial liquid velocity is in the range of 1100–2200.
5. Surface tension is constant. The absence of temperature gradients and surface active agents allows surface tension to be assumed a constant.

6. No heat or mass transfer. There is no heat or mass transfer across the interface or through the pipe wall.
7. Inertial and viscous forces in the gas are negligible. The density ratio between air and water is 833.3, which indicates that the inertia of the water is highly dominant. The dynamic (absolute) viscosity ratio between the water and air is 55.6, which indicates again that the water phase viscous forces dominate.

2. GOVERNING EQUATIONS

The governing equations are the radial and axial momentum equations in two-dimensional axisymmetric form:

$$\frac{\partial u}{\partial t}+u\frac{\partial u}{\partial r}+v\frac{\partial u}{\partial z}=-\frac{1}{\rho}\frac{\partial p}{\partial r}+\nu\left[\frac{1}{r}\frac{\partial}{\partial r}\left(r\frac{\partial u}{\partial r}\right)+\frac{\partial^2 u}{\partial z^2}-\frac{u}{r^2}\right]+\frac{1}{\rho}g, \tag{1}$$

$$\frac{\partial v}{\partial t}+u\frac{\partial v}{\partial r}+v\frac{\partial v}{\partial z}=-\frac{1}{\rho}\frac{\partial p}{\partial z}+\nu\left[\frac{\partial^2 v}{\partial r^2}+\frac{\partial^2 v}{\partial z^2}+\frac{1}{r}\frac{\partial v}{\partial r}\right]+\frac{1}{\rho}g, \tag{2}$$

and the continuity equation,

$$\frac{\partial u}{\partial r}+\frac{\partial v}{\partial z}+\frac{u}{r}=0. \tag{3}$$

In these equations, u and v are the radial and axial velocity components respectively, t is time, r and z are the radial and axial co-ordinates, and p is the pressure.

3. INTERFACE-TRACKING MODEL

The interface position is tracked using the VOF method pioneered by Hirt and Nichols (1981). The VOF method introduces an additional scalar field variable, the VOF function, F, which is defined as unity in the liquid and zero in the gas. This F function acts as a marker to differentiate gas regions from liquid regions. The interface is advected according to:

$$\frac{\partial F}{\partial t}+u\frac{\partial F}{\partial r}+v\frac{\partial F}{\partial z}=0. \tag{4}$$

Since the VOF method only requires the storage of one piece of information about the interface (the F function), it provides a simple and compact method for tracking an arbitrary interface as it moves, deforms, and even undergoes topology changes.

4. BOUNDARY CONDITIONS

The experimental data obtained by Elkow (1996), that is used for model verification, shows long sequences of similarly sized gas bubbles – an indicator of a highly periodic flow. To simplify the model, these bubbles are modeled by applying a set of periodic boundary conditions at the inflow/outflow boundaries of the solution domain. These periodic boundary conditions reduce the size of the solution domain required to model the development of the bubble. The periodic solution domain becomes representative of a unit wavelength of the periodic cycle.

Since the application of periodic boundary conditions makes the velocity and pressure fields identical at the inlet and the outlet, the actual pressure field, p, must be represented by a combination of an overall pressure gradient, $\Delta p/L$, and a pressure deviation field, $\hat{p}$, as shown below:

$$p(r,z,t) = \frac{\Delta p}{L} z + \hat{p}(rz,t). \tag{5}$$

Differentiating with respect to r yields

$$\frac{\partial p}{\partial r} = \frac{\partial \hat{p}}{\partial r}. \tag{6}$$

Differentiating with respect to z yields

$$\frac{\partial p}{\partial z} = \frac{\Delta p}{L} + \frac{\partial \hat{p}}{\partial z}. \tag{7}$$

The periodic form of the momentum equations is arrived at by substituting the differentiated equations for the pressure derivatives in the momentum equations. The overall pressure drop, $\Delta p/L$, is then treated as a body force in the axial momentum equation.

$$\frac{\partial u}{\partial t} + u\frac{\partial u}{\partial r} + v\frac{\partial u}{\partial z} = -\frac{1}{\rho}\frac{\partial \hat{p}}{\partial r} + \nu\left[\frac{1}{r}\frac{\partial}{\partial r}\left(r\frac{\partial u}{\partial r}\right) + \frac{\partial^2 u}{\partial z^2} - \frac{u}{r^2}\right] + \frac{1}{\rho}g, \tag{8}$$

$$\frac{\partial v}{\partial t} + u\frac{\partial v}{\partial r} + v\frac{\partial v}{\partial z} = -\frac{1}{\rho}\left[\frac{\Delta p}{L} + \frac{\partial \hat{p}}{\partial z}\right] + \nu\left[\frac{\partial^2 v}{\partial r^2} + \frac{\partial^2 v}{\partial z^2} + \frac{1}{r}\frac{\partial v}{\partial r}\right] + \frac{1}{\rho}g, \tag{9}$$

The boundary conditions along the tube wall are the standard no-slip and no-penetration conditions. Since the model is axisymmetric, the centerline boundary condition is a free slip condition.

The only boundary condition which remains unaddressed is the interfacial boundary condition. Since the inertial and viscous effects of the gas are neglected, this leaves a solution domain where the governing equations are solved only in the liquid phase, coupled with a boundary condition at the interface which provides the link between the two phases. The interfacial boundary condition consists of two components: a kinematic boundary condition and a stress boundary condition. Since the gas inertia is neglected, the velocity in the gas is not solved for, rendering the kinematic condition irrelevant. The stress boundary condition is composed of two components: a normal and a tangential. The normal component is

$$\left(p - 2\mu\frac{\partial u_n}{\partial n}\right)_{\text{gas}} = \left(p - 2\mu\frac{\partial u_n}{\partial n}\right)_{\text{liquid}} - \sigma\kappa \tag{10}$$

and the tangential component is

$$\left[\mu\left(\frac{\partial u_n}{\partial s}+\frac{\partial u_s}{\partial n}\right)\right]_{\text{liquid}}=\left[\mu\left(\frac{\partial u_n}{\partial s}+\frac{\partial u_s}{\partial n}\right)\right]_{\text{gas}}+\frac{\partial \sigma}{\partial s}, \tag{11}$$

where u_n and u_s are the normal and tangential velocities respectively, σ is the surface tension, κ is the interface curvature, and μ is the viscosity.

By neglecting the viscosity of the gas and assuming surface tension to be constant, the tangential stress in the liquid is forced to be zero. The viscous contribution to the normal stress is also neglected, yielding:

$$p_{\text{liquid}} - p_{\text{gas}} = \sigma\kappa \tag{12}$$

Thus, the pressure jump at the interface is due solely to surface tension.

5. SUPERFICIAL TWO-PHASE FLOW PARAMETERS

In general, conventional two-phase flow experimental measurements are made of bulk quantities such as superficial gas and liquid velocities, the pressure drop between different locations, and the average void fraction of the flow. On the other hand, numerical modeling yields velocities, pressures, and fluid phases for discrete control volumes that comprise the numerical grid. Since the model uses primitive field variables, the variables need to be related to the measured quantities before model performance can be assessed.

Through manipulation of the more detailed numerical results, bulk averaged quantities typical of experimental measurements can be obtained. By taking advantage of the assumed periodic nature of the flow, it is a simple matter to obtain the averaged quantity. The void fraction, α, and the superficial gas and liquid velocities (v_{sg}, v_{sl}) can be calculated from volume integrals over the periodic solution domain.

$$v_{sl} = \frac{1}{V_{\text{cell}}}\iiint_{V_{cell}}\left(V\cdot\hat{k}\right)F dV, \tag{13}$$

$$v_{sg} = \frac{1}{\text{Ɏ}_{cell}}\iiint_{\text{Ɏ}_{cell}}\left(V\cdot\hat{k}\right)(1-F)d\text{Ɏ}, \tag{14}$$

$$\alpha = \frac{1}{\text{Ɏ}_{cell}}\iiint_{\text{Ɏ}_{cell}}(1-F)d\text{Ɏ}. \tag{15}$$

Recall that the model does not calculate any velocity components in the gas. Since F equals zero in the gas, this means the (1-F) term equals one in the superficial gas velocity equation. The equation for the superficial gas velocity can then be rewritten as:

$$v_{sg} = \frac{1}{\text{Ɏ}_{cell}}\iiint_{\text{Ɏ}_{bubble}}\left(V\cdot\hat{k}\right)d\text{Ɏ}, \tag{16}$$

$$v_{sg} = \frac{\text{Ɏ}_{bubble}}{\text{Ɏ}_{cell}}\left[\frac{1}{\text{Ɏ}_{bubble}}\iiint_{\text{Ɏ}_{bubble}}\left(V\cdot\hat{k}\right)d\text{Ɏ}\right]. \tag{17}$$

When this triple integral is evaluated only in the gas, the integral simplifies to v_b, the bubble velocity, so that

$$v_{sg} = \alpha v_b. \tag{18}$$

6. NUMERICAL IMPLEMENTATION

The governing equations are solved using a finite difference method on a staggered grid. The solution procedure is essentially a SOLution Algorithm (SOLA) type method as described by Hirt and Nicholas (1981). This method requires the solution of a Poisson equation for pressure at every time step to obtain a proper time-advanced velocity field. The Poisson pressure equation (PPE) is used to obtain a pressure correction which is used to correct the provisional velocity field to ensure that zero divergence is satisfied. The basic solution procedure can be summarized as:

- The explicit approximations to the momentum equations are used to obtain an initial guess for the $n+1$ time level velocities using time level n values for all advection, pressure, and viscosity terms.
- To satisfy continuity, the pressures are iteratively adjusted in each control volume and velocity changes caused by each pressure correction change are added to the velocities calculated in the first step.

The PPE has the form:

$$\nabla^2 \delta p = \frac{\rho}{\Delta t} \nabla \cdot \tilde{V} \tag{19}$$

The solution of the PPE yields pressure corrections which are applied to the provisional velocity field to yield a divergence free time-advanced velocity field. These corrections are applied using:

$$u_{i,j}^{n+1} = \tilde{u}_{i,j} + \frac{\Delta t \Delta r}{\rho}(\delta p_{i,j} - \delta p_{i+1,j}), \tag{20}$$

$$v_{i,j}^{n+1} = \tilde{v}_{i,j} + \frac{\Delta t \Delta z}{\rho}(\delta p_{i,j} - \delta p_{i,j+1}). \tag{21}$$

7. VOLUME OF FLUID INTERFACE-TRACKING METHOD

The Volume of Fluid (VOF) interface-tracking method was originally developed by Hirt and Nichols (1981). Volume Of Fluid allows free surfaces to be represented by discrete line segments within the computational grid. The VOF method discards explicit information about interface location in favor of discrete fractional fluid volume data, F. The VOF procedure is composed of two parts – the first is the reconstruction of the interface and the second is the advection of the interface. During each time step, the precise location of the interface is unknown. When it is required, the free surface is reconstructed using the local VOF data. This reconstructed interface is approximate and not necessarily continuous – in general, it is a set of discrete, discontinuous line segments. The properties of the F function used in the VOF algorithm are

$F = 1$	liquid
$0 < F < 1$	interface
$F = 0$	gas

8. MODEL GEOMETRY

The experimental conditions provide a set of specifications used to construct the solution domain as well as to determine the final desired flow conditions. Based on the dimensions of the experimental apparatus, the unit cell has a known diameter of 9.525 mm. Image analysis performed on the video recording of the experimental flow indicate that the unit cell length is 55.15 mm. The experimentally measured flow conditions such as void fraction and superficial gas and liquid velocities comprise the desired final flow conditions. The experimentally measured void fraction is used to specify the size of the initial bubble, while the superficial velocities are used to specify target flow conditions. The initial geometry of the gas bubble in the solution domain is represented by two hemispheres capping a cylinder.

The solution domain was discretized into a uniform computational grid whose size was limited by computing constraints. To ensure reasonable run times, the grid size was limited to 50×579. With the 50×579 grid, the typical CPU time required to simulate 0.25 sec on a Sun UltraSparc 1/170 was about 10 h.

9. SIMULATION RESULTS

The experiments performed by Elkow (1996) were conducted using air–water flows under simulated microgravity conditions attained aboard the NASA KC-135 aircraft. Although there is a difference in gravity level between the experiments and the numerical model, the quality of the microgravity environment aboard the KC-135 aircraft during the experiment was typically on the order of $10^{-3}g$ and assumed to be negligible. The periodic slug flows observed in the experimental data provides the basis for the flow conditions modeled in the zero-gravity simulations. Additional modeling was performed to examine the differences between zero gravity and normal gravity flows under similar flow conditions.

9.1 Zero-Gravity Slug Flow

The zero-gravity slug flow simulation was performed for conditions similar to what was previously reported for the experimental data of Elkow (1996). The VOF modeling technique used is able to provide detailed interface information, allowing comparisons to be made with video recordings taken during the experiment. In the simulation, the superficial liquid velocity was matched to the experimentally observed superficial liquid velocity. Looking only at the steady-state condition, the numerical bubble velocity was found to be 0.698 m/s. This is very close to the experimental velocity of 0.708 m/s extracted from the video images. Knowing the numerical bubble velocity, the numerical superficial gas velocity was calculated from the product αv_b. The numerical superficial gas velocity was estimated to be 0.384 m/s. This is within 13% of the experimental superficial gas velocity of 0.341 m/s. A side-by-side illustration of the numerical and experimental bubble profiles is shown in Fig. 100. This figure shows how closely the model predicts the bubble shape.

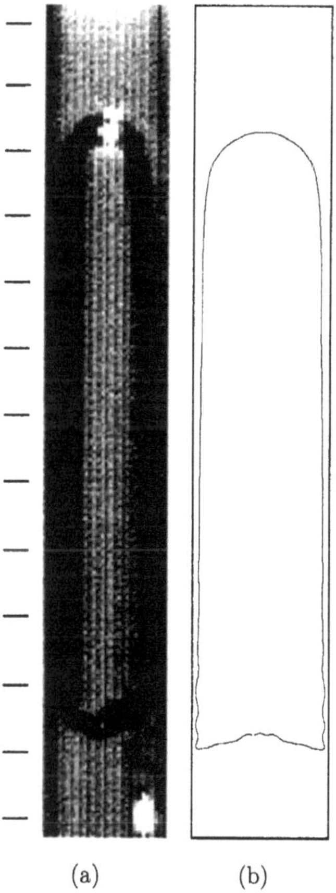

Figure 100. Comparison of (a) experimental video image and (b) numerical result at 4.880s for zero-gravity case

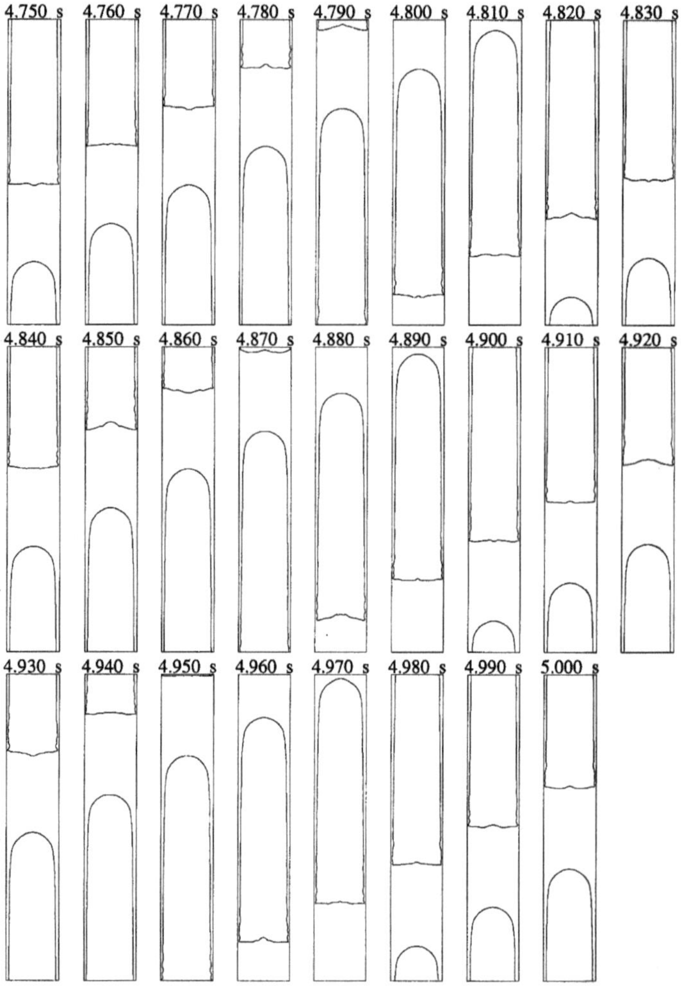

Figure 101. Sequence illustrating time variation of bubble shape from 4.75 s to 5.00 s in 0.01 s increments

A sequential progression of the quasi-steady state bubble shape is shown in Fig. 101 for a period of 0.25 s (in increments of 0.01 s).

Besides bubble shapes, the simulation can also provide other details of interest, such as the velocity field and streamlines. Figure 102 shows the velocity profiles in the liquid slug at 4.861 s.

The primary features noted in this figure are the velocity peak away from the centerline in the liquid slug and the flatness of the velocity profile in the liquid core. The cause of this particular distribution of the velocity profiles can be explained by examining the motion of the gas bubble from a different frame of reference.

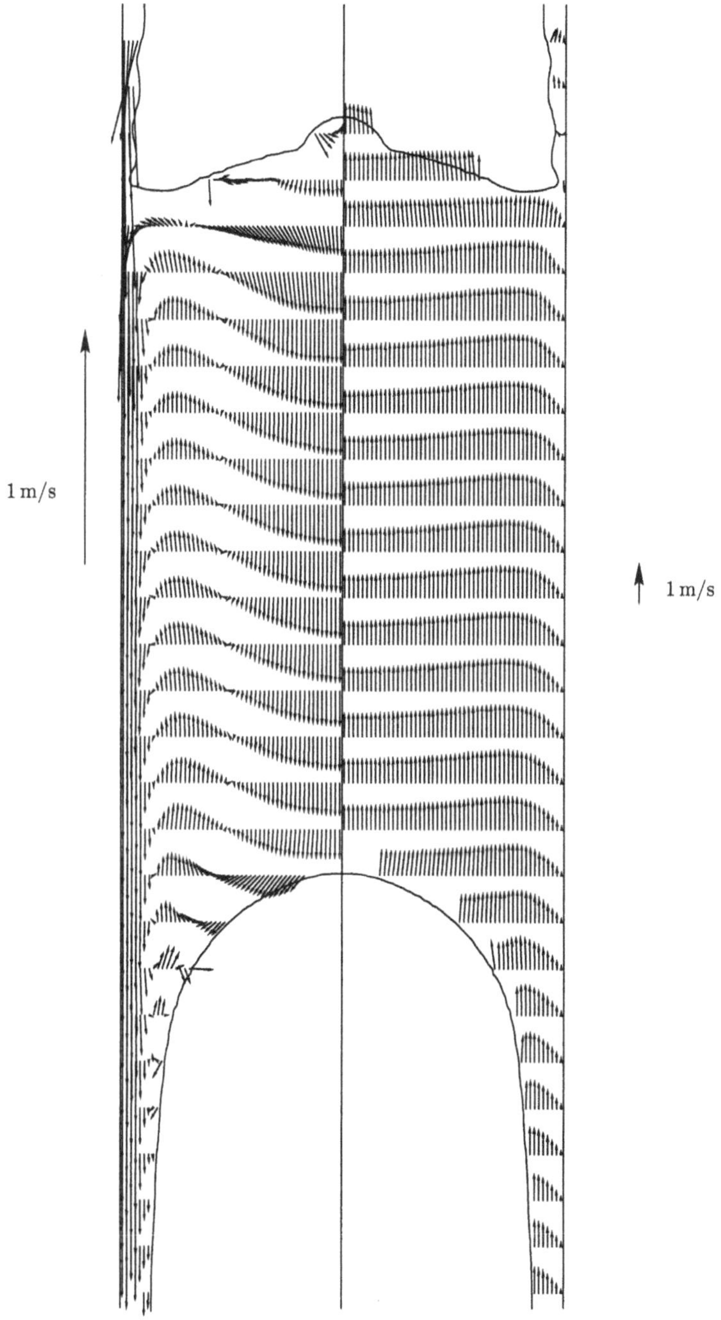

Figure 102. The velocity field in the liquid slug at 4.861 s for the zero-gravity case. Vectors are shown at every node radially and at every eighth node axially

Figure 103 shows the streamlines and velocity vectors for the liquid slug plotted relative to an observer placed on the "nose" of the bubble. This frame of reference provides a much more interesting view of the flow field since it reveals more of the flow structure. The figure shows the streamlines on the left half and the velocity field on the right half.

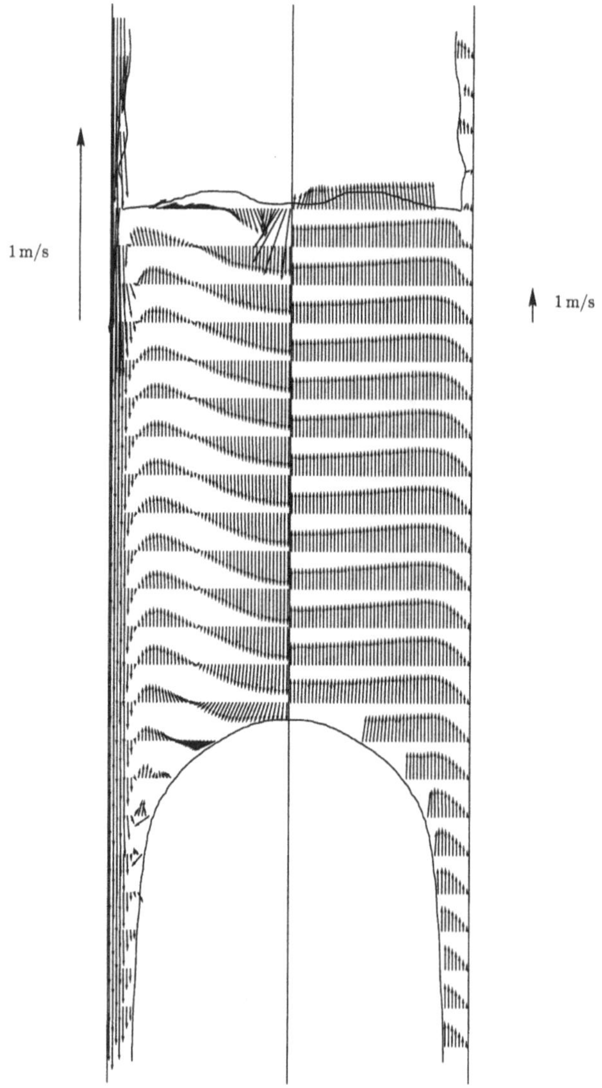

Figure 103. A combined streamline-vector field at 4.861 s

As can be seen from Fig. 103, there are two distinct flow circulation zones. Within the liquid slug itself, the core region is dominated by a counterclockwise circulation, while the wall region is dominated by a clockwise circulation. Although from a cursory observation of the general shape of the bubble, it appears that there is a strong resemblance to a typical single 1-*g* Taylor bubble rising in a quiescent liquid, the details of the flow field are quite different from those reported earlier for 1-*g* flows (Bugg et al., 1998). In that study, it was found that the Taylor bubbles possess a wake that consists exclusively of a single clockwise circulation zone. The flow pattern that develops in that case makes perfect sense. When the flow field is examined relative to the velocity of the nose, the wake is a stationary plug of liquid beside a moving wall. The flow is essentially the same as a lid-driven cavity flow where a moving lid induces a circulation within the cavity due purely to viscous drag. The reason the 0-*g* slug flow is somewhat different is because of the presence of a trailing bubble. As the near wall flow approaches the nose of the bubble and begins to enter the film region, a large amount of liquid appears to reverse and flow upwards. It is this upward flow that is responsible for the formation of the two circulation regions in the liquid slug. The reason for the flow reversal is attributed to the film of the trailing bubble being unable to accommodate the liquid flow, forcing the remainder to reverse. The reason the film is unable to accommodate the liquid flow is due to the entrainment of additional liquid by the wall, much like a lid-driven cavity flow where the moving lid drags the fluid in the cavity. It is this newly entrained liquid that ends up reversing.

9.2 Normal Gravity Periodic Slug Flow

A simulation was performed for a virtually identical configuration as the zero-gravity case. The only differences being the presence of gravity and a corresponding change in the overall pressure drop to maintain the same superficial liquid velocity. The results shed light on the similarities and the differences between the two cases of zero gravity and normal gravity.

Looking only at the steady-state condition, the numerical bubble velocity was found to be 0.769 m/s. As expected, the bubble velocity is higher in the 1-*g* case than the 0-*g* case. This is due to the density difference in the presence of gravity causing "slip" to occur between the two phases. The end result of this slip is a faster moving bubble under normal gravity.

The progression of the quasi-steady bubble shape results for the 1-*g* simulation are shown in Fig. 104 for a period of 0.25 s.

Again, from a cursory inspection, the differences between the 0-*g* and the 1-*g* bubble shapes appear relatively minor. They appear to exhibit the same characteristics: smooth, round nose transitioning to a wavy tail. Ideally, a side-by-side comparison would be useful in determining the extent of the similarities between the two cases. Unfortunately, due to algorithm deficiencies, the side-by-side comparison

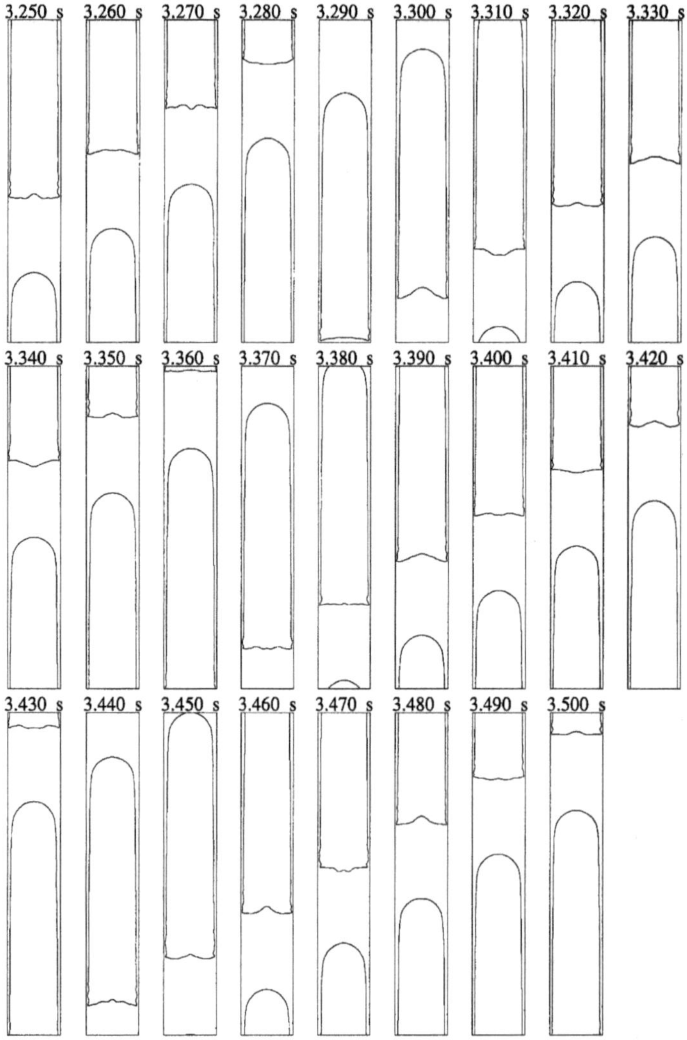

Figure 104. Progression of 1-g quasi-steady bubble shapes from 3.25 s to 3.50 s in 0.01 s increments.

(Fig. 105) instead showcases the size difference between the two bubbles. While the original simulation initial conditions set the gas volume for the two bubbles to be the same, the gas volume loss inherent with a low order interface-tracking algorithm over an extended simulation time make the resulting bubble volumes different. Since the normal gravity simulation took less time to reach a quasi-steady state, its gas volume loss is less, with the result being a larger bubble. The initial condition

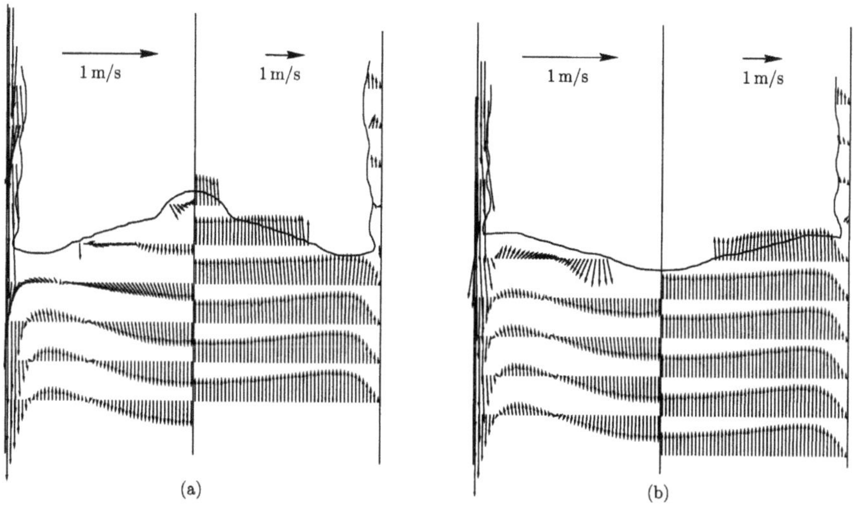

Figure 105. Comparison of (a) normal gravity and (b) zero-gravity bubbles

specifies a gas void fraction of 63.5%. The zero-gravity simulation concludes with the gas void fraction dropping to 55.5%. The normal gravity simulation finishes with a gas void fraction of 58.1%. Certainly, a higher order tracking algorithm would provide a better result, but at the cost of significantly greater computation resources.

The velocity profiles at 3.40 s in Fig. 106 show similar features to the 0-*g* case; the velocity peaks near the wall and core profile is relatively flat. This is expected since the flow configurations are essentially identical.

The combined streamline-vector field in Fig. 107 would also appear to confirm the lid-driven cavity analogy. The two zone circulation is evident, strongly resembling the 0-*g* result. It would appear that the two zone circulation is an artifact of geometry and not of gravity.

10. CONCLUSIONS

The VOF method was applied to the modeling of zero gravity and normal gravity periodic slug flows consisting of air and water. In the comparisons of the zero-gravity results with experimental measurements, good agreement was found in general bubble shape and gas velocities. The model predicts that a two-zone recirculation pattern develops in the wake of the bubble. The normal gravity results confirm that the two-zone recirculation observed in the wake of the bubble is an

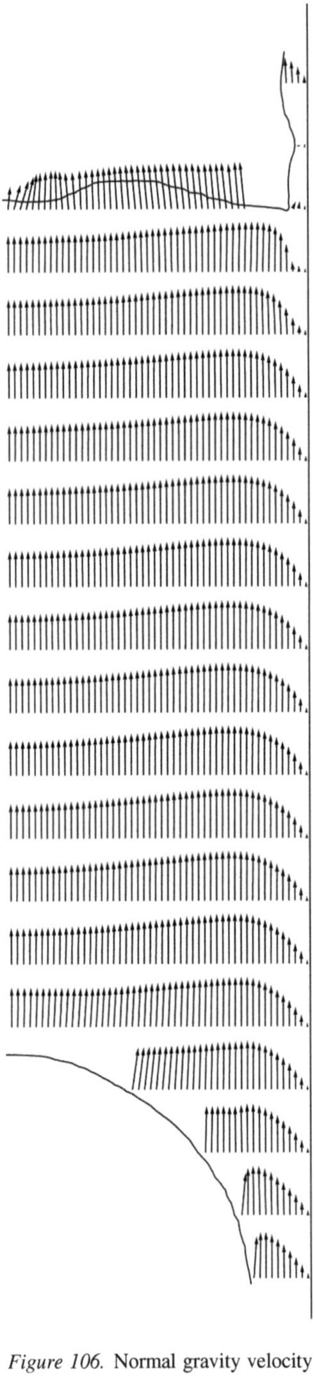

Figure 106. Normal gravity velocity profile in the liquid slug at 3.40 s

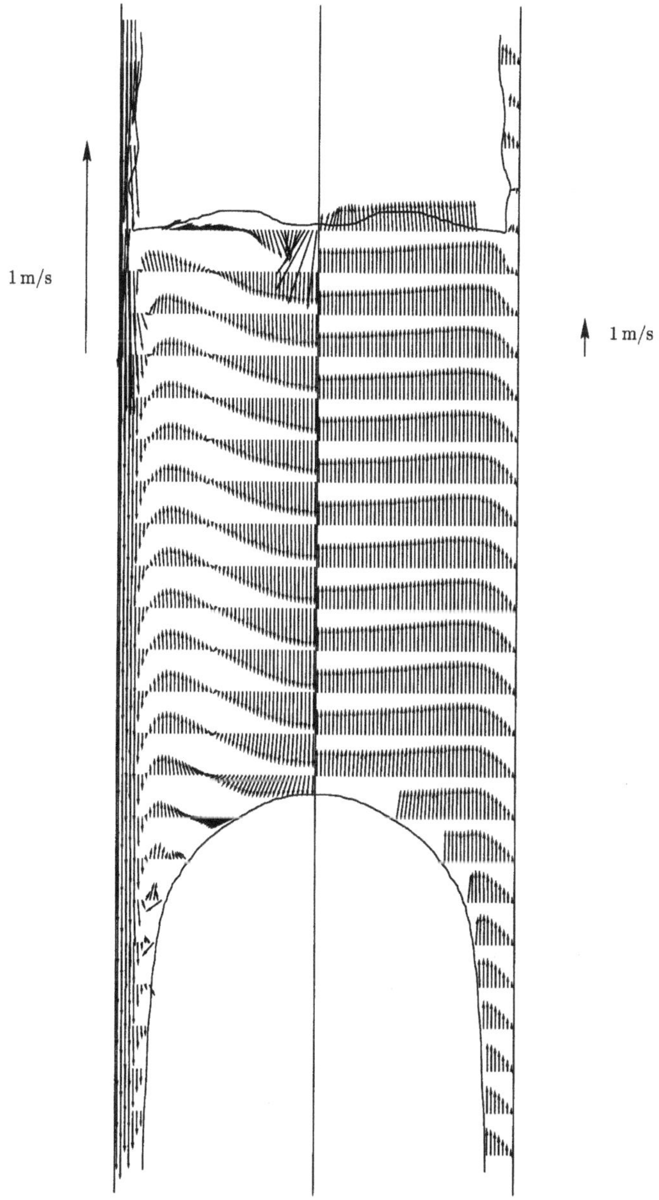

Figure 107. Normal gravity liquid slug streamline-vector field at 3.40 s

artifact caused by the presence of a trailing bubble, and not an artifact of gravity level. Also as expected, the main difference that arises between zero gravity and normal gravity flows maintained under similar flow conditions is the presence of slip between the phases.

CHAPTER 8

SUMMARY AND CONCLUSION

It should be pointed out that the study of two-phase flow at microgravity conditions is far from complete. (The same can be said for the study at 1-g conditions.) The predictions of flow behavior is characterized by empiricism due to its complexity. Moreover, it is somewhat restricted by the fact that only very limited experimental data have been collected at actual microgravity conditions. Flow conditions were usually restricted to small diameter tube and low fluid flow rates.

Thanks to the efforts of many researchers all over the world, some progress has been made toward the understanding of the phenomena. It was found that gas–liquid flow at microgravity conditions can be classified as bubble flow, slug flow, frothy slug-annular flow (transitional flow), and annular flow. Stratification was not observed at $0.01g_0$ regardless of the tube orientation. For the transition between various flow patterns, transitions based on the Weber number criteria seem to give the best overall prediction for the available experimental data.

It was found that the values of the frictional pressure drop of gas–liquid flow change very little when gravity is reduced while keeping the flow rates unchanged. The difference is within 14% with those at μ-g being slightly higher. The homogeneous model, Chisholm's correlation, and Friedel's model all gave reasonable predictions.

Based on the large pool of two-phase heat transfer data points that have been collected both in-flight and on-ground, several new and interesting conclusions have been reached. These results may be summarized as follows:

- For superficial liquid Reynolds numbers (Re_{SL}) less than 2300, μ-g reduces the heat transfer coefficient by up to 50% at the lowest gas flow rates in the slug flow regime compared with 1-g. As the gas quality is increased, the difference between 1-g and μ-g data becomes smaller.
- At higher liquid flow rates in "transitional or turbulent" flow ($\mathrm{Re}_{SL} > 2300$) and very low gas qualities there is again a tendency for the 1-g heat transfer coefficients to be greater than those at μ-g. The two approach one another until above $x \approx 0.002$, the μ-g heat transfer data are greater than the 1-g. However, the magnitude of the difference is on the order of 10–15%, which is the same order of magnitude as the uncertainty of the measurements.

- Through an examination of local heat transfer coefficients, the influence of gravity was determined to be both a single-and a two-phase effect. It can be considered a single-phase effect because mixed convection in the liquid phase is significant under 1-g conditions for the air–water data that were recorded up to a superficial liquid Reynolds number of approximately 8000. The two-phase effect manifests itself in that the addition of gas bubbles tends to promote heat transfer more under 1-g conditions than low gravity due to increased turbulence caused by the buoyancy forces acting on the gas.

Through ground and low-gravity flight experiments with two-phase gas–liquid flows having liquid viscosities and Prandtl numbers that varied by more than an order of magnitude, it was found that unlike single-phase flows, Re_{SL} and Pr_{L} numbers and fluid properties are not the only factors governing heat transfer. New dimensionless groups need to be considered involving gravitational acceleration, flow regime as well as some properties previously ignored such as surface tension. These dimensionless groups were determined to be the Weber, Froude, and Morton numbers. The Weber number was found to provide a means of incorporating flow regime while the force balance between body and surfaces tension forces was represented by the Froude and Morton numbers.

Eight new correlations have been developed to model two-phase gas–liquid flow over a wide range of flow rates and gravity levels from $g/g_0 = 0.01$–1.0. Through these correlations, it can be seen that gravity does not have a profound effect on the heat transfer capability of the flows, if the effects of mixed convection are eliminated. Rather, the influence of gravity is felt subtly in regard to film thickness and bubble size. The new correlations address the effect of flow regime and the influence of gravity on the interface between the two phases. For $Re_{\mathrm{SL}} < 2300$, the correlations fit the new data within 25%. Above $Re_{\mathrm{SL}} = 2300$, the empirical correlations are within 15% of all of the data that were generated.

APPENDIX A

TABLE KEY

Code:	Data Set – date (MMDDYY), Flight (AM/PM) or Ground (A/B/C), Data Point (#)
V_{sl}:	Superficial liquid velocity (m/s)
V_{sg}:	Superficial gas velocity (m/s)
α:	Void fraction
gz/g0:	Ratio of gravitational acceleration to normal gravity in z-axis
Tf,avg.:	Average Fluid temperature (°C)
Ts,avg.:	Average heated test section surface temperature (°C)
Pr_L:	Liquid Prandtl number
x:	Gas mass quality
q'':	Heat flux at heated test section (W/m^2)
h_L:	Single-phase convective heat transfer coefficient (W/m^2–K)
$\psi 2$:	Two-phase to single-phase heat transfer coefficient ratio
Pin:	Absolute test section pressure (Pa)
dp/dzf:	Pressure gradient (Pa/m)
FP:	Flow pattern B: bubbly
	BS: bubbly-slug transition
	S: slug
	SC: slug-churn transition
	C: churn
	CA: churn-annular transition
	FSA: frothy slug-annular transition
	A: annular

February 1993, Ground Data
65% Glycol/Water-Air, 9.525 mm i.d., L (pressure taps): dz1 = 16 cm, dz2 = 39 cm

Code	V_{sl}	V_{sg}	α	gz/g0	Tf,avg	Ts,avg	Pr_L	x	q″	h_L	$\psi 2$	Pin	dp/dzf	FP
020693C1	0,81	0,24	–	1,0	30,8	52,0	77,26	3,10E-4	22934,4	661,8	1,63	103694	–	S
020693C2	0,78	0,35	–	1,0	31,2	50,1	76,05	4,76E-4	23128,6	652,2	1,88	105714	–	S
020693C3	0,81	0,55	–	1,0	31,5	48,2	75,24	7,04E-4	23173,9	655,3	2,11	106052	–	S
020693C4	0,80	1,04	–	1,0	31,8	46,7	74,35	1,36E-3	23407,3	649,7	2,41	107059	–	S
020693C5	0,80	1,72	–	1,0	32,0	46,3	73,81	2,31E-3	23345,9	646,8	2,53	109417	–	S
020693C6	0,83	3,30	–	1,0	32,6	45,5	71,97	4,49E-3	23317,7	649,4	2,80	115140	–	SC
020693C7	0,81	5,33	–	1,0	33,1	45,3	70,49	7,67E-3	23483,8	643,8	3,01	119856	–	SC
020693C8	0,84	7,74	–	1,0	33,5	45,3	69,40	1,18E-2	23514,7	648,2	3,12	131639	–	CA
020693C9	0,83	10,65	–	1,0	33,9	45,2	68,45	1,70E-2	23482,8	644,4	3,25	137024	–	A
020693C10	0,82	16,30	–	1,0	33,8	44,6	68,64	2,86E-2	23582,7	642,3	3,43	151504	–	A
020693C11	1,11	0,23	–	1,0	32,4	51,7	70,77	4,46E-4	23926,5	709,1	2,01	109079	4302	BS
020693C12	1,11	0,46	–	1,0	33,0	49,8	69,46	6,82E-4	23601,5	703,5	2,18	109755	6073	S
020693C13	1,10	0,69	–	1,0	33,5	48,9	69,46	6,84E-4	23601,5	703,5	2,18	110092	7065	S
020693C14	1,09	0,91	–	1,0	33,8	48,3	68,72	9,14E-4	23265,4	700,8	2,29	111437	7854	S
020693C15	1,08	1,14	–	1,0	34,0	48,2	68,25	1,16E-3	23293,6	697,3	2,35	111437	8395	S
020693C16	1,10	1,72	–	1,0	34,5	48,0	66,92	1,76E-3	23267,4	698,7	2,47	114802	10154	S
020693C17	1,14	3,21	–	1,0	35,3	47,6	64,83	3,38E-3	23260,4	701,6	2,70	122545	–	SC
020693C18	1,12	5,03	–	1,0	36,2	47,6	62,49	5,68E-3	23221,2	696,3	2,95	130626	–	SC
020693C19	1,14	7,37	–	1,0	36,7	47,7	61,45	8,85E-3	23314,0	697,3	3,05	141065	–	SC
020693C20	1,13	10,38	–	1,0	36,9	47,7	60,81	1,27E-2	23298,6	695,2	3,13	143085	–	CA
020693C21	1,59	0,29	–	1,0	35,0	51,5	65,52	2,03E-4	23131,1	996,7	1,40	113457	8247	BS

020693C22	1,60	0,55	–	1,0	35,0	49,3	65,42	3,90E-4	23154,8	999,8	1,63	115140	10328	S
020693C23	1,62	0,86	–	1,0	35,0	48,4	65,54	6,17E-4	23279,5	1007,0	1,74	118842	12621	S
020693C24	1,60	1,07	–	1,0	35,1	48,1	65,30	7,84E-4	23234,2	995,3	1,80	119856	13288	S
020693C25	1,61	1,57	–	1,0	35,2	47,8	64,94	1,19E-3	23225,4	1000,7	1,86	124565	15271	S
020693C26	1,61	2,50	–	1,0	35,7	47,5	63,74	1,99E-3	23265,4	1008,4	1,96	131301	–	SC
020693C27	1,63	4,43	–	1,0	36,5	47,3	61,84	3,87E-3	23260,0	1026,7	2,12	146450	–	SC
020693C28	1,60	6,64	–	1,0	37,1	47,5	60,45	6,29E-3	23309,7	1023,1	2,20	156889	–	CA
020693C29	1,65	7,80	–	1,0	37,5	47,7	59,51	7,66E-3	23335,6	1051,0	2,20	167328	–	CA
020693C30	1,60	8,72	–	1,0	37,9	48,0	58,67	8,94E-3	23337,6	1035,7	2,25	170693	–	CA
020693C31	1,92	0,22	–	1,0	35,6	51,7	54,03	1,27E-4	23089,3	1165,5	1,23	115140	9861	B
020693C32	1,93	0,32	–	1,0	35,7	50,8	53,68	1,93E-4	23099,6	1170,0	1,31	117160	10375	BS
020693C33	1,88	0,47	–	1,0	35,6	49,9	53,93	2,91E-4	23382,1	1143,5	1,44	118511	11417	BS
020693C34	1,92	0,91	–	1,0	35,9	48,5	53,29	5,75E-4	23418,3	1160,5	1,61	124234	14661	S
020693C35	1,91	1,47	–	1,0	36,0	48,1	52,95	9,75E-4	23370,1	1155,8	1,68	129957	18322	S
020693C36	1,93	2,69	–	1,0	36,7	47,9	51,43	1,94E-3	23366,9	1175,0	1,78	143085	–	S
020693C37	1,91	4,25	–	1,0	37,4	48,0	59,66	3,31E-3	23341,8	1178,3	1,90	153186	–	SC
020693C38	1,91	6,23	–	1,0	38,2	48,2	58,05	5,28E-3	23343,9	1191,6	1,97	167659	–	CA
020693C39	1,92	8,36	–	1,0	38,8	48,6	56,67	7,81E-3	23277,5	1209,2	2,00	186855	–	CA
020693C40	2,02	9,16	–	1,0	37,9	46,7	58,66	8,70E-3	20916,4	1231,9	1,95	198638	–	A

February 1993, Ground Data
Water-Air, 9.525 mm i.d.

Code	V_{sl}	V_{sg}	α	$gz/g0$	Tf,avg.	Ts,avg.	Pr_L	x	q″	h_L	$\psi 2$	Pin	dp/dzf	FP
051993A1	0,37	0,11	–	1,0	27,8	37,8	5,68	3,46E-4	23489,3	1188,9	1,98	102349	–	S
051993A2	0,36	0,21	–	1,0	28,4	38,0	5,60	7,08E-4	23439,0	1179,4	2,07	102349	–	S
051993A3	0,24	0,61	–	1,0	28,8	38,3	5,54	3,03E-3	23456,9	959,5	2,57	102349	–	S
051993A4	0,36	0,38	–	1,0	29,2	38,7	5,49	1,25E-3	23403,3	1181,2	2,08	102012	–	S
051993A5	0,36	0,54	–	1,0	29,4	39,1	5,46	1,76E-3	23461,5	1187,9	2,04	101336	–	S
051993A6	0,36	0,82	–	1,0	29,6	39,4	5,43	2,64E-3	23293,4	1190,3	2,01	100998	–	SC
051993A7	0,36	1,10	–	1,0	30,1	39,9	5,37	3,51E-3	23360,3	1195,7	1,98	100998	–	SC
051993A8	0,36	1,66	–	1,0	30,3	40,0	5,34	5,28E-3	23288,8	1193,3	2,01	99991	–	SC
051993A9	0,36	2,70	–	1,0	30,9	39,8	5,27	8,87E-3	23360,9	1187,1	2,20	103018	–	C
051993A10	0,36	4,31	–	1,0	31,2	39,5	5,23	1,39E-2	23347,7	1198,5	2,35	103356	–	CA
051993A11	0,13	0,11	–	1,0	31,8	39,5	5,15	9,68E-4	16931,1	611,0	3,56	100329	–	S
051993A12	0,13	0,23	–	1,0	31,8	39,3	5,16	1,93E-3	16915,8	616,4	3,58	99316	–	S
051993A13	0,13	0,27	–	1,0	31,9	39,3	5,15	2,26E-3	16935,0	615,4	3,60	98647	–	S
051993A14	0,13	0,40	–	1,0	31,8	39,7	5,16	3,35E-3	16972,8	616,9	3,41	97971	–	SC
051993A15	0,13	0,57	–	1,0	31,7	40,0	5,16	4,78E-3	16822,5	615,1	3,27	97971	–	SC
051993A16	0,13	0,86	–	1,0	31,7	40,3	5,17	7,29E-3	16875,5	611,1	3,16	97357	–	SC
051993A17	0,13	1,14	–	1,0	31,5	40,3	5,19	9,66E-3	16823,8	613,1	3,10	97633	–	SC
051993A18	0,13	1,71	–	1,0	31,5	40,2	5,19	1,45E-2	16920,5	611,0	3,13	97633	–	SC
051993A19	0,13	2,86	–	1,0	31,4	38,8	5,20	2,45E-2	16955,5	601,8	3,71	97295	–	C
051993A20	0,13	4,50	–	1,0	31,1	38,0	5,24	3,78E-2	16938,3	607,5	3,98	98978	–	CA
051993A21	0,23	0,11	–	1,0	30,2	40,3	5,35	5,49E-4	23520,6	951,9	2,44	101929	–	S
051993A22	0,23	0,22	–	1,0	30,8	40,7	5,28	1,08E-3	23528,9	966,8	2,44	101088	–	S

051993A23	0,22	0,26	–	1,0	31,2	41,2	5,23	1,34E-3	23375,0	948,0	2,45	100329	–	S
051993A24	0,22	0,39	–	1,0	31,5	41,3	5,18	2,00E-3	23727,5	945,7	2,42	99233	–	S
051993A25	0,23	0,56	–	1,0	31,7	42,3	5,16	2,78E-3	23668,8	957,4	2,32	99488	–	S
051993A26	0,23	0,84	–	1,0	32,0	42,3	5,13	4,15E-3	23636,5	963,9	2,25	99067	–	SC
051993A27	0,22	1,11	–	1,0	32,2	43,3	5,10	5,73E-3	23724,2	949,2	2,25	100495	–	SC
051993A28	0,22	1,69	–	1,0	32,6	43,5	5,06	8,58E-3	23605,0	949,8	2,26	99067	–	C
051993A29	0,22	2,77	–	1,0	32,8	42,6	5,03	1,44E-2	23725,8	941,5	2,55	100667	–	C
051993A30	0,23	4,52	–	1,0	32,9	42,0	5,03	2,20E-2	23650,5	958,6	2,69	98812	–	CA
051993A31	0,52	0,11	–	1,0	34,9	48,3	4,80	2,42E-4	31792,1	1468,1	1,63	101591	–	S
051993A32	0,52	0,22	–	1,0	35,5	48,4	4,73	4,88E-4	31685,4	1475,5	1,68	101929	–	S
051993A33	0,53	0,26	–	1,0	36,1	48,3	4,67	5,73E-4	31640,7	1482,3	1,68	100998	–	S
051993A34	0,52	0,39	–	1,0	36,5	49,1	4,62	8,59E-4	31555,5	1479,6	1,60	100998	–	S
051993A35	0,52	0,55	–	1,0	36,7	49,4	4,60	1,22E-3	31571,2	1482,3	1,69	101591	–	S
051993A36	0,52	0,83	–	1,0	36,9	49,7	4,58	1,83E-3	31478,5	1484,9	1,67	101508	–	S
051993A37	0,52	1,13	–	1,0	37,3	49,9	4,54	2,44E-3	31425,5	1487,7	1,68	100074	–	S
051993A38	0,52	1,69	–	1,0	37,8	50,1	4,50	3,65E-3	31563,7	1490,8	1,72	100915	–	SC
051993A39	0,52	2,79	–	1,0	38,5	49,9	4,43	6,11E-3	31726,8	1486,6	1,88	101929	–	C
051993A40	0,52	4,22	–	1,0	38,9	49,5	4,39	9,70E-3	31596,0	1489,5	2,00	107989	–	A

Colin and Fabre (1994) Flight Data
Water-Air, 40 mm i.d.

V_{sl}	V_{sg}	FP
0,49	0,08	B
0,55	0,12	B
0,53	0,16	B
0,69	0,04	B
0,65	0,11	B
0,63	0,16	B
0,61	0,19	B
0,59	0,25	B
0,88	0,05	B
0,85	0,12	B
0,82	0,17	B
0,77	0,25	B
1,10	0,09	B
1,06	0,15	B
1,00	0,19	B
0,97	0,27	B
0,33	0,05	B
0,31	0,06	B
0,31	0,05	B
0,32	0,07	B
0,31	0,12	B
0,45	0,05	B
0,42	0,06	B
0,88	0,16	B
0,27	0,05	B
0,26	0,06	B
0,49	0,05	B
0,96	0,05	B
0,93	0,07	B
0,95	0,13	B
0,94	0,22	B
0,34	0,13	B
0,87	0,13	B
0,84	0,13	B
0,84	0,13	B
0,82	0,13	B
0,81	0,13	B
0,94	0,06	B
0,92	0,06	B
0,90	0,06	B
0,89	0,06	B
0,88	0,06	B
0,87	0,06	B

V_{sl}	V_{sg}	FP
1,49	0,12	B
0,24	0,04	B
0,23	0,02	B
0,26	0,02	B
0,25	0,02	B
0,24	0,02	B
0,38	0,11	B
0,26	0,11	B
0,41	0,11	B
0,97	0,04	B
0,99	0,05	B
0,99	0,04	B
0,89	0,11	B
0,85	0,05	B
0,84	0,04	B
0,79	0,13	B
0,76	0,13	B
0,45	0,06	B
0,45	0,08	B
0,88	0,07	B
0,87	0,09	B
0,86	0,09	B
0,84	0,13	B
0,83	0,16	B
0,83	0,17	B
0,54	0,22	B
0,54	0,22	B
0,47	0,13	B
0,86	0,12	B
0,84	0,22	B
0,82	0,38	B
0,80	0,47	B
0,82	0,56	B
0,80	0,56	B
0,81	0,62	B
0,67	0,18	B
0,68	0,28	B
0,67	0,47	B
0,67	0,46	B
0,49	0,19	S
0,29	0,30	S
0,56	0,29	S
0,49	0,32	S

V_{sl}	V_{sg}	FP
0,30	0,23	S
0,28	0,47	S
0,29	0,49	S
0,30	0,23	S
0,28	0,12	S
0,27	0,20	S
0,28	0,44	S
0,48	0,07	S
0,46	0,12	S
0,91	0,35	S
0,95	0,55	S
0,26	0,13	S
0,33	0,13	S
0,30	0,13	S
0,47	0,13	S
0,46	0,13	S
0,46	0,13	S
0,54	0,13	S
0,24	0,10	S
0,32	0,21	S
0,44	0,13	S
0,44	0,12	S
0,43	0,19	S
0,43	0,19	S
0,41	0,19	S
0,47	0,23	S
0,46	0,23	S
0,46	0,23	S
0,45	0,28	S
0,54	0,25	S
0,53	0,32	S
0,21	0,14	S
0,22	0,14	S
0,20	0,21	S
0,21	0,29	S
0,21	0,39	S
0,18	0,31	S
0,17	0,26	S
0,47	0,18	S
0,47	0,23	S
0,46	0,32	S
0,46	0,41	S

Bousman (1995) Flight Data
50% Glycerin/Water-Air, 12.7 mm i.d.

Code	V_{sl}	V_{sg}	α	FP
52,1	0,31	10,12	–	A
52,12	0,30	10,35	–	A
52,2	0,09	0,12	0,434	S
53,13	0,49	0,78	0,458	S
53,2	0,09	2,11	–	FSA
54,1	0,48	3,02	–	FSA
54,2	0,10	0,78	0,621	S
55,2	0,31	1,11	0,574	S
62,1	0,33	6,50	–	A
62,2	0,34	0,33	0,420	BS
62,32	0,14	2,16	–	FSA
62,4	0,14	4,29	–	FSA
63,1	0,30	3,47	–	FSA
63,2	0,54	8,20	–	A
71,12	0,27	0,57	0,526	S
71,22	0,13	0,46	0,548	S
71,32	0,21	2,28	–	FSA
71,42	0,44	5,55	–	FSA
72,1	0,07	3,11	–	FSA
72,2	0,32	4,56	–	FSA
72,3	0,49	6,52	–	FSA
73,1	0,18	10,93	–	A
73,2	0,32	10,67	–	A
73,3	0,49	1,02	0,609	S
92,12	0,09	0,33	0,539	S
92,2	0,20	0,31	0,546	S
92,32	0,71	9,64	–	A
93,1	0,08	1,08	0,703	S
93,2	0,20	1,05	0,652	S
93,3	0,82	3,80	–	S
94,1	0,21	0,11	0,304	BS
94,2	0,49	9,09	–	A
94,3	0,51	0,30	0,291	B
95,13	0,08	9,77	–	A
95,23	0,53	0,12	0,137	B
95,33	0,89	0,12	0,079	B
96,1	0,09	4,04	–	FSA
96,22	0,21	4,46	–	FSA
96,32	0,90	0,30	0,172	B
97,13	0,51	1,05	0,558	S
97,22	0,87	1,03	0,43	BS
98,1	0,53	2,46	0,657	S
98,2	0,82	2,30	0,594	S
99,12	0,09	2,30	–	FSA
99,22	0,20	2,46	–	FSA

Code	V_{sl}	V_{sg}	α	FP
99,3	0,50	3,99	–	FSA
145,1	0,11	25,20	0,855	A
145,2	0,11	10,27	0,772	A
145,3	0,20	10,19	0,739	A
145,4	0,12	5,52	–	A
145,5	0,21	5,21	0,637	A
145,6	0,51	9,80	0,665	A
146,1	0,11	25,42	0,875	A
146,2	0,20	24,86	0,86	A
146,3	0,12	15,44	0,834	A
146,4	0,21	15,28	0,817	A
146,5	0,12	5,46	0,745	A
146,6	0,53	13,97	0,772	A
147,1	0,06	25,36	0,878	A
147,2	0,07	16,02	0,849	A
147,3	0,07	10,71	0,809	A
147,4	0,06	5,43	0,769	A
147,5	0,50	22,11	0,817	A
147,6	0,55	4,89	–	FSA
148,1	0,11	15,29	0,829	A
148,2	0,12	10,64	0,803	A
148,3	0,07	25,98	0,884	A
148,4	0,07	15,95	0,849	A
148,5	0,50	22,02	0,834	A
148,6	0,55	4,95	–	FSA
157,1	0,11	2,09	–	FSA
157,2	0,06	2,13	–	FSA
157,3	0,07	1,13	–	FSA
157,4	0,06	0,53	0,610	S
157,5	0,51	1,07	0,547	S
157,6	0,52	0,54	–	BS
158,1	0,20	2,14	–	FSA
158,2	0,20	1,09	0,682	S
158,3	0,21	0,56	0,597	S
158,4	0,11	1,07	0,694	S
158,5	0,11	0,54	0,627	S
158,6	0,53	2,09	0,651	S

Bousman (1995) Flight Data
Water-Air, 12.7 mm i.d.

Code	V_{sl}	V_{sg}	α	FP	dp/dzf
31,100	0,07	2,19	–	FSA	–
32,100	0,07	0,43	0,617	S	–
33,100	0,08	1,77	–	FSA	–
34,300	0,07	0,15	–	S	–
35,200	0,07	5,30	–	A	–
36,100	0,19	4,40	–	A	–
37,220	0,41	7,50	–	A	–
38,100	0,19	11,57	–	A	–
45,100	0,18	0,22	0,402	BS	–
58,100	0,20	1,11	0,660	S	–
58,220	0,08	0,40	0,652	S	–
59,100	0,21	7,67	–	A	–
59,200	0,21	0,51	0,610	S	–
60,100	0,10	0,11	0,431	BS	–
60,200	0,10	1,67	–	FSA	–
64,200	0,34	0,18	0,297	BS	–
64,320	0,90	0,23	–	B	–
65,100	0,14	0,36	0,554	S	–
65,200	0,34	0,59	0,506	S	–
65,300	0,86	0,60	0,375	BS	–
66,130	0,16	1,43	–	S	–
66,200	0,34	1,15	0,600	S	–
66,300	0,86	4,29	–	FSA	–
67,120	0,10	0,90	0,652	S	–
67,200	0,06	0,20	0,546	S	–
67,300	0,88	1,12	0,491	S	–

Code	V_{sl}	V_{sg}	α	FP	dp/dzf
84,220	0,51	4,43	–	FSA	–
85,100	0,20	2,01	–	FSA	–
85,200	0,50	0,11	0,162	B	752
85,300	0,79	2,03	0,603	S	4099
86,120	0,20	1,24	0,690	S	954
86,220	0,20	0,62	0,704	S	411
86,320	0,79	1,21	0,552	S	2386
87,100	0,06	4,69	0,809	A	–
87,220	0,53	0,31	0,313	B	1215
87,300	0,81	0,36	0,283	B	2037
88,200	0,77	10,24	–	A	–
89,100	0,52	0,58	0,488	BS	2009
89,200	0,80	4,43	–	FSA	–
90,100	0,20	3,99	0,731	A	–
90,200	0,81	0,12	0,113	B	1237
91,100	0,52	2,18	0,668	S	–
141,100	0,20	24,75	0,875	A	9716
141,200	0,08	10,83	0,849	A	1774
141,300	0,12	10,40	0,803	A	2321
141,400	0,21	10,33	0,772	A	3513
141,500	0,12	15,67	0,840	A	3638
141,600	0,55	9,26	–	A	5434
142,100	0,07	15,66	0,875	A	2411
142,200	0,11	15,68	0,860	A	3307
142,300	0,20	15,18	0,826	A	4854
142,400	0,13	10,30	0,823	A	2349

68,100	0,10	2,21	–	FSA	–
68,200	0,61	0,25	–	B	–
68,300	0,60	0,77	0,497	S	–
69,120	0,10	10,83	–	A	–
69,200	0,33	4,60	–	FSA	–
69,320	0,58	1,67	0,586	S	–
70,120	0,15	2,49	–	FSA	–
70,200	0,35	10,14	–	A	–
70,300	0,60	5,78	–	FSA	–
81,120	0,20	0,12	0,312	BS	388
81,220	0,51	9,23	0,742	A	–
81,320	0,53	0,98	0,600	S	2683
82,120	0,07	10,07	0,846	A	–
82,220	0,07	1,12	–	FSA	–
83,120	0,07	2,25	–	FSA	–
83,220	0,20	0,29	0,504	S	584
83,320	0,81	0,49	0,303	B	2434
84,120	0,07	0,35	0,677	S	–

142,500	0,13	5,28	0,758	A	1346
142,600	0,50	22,43	0,834	A	7629
143,100	0,07	26,18	0,923	A	4884
143,200	0,11	25,50	0,911	A	6578
143,300	0,07	5,48	0,797	A	810
143,400	0,12	5,38	0,781	A	1270
143,500	0,21	5,29	0,734	A	2028
143,600	0,55	5,11	0,684	A	8096
144,100	0,07	25,08	0,905	A	4769
144,200	0,11	25,45	0,893	A	6507
144,300	0,40	23,52	0,849	A	17194
144,400	0,07	10,44	0,849	A	1566
144,500	0,12	0,99	0,728	S	559
144,600	0,54	10,01	0,747	A	–
155,100	0,11	2,10	0,760	S	465
155,200	0,07	2,10	0,770	S	284
155,300	0,07	1,09	0,674	S	234
155,400	0,07	0,53	0,705	S	119
155,500	0,53	1,07	0,618	S	1973
155,600	0,53	0,54	0,509	S	1453
156,100	0,19	2,10	0,727	S	1068
156,200	0,19	1,08	0,708	S	684
156,300	0,20	0,54	0,648	S	413
156,400	0,11	1,06	0,741	S	355
156,500	0,11	5,46	0,703	S	123
156,600	0,53	2,06	0,667	S	3385

February 1994, Ground Data
Water-Air, 9.525 mm i.d., L (pressure taps): dz1 = 69.0 cm, dz2 = 38.5 cm, and dz3 = 30.5 cm

Code	V_{sl}	V_{sg}	α	$gz/g0$	Tf,avg	Ts,avg	Pr_L	x	q''	h_L	$\psi 2$	Pin	dp/dzf	FP
031894A2	0,05	0,09	–	1,0	26,3	28,8	5,90	2,37E-3	9997,0	509,4	8,02	99021	–	S
031894A3	0,04	0,19	–	1,0	28,1	32,3	5,64	5,25E-3	13730,5	497,1	6,61	96650	–	S
031894A4	0,03	0,39	–	1,0	28,2	34,2	5,63	1,53E-2	13573,3	440,0	5,26	95202	–	S
031894A6	0,10	0,14	0,405	1,0	26,9	30,5	5,80	1,53E-3	15421,9	665,6	6,70	98348	–	S
031894A7	0,10	0,19	0,483	1,0	27,9	31,9	5,67	2,16E-3	18967,0	667,6	7,25	97455	–	S
031894A8	0,10	0,39	0,628	1,0	28,0	32,7	5,65	4,26E-3	18928,7	672,2	6,08	95845	–	S
031894A9	0,10	0,59	0,676	1,0	28,7	34,0	5,56	6,55E-3	18857,7	666,5	5,41	95019	–	S
031894A10	0,10	0,79	0,708	1,0	29,0	34,6	5,52	8,63E-3	18737,9	669,2	5,03	94696	–	S
031894A12	0,25	0,09	0,205	1,0	29,1	34,2	5,49	4,40E-4	22862,5	1556,2	2,95	100436	–	S
031894A13	0,24	0,19	0,363	1,0	29,5	34,4	5,44	9,13E-4	22826,4	1542,4	3,11	99384	–	S
031894A14	0,23	0,34	0,505	1,0	30,4	36,2	5,33	1,69E-3	28127,4	1492,1	3,37	97931	–	S
031894A15	0,24	0,64	0,636	1,0	31,0	36,8	5,26	3,00E-3	28098,3	1545,8	3,21	96738	–	S
031894A16	0,12	0,09	–	1,0	32,6	38,8	5,06	9,17E-4	27842,2	710,5	6,42	99431	–	S
031894A17	0,37	0,19	0,288	1,0	32,8	39,2	5,03	5,99E-4	32213,8	2271,0	2,30	100746	–	S
031894A18	0,43	0,34	0,391	1,0	33,4	40,2	4,96	9,11E-4	37934,9	2559,9	2,24	99900	–	S
031894A19	0,29	0,49	0,567	1,0	34,0	41,0	4,89	1,90E-3	37797,4	1875,9	2,98	97828	–	S
031894A20	0,27	0,69	0,649	1,0	34,3	41,6	4,86	2,79E-3	37713,7	1806,5	2,95	97040	–	S
031894A22	0,84	0,09	0,141	1,0	32,8	38,5	5,04	1,29E-4	29736,1	4340,2	1,23	101185	–	S
031894A23	0,67	0,13	0,177	1,0	32,9	38,8	5,02	2,29E-4	29629,5	3613,7	1,44	99841	–	S
031894A24	0,75	0,19	0,185	1,0	33,0	38,4	5,01	2,97E-4	29604,5	3975,2	1,43	101019	–	S
031894A25	0,89	0,39	0,282	1,0	33,4	38,5	4,96	5,20E-4	32780,4	4545,6	1,48	103023	–	S
031894A26	1,11	0,10	0,142	1,0	33,8	39,4	4,92	1,04E-4	32763,6	5453,2	1,12	104459	–	B
031894A27	1,51	0,14	0,132	1,0	34,4	39,5	4,85	1,08E-4	37715,6	7018,2	1,10	105198	3178	B
031894A28	1,36	0,19	0,144	1,0	34,6	39,8	4,83	1,71E-4	37659,7	6461,9	1,17	104459	2866	S

031894A29	1,50	0,29	0,156	1,0	34,8	39,5	4,81	2,36E-4	37716,2	7027,7	1,19	106065	3700	S
031894A30	1,25	0,39	0,192	1,0	34,9	39,8	4,80	3,75E-4	37846,8	6080,1	1,31	104839	3106	S
031894A32	2,49	0,10	0,099	1,0	33,1	39,2	5,00	4,95E-5	66023,0	10343,2	1,11	113115	6929	B
031894A33	2,49	0,13	0,097	1,0	34,9	40,9	4,80	7,02E-5	65727,2	10540,7	1,11	113764	7104	B
031894A34	2,48	0,19	0,098	1,0	36,3	42,2	4,65	1,01E-4	65223,6	10674,6	1,11	114567	7356	B
031894A35	2,50	0,23	0,095	1,0	37,6	43,5	4,51	1,22E-4	65354,0	10882,0	1,10	115719	7572	B
031894A36	2,49	0,29	0,112	1,0	39,3	45,5	4,35	1,55E-4	71536,4	11052,7	1,12	116337	7719	S
031894A37	2,50	0,39	0,136	1,0	40,3	46,4	4,26	2,08E-4	71279,0	11196,4	1,12	117596	8151	S
031894B12	0,10	0,96	0,713	1,0	28,1	32,2	5,63	1,03E-2	13727,3	671,9	5,13	95603	–	S
031894B13	0,10	1,96	0,689	1,0	34,2	46,7	4,87	2,13E-2	47053,6	681,6	5,71	95973	–	S
031894B14	0,13	3,96	0,665	1,0	32,1	40,4	5,12	3,28E-2	46366,0	737,7	7,78	97160	–	C
031894B15	0,06	6,94	–	1,0	36,0	45,0	4,68	1,12E-1	44693,9	574,2	8,85	95982	–	A
031894B16	0,10	9,92	–	1,0	32,7	40,6	5,05	1,01E-1	45153,6	672,3	8,79	97060	–	A
031894B17	0,09	11,92	–	1,0	33,5	42,3	4,95	1,25E-1	49058,7	661,2	8,71	97588	1932	A
031894B18	0,10	13,93	0,881	1,0	33,4	42,0	4,96	1,41E-1	49172,0	666,7	8,86	98681	2421	A
031894B19	0,06	15,92	–	1,0	33,5	43,5	4,95	2,26E-1	44931,5	576,8	8,12	97883	1983	A
031894B20	0,09	17,92	0,925	1,0	27,5	–	–	1,81E-1	–	–	–	99859	3102	A
031894B22	0,20	0,96	0,704	1,0	34,0	45,1	4,89	5,35E-3	51638,4	1394,6	3,41	96820	–	S
031894B23	0,20	1,47	0,739	1,0	35,2	46,7	4,76	8,17E-3	52710,1	1408,0	3,34	96964	–	S
031894B24	0,22	2,47	0,675	1,0	35,8	46,3	4,70	1,25E-2	52792,8	1529,3	3,37	97969	–	S
031894B25	0,22	3,45	0,673	1,0	32,5	–	–	1,78E-2	–	–	–	98805	2081	C
031894B26	0,13	6,94	–	1,0	38,2	47,8	4,46	5,48E-2	59535,5	746,5	8,54	97356	1863	A
031894B27	0,16	9,93	0,833	1,0	31,8	–	–	6,74E-2	–	–	–	98874	2466	A
031894B28	0,07	11,92	–	1,0	38,6	49,7	4,41	1,49E-1	56385,9	621,4	8,39	97116	–	A
031894B29	0,13	13,92	0,896	1,0	30,5	–	–	1,09E-1	–	–	–	99909	2924	A
031894B30	0,06	17,85	–	1,0	29,8	–	–	2,40E-1	–	–	–	98736	2320	A
032294A2	0,08	0,56	0,687	1,0	32,2	40,6	5,10	8,08E-3	28072,9	618,3	5,55	96309	–	S
032294A3	0,08	0,96	0,742	1,0	32,3	41,7	5,09	1,32E-2	27956,8	629,4	4,85	95804	–	S
032294A4	0,08	1,41	–	1,0	32,7	42,6	5,05	1,98E-2	27786,9	624,1	4,63	95831	–	S
032294A5	0,07	1,96	0,669	1,0	33,4	42,9	4,97	2,80E-2	27836,4	617,8	4,81	96071	–	S
032294A6	0,07	4,96	–	1,0	34,3	42,3	4,86	7,09E-2	37280,8	606,4	7,83	96381	–	C

(Continued)

February 1994, Ground Data
Water-Air, 9.525 mm i.d., L (pressure taps): dz1 = 69.0 cm, dz2 = 38.5 cm, and dz3 = 30.5 cm (*Continued*)

Code	V_{sl}	V_{sg}	α	*gz/g0*	Tf,avg	Ts,avg	Pr_L	x	q″	h_L	$\psi 2$	Pin	dp/dzf	FP
032294A7	0,05	9,93	–	1,0	33,9	42,8	4,91	1,83E-1	37388,2	535,7	8,02	96512	–	A
032294A9	0,07	13,92	0,887	1,0	31,7	38,7	5,17	1,74E-1	35288,9	611,7	8,39	98138	–	A
032294A10	0,06	17,95	–	1,0	31,0	39,5	5,26	2,48E-1	37740,6	583,6	7,93	101731	2165	A
032294A12	0,40	0,97	0,628	1,0	30,7	36,0	5,29	2,78E-3	26815,9	2330,9	2,26	100195	–	S
032294A13	0,37	1,43	–	1,0	32,3	40,6	5,10	4,38E-3	46593,3	2255,0	2,54	99918	–	S
032294A14	0,29	2,41	0,667	1,0	33,4	41,8	4,97	9,40E-3	46457,3	1876,0	3,03	100081	–	S
032294A15	0,53	3,40	0,687	1,0	34,3	40,1	4,87	7,60E-3	43199,4	3029,3	2,53	104299	4182	C
032294A16	0,40	6,92	0,721	1,0	34,4	39,8	4,85	2,02E-2	43035,4	2414,3	3,46	103951	4405	C
032294A17	0,28	9,92	–	1,0	35,1	41,7	4,77	3,96E-2	52104,3	1852,2	4,47	102846	4175	A
032294A18	0,25	11,93	–	1,0	35,4	41,9	4,74	5,31E-2	51844,8	1688,1	4,94	103208	4328	A
032294A19	0,27	13,88	0,886	1,0	35,5	41,6	4,73	5,80E-2	52034,5	1796,2	4,96	105336	5366	A
032294A20	0,07	15,91	–	1,0	34,4	41,1	4,85	2,03E-1	30593,3	601,5	7,86	98727	2076	A
032294A22	0,92	0,97	0,454	1,0	32,7	39,4	5,05	1,27E-3	49579,1	4675,8	1,66	105545	2963	S
032294A23	0,82	1,43	0,580	1,0	33,9	40,5	4,91	2,09E-3	49261,2	4302,7	1,81	105225	3083	S
032294A24	0,90	1,97	0,615	1,0	35,4	42,6	4,75	2,68E-3	59619,8	4706,6	1,84	107755	4702	S
032294A25	0,70	3,97	0,665	1,0	35,7	42,4	4,71	7,00E-3	59912,0	3844,8	2,45	108691	6446	C
032294A26	0,51	5,93	–	1,0	36,6	43,4	4,62	1,40E-2	59970,0	3009,4	3,08	107108	5537	C
032294A27	0,40	7,93	–	1,0	36,3	41,1	4,65	2,31E-2	37705,0	2465,5	3,32	104765	4951	A
032294A28	0,13	9,93	–	1,0	36,7	41,9	4,61	8,02E-2	30351,6	730,7	8,24	98789	2024	A
032294A29	0,54	11,96	0,827	1,0	36,2	40,4	4,66	2,79E-2	41790,9	3137,3	3,40	114230	8990	A
032294A30	0,61	8,78	0,753	1,0	32,1	33,8	5,12	1,83E-2	13378,4	3291,3	2,39	113129	8665	A
032294A32	1,50	0,77	0,267	1,0	33,2	38,0	4,99	6,54E-4	42049,9	6896,7	1,33	111054	5246	S
032294A33	1,40	1,47	0,446	1,0	33,8	38,3	4,92	1,36E-3	42096,7	6544,3	1,51	113582	6670	S
032294A34	1,60	1,97	0,474	1,0	34,3	37,9	4,87	1,67E-3	35737,7	7309,3	1,44	119390	9688	S
032294A35	1,50	3,98	0,653	1,0	34,1	38,7	4,88	3,88E-3	54228,6	6936,0	1,85	128331	15467	S
032294A36	0,75	6,93	0,713	1,0	35,2	41,2	4,77	1,20E-2	62086,9	4047,5	2,70	114862	9894	C
032294A38	0,77	5,93	0,686	1,0	33,0	–	–	9,91E-3	–	–	–	113492	9652	C
032294A40	1,52	8,16	0,792	1,0	32,6	35,3	5,06	9,22E-3	33072,2	6871,3	1,95	151114	25671	A

October 1991, Flight Data
Water-Air, 9.525 mm i.d.

Code	V_{sl}	V_{sg}	α	$gz/g0$	Tf,avg	Ts,avg	Pr_L	x	q″	h_L	$\psi 2$	Pin	dp/dzf	FP
100391AM1	0,92	0,47	–	1,84E-2	35,3	48,9	4,75	2,90E-4	54222,7	2718,8	1,51	48803	–	S
100391AM2	0,89	0,72	–	1,42E-2	35,9	49,5	4,69	4,39E-4	54035,7	2697,0	1,51	49616	–	S
100391AM3	0,95	0,95	–	1,02E-2	36,8	50,0	4,59	6,76E-4	53980,2	2788,0	1,51	62069	–	S
100391AM4	0,86	1,56	–	1,40E-2	37,0	50,3	4,57	1,14E-3	53962,4	2695,8	1,56	58945	–	S
100391AM5	0,88	2,76	–	1,86E-2	37,9	50,3	4,48	2,86E-3	54001,2	2688,6	1,67	82375	–	S
100391AM6	0,92	5,36	–	1,40E-2	37,0	50,3	4,57	1,14E-3	53962,4	2695,8	1,56	85436	–	FSA
100391AM7	0,85	7,50	–	2,22E-2	37,6	48,8	4,51	8,98E-3	54084,2	2627,9	1,92	91386	–	FSA
100391AM8	0,90	9,95	–	2,59E-2	38,3	49,1	4,44	1,15E-2	53836,5	2681,3	1,95	91924	–	FSA
100391AM9	0,85	12,25	–	2,53E-2	38,8	50,3	4,39	1,46E-2	56784,6	2664,8	1,94	93317	–	FSA
100391AM10	0,75	14,70	–	2,07E-2	39,5	50,8	4,33	1,92E-2	56777,0	2558,8	2,05	92462	–	FSA
100391AM19	0,56	11,53	–	1,24E-2	35,9	–	–	2,25E-2	–	–	–	99005	–	FSA
100391AM20	1,38	13,22	–	2,17E-2	35,9	–	–	1,11E-2	–	–	–	103770	–	FSA
100391AM21	2,27	0,30	–	2,14E-2	36,9	49,7	4,58	1,20E-4	63264,0	4077,4	1,27	75997	–	B
100391AM22	2,24	0,62	–	1,63E-2	38,2	56,4	4,45	1,59E-4	94306,0	4242,0	1,27	57608	–	BS
100391AM23	2,26	0,65	–	1,64E-2	39,0	57,3	4,37	2,85E-4	93941,6	4153,8	1,29	88552	–	S
100391AM24	2,35	0,81	–	2,03E-2	40,4	58,4	4,24	3,21E-4	93257,3	4241,9	1,28	84781	–	S
100391AM25	2,33	1,13	–	2,02E-2	41,6	59,5	4,14	4,37E-4	93086,1	4238,4	1,29	82044	–	S
100391AM26	2,33	2,49	–	1,95E-2	42,9	60,5	4,03	1,09E-3	94764,4	4252,2	1,33	93613	–	S
100391AM27	2,29	4,86	–	2,22E-2	44,1	61,1	3,93	2,24E-3	94393,7	4232,0	1,38	95972	–	FSA
100391AM28	2,18	6,72	–	1,76E-2	43,8	60,3	3,95	3,55E-3	94182,8	4115,2	1,47	104776	–	FSA
100391AM29	2,13	9,08	–	2,36E-2	45,3	61,6	3,84	4,80E-3	93664,2	4107,5	1,47	104066	–	FSA
100391AM30	2,08	10,46	–	2,25E-2	43,3	59,2	3,99	6,30E-3	94012,4	3986,9	1,57	111878	–	FSA
100391AM31	3,46	0,29	–	1,21E-2	43,3	60,6	3,99	7,64E-5	94151,5	5078,4	1,13	82485	–	B
100391AM32	3,53	0,48	–	7,72E-3	45,3	62,6	3,84	1,06E-4	93433,0	5180,7	1,09	73377	–	B
100391AM33	3,45	0,66	–	1,48E-2	47,4	64,5	3,68	1,41E-4	92955,0	5155,5	1,10	71094	–	B
100391AM34	3,73	0,75	–	1,53E-2	47,5	64,2	3,68	2,07E-4	92903,4	5269,3	1,11	93958	–	B
100391AM35	3,71	1,00	–	1,84E-2	47,7	64,3	3,66	2,82E-4	93170,4	5249,8	1,12	95220	–	B
100391AM36	3,34	2,42	–	1,56E-2	47,9	64,2	3,65	7,79E-4	92801,9	5007,5	1,20	97654	–	S
100391AM37	3,06	4,65	–	1,41E-2	47,3	63,4	3,69	1,74E-3	93598,2	4768,7	1,29	102232	–	FSA

(Continued)

October 1991, Flight Data
Water-Air, 9.525 mm i.d. (*Continued*)

Code	V_{sl}	V_{sg}	α	gz/g0	Tf,avg	Ts,avg	Pr_L	x	q″	h_L	$\psi 2$	Pin	dp/dzf	FP
100391AM38	2,83	6,62	–	2,02E-2	46,3	62,0	3,76	2,82E-3	93505,0	4588,5	1,37	107100	–	FSA
100391AM39	2,59	8,56	–	1,43E-2	45,4	61,0	3,83	4,20E-3	93474,3	4355,1	1,46	110354	–	FSA
100391AM40	2,39	10,42	–	1,37E-2	44,6	60,5	3,89	5,72E-3	93877,7	4178,7	1,49	112823	–	FSA
100391PM1	0,09	0,33	–	2,53E-2	37,5	–	–	2,92E-3	–	–	–	71563	–	S
100391PM2	0,10	0,61	–	1,71E-2	41,0	57,9	4,19	4,44E-3	18687,3	622,2	1,80	56760	–	S
100391PM3	0,09	0,67	–	2,08E-2	41,9	58,2	4,12	5,71E-3	18693,3	625,9	1,88	65440	–	S
100391PM4	0,09	1,26	–	1,05E-2	41,6	54,7	4,14	1,17E-2	18808,4	629,6	2,37	70129	–	FSA
100391PM5	0,09	1,68	–	1,66E-2	42,7	56,2	4,05	1,44E-2	18779,7	629,7	2,32	67364	–	FSA
100391PM6	0,09	2,63	–	1,91E-2	42,8	53,5	4,04	2,85E-2	18913,5	621,0	3,13	84064	–	FSA
100391PM7	0,09	4,10	–	1,78E-2	42,0	51,7	4,11	4,25E-2	18927,3	615,4	3,49	81237	–	FSA
100391PM8	0,09	6,58	–	1,94E-2	38,0	–	–	6,49E-2	–	–	–	84657	–	FSA
100391PM9	0,09	8,96	–	2,67E-2	38,2	–	–	8,83E-2	–	–	–	86960	–	FSA
100391PM10	0,09	12,48	–	1,86E-2	37,8	–	–	1,21E-1	–	–	–	88690	–	FSA
100391PM11	0,13	0,39	–	8,25E-3	35,7	50,7	4,71	2,11E-3	25478,1	766,2	2,23	60545	–	S
100391PM12	0,11	0,47	–	2,07E-2	40,2	–	–	3,45E-3	–	–	–	73335	–	S
100391PM13	0,12	0,66	–	1,72E-2	41,2	57,7	4,17	3,21E-3	27781,5	877,8	1,93	67371	–	S
100391PM14	0,13	1,05	–	1,60E-2	41,0	57,2	4,19	7,86E-3	27662,6	793,1	2,17	86250	–	S
100391PM15	0,13	1,41	–	1,10E-2	41,8	56,1	4,13	9,94E-3	27845,8	787,8	2,48	81506	–	S
100391PM16	0,13	2,63	–	1,81E-2	41,9	53,7	4,12	1,92E-2	28091,0	796,2	3,01	85395	–	FSA
100391PM17	0,13	3,85	–	2,13E-2	42,1	53,8	4,10	2,95E-2	28149,4	782,6	3,10	88863	–	FSA
100391PM18	0,13	6,51	–	1,54E-2	41,7	52,5	4,13	4,61E-2	28684,9	802,5	3,38	87284	–	FSA
100391PM19	0,13	8,99	–	1,16E-2	41,7	51,6	4,13	6,35E-2	28682,5	798,5	3,69	88173	–	FSA
100391PM20	0,13	12,29	–	1,07E-2	40,0	49,0	4,29	8,72E-2	28844,0	799,2	4,10	91276	–	FSA
100391PM21	0,21	0,14	–	1,52E-2	40,2	55,0	4,27	5,92E-4	37565,8	1169,8	2,19	91593	–	S
100391PM22	0,21	0,21	–	1,61E-2	40,9	56,0	4,20	1,02E-3	37480,4	1421,8	1,78	91779	–	S
100391PM23	0,21	0,27	–	2,15E-2	41,3	56,4	4,17	1,32E-3	37421,6	1426,3	1,77	91359	–	S
100391PM24	0,20	0,65	–	2,19E-2	41,3	56,5	4,17	1,87E-3	37415,3	1422,7	1,76	53105	–	S
100391PM25	0,20	0,97	–	2,18E-2	41,1	56,9	4,19	3,24E-3	37484,8	1418,1	1,69	60669	–	S

100391PM26	0,20	1,82	–	1,98E-2	41,8	–	–	6,34E-3	–	–	–	63351	–	S
100391PM27	0,20	3,15	–	2,17E-2	41,0	53,7	4,19	1,27E-2	37709,0	1387,3	2,17	71453	–	FSA
100391PM28	0,20	5,53	–	1,46E-2	40,6	52,8	4,22	2,56E-2	37753,2	1390,2	2,29	82085	–	FSA
100391PM29	0,19	9,11	–	1,34E-2	40,8	51,8	4,21	4,41E-2	37709,0	1375,6	2,54	86967	–	FSA
100391PM30	0,21	11,89	–	1,18E-2	39,1	49,1	4,37	5,84E-2	37892,9	1383,8	2,82	94917	–	FSA
100391PM31	0,30	4,61	–	1,43E-2	39,7	52,4	4,31	8,74E-4	37446,7	1405,3	2,14	52526	–	S
100391PM32	0,30	0,49	–	1,25E-2	39,8	52,7	4,30	1,07E-3	37554,2	1673,8	1,77	59800	–	S
100391PM33	0,30	0,52	–	2,02E-2	40,1	53,0	4,27	1,26E-3	36854,3	1674,7	1,74	60690	–	S
100391PM34	0,30	0,77	–	1,55E-2	40,3	53,2	4,25	2,12E-3	36757,1	1671,7	1,73	73728	–	S
100391PM35	0,30	1,75	–	8,46E-3	40,5	52,7	4,24	4,29E-3	36742,6	1665,0	1,84	65978	–	S
100391PM36	0,29	3,14	–	9,90E-3	40,6	51,8	4,22	8,93E-3	37156,4	1656,0	2,06	71853	–	FSA
100391PM37	0,29	5,43	–	1,40E-2	40,5	50,9	4,24	1,66E-2	37268,9	1620,3	2,25	83823	–	FSA
100391PM38	0,29	7,35	–	1,64E-2	39,8	49,5	4,30	2,59E-2	37523,8	1674,7	2,38	92793	–	FSA
100391PM39	0,30	9,98	–	4,55E-3	39,2	48,2	4,35	3,32E-2	37588,7	1631,1	2,67	90635	–	FSA
100391PM40	0,29	12,37	–	1,53E-2	39,1	48,3	4,37	4,21E-2	37681,8	1650,2	2,57	91200	–	FSA
100491AM1	0,09	0,54	–	1,74E-2	32,4	–	–	3,09E-3	–	–	–	45604	–	S
100491AM2	0,09	0,81	–	2,35E-2	34,5	–	–	5,42E-3	–	–	–	53098	–	S
100491AM3	0,09	1,26	–	1,62E-2	34,3	–	–	8,17E-3	–	–	–	51795	–	S
100491AM4	0,09	1,91	–	1,11E-2	36,1	–	–	1,36E-2	–	–	–	57615	–	FSA
100491AM5	0,09	2,26	–	2,60E-2	38,2	50,3	4,46	2,12E-2	14217,0	627,2	1,89	72908	–	FSA
100491AM6	0,09	3,39	–	2,23E-2	35,2	–	–	3,31E-2	–	–	–	80361	–	FSA
100491AM7	0,09	4,58	–	2,39E-2	35,8	–	–	4,58E-2	–	–	–	83595	–	FSA
100491AM8	0,09	6,42	–	2,61E-2	35,1	–	–	6,44E-2	–	–	–	85319	–	FSA
100491AM9	0,09	8,90	–	2,21E-2	34,9	–	–	8,78E-2	–	–	–	85994	–	FSA
100491AM10	0,09	12,52	–	2,63E-2	34,4	–	–	1,21E-1	–	–	–	87015	–	FSA
100491AM11	0,09	0,37	–	2,78E-2	37,0	–	–	2,90E-3	–	–	–	62786	–	S
100491AM12	0,09	0,71	–	2,70E-2	36,2	–	–	5,24E-3	–	–	–	58856	–	S
100491AM13	0,09	1,05	–	3,12E-2	35,0	–	–	8,14E-3	–	–	–	62131	–	S
100491AM14	0,09	1,60	–	4,05E-2	34,8	–	–	1,36E-2	–	–	–	68336	–	FSA
100491AM15	0,09	2,20	–	9,42E-3	39,0	52,2	4,37	2,10E-2	16055,1	634,2	2,02	74818	–	FSA

(Continued)

October 1991, Flight Data
Water-Air, 9.525 mm i.d. (*Continued*)

Code	V_{sl}	V_{sg}	α	$gz/g0$	Tf,avg	Ts,avg	Pr_L	x	q''	h_L	$\psi 2$	Pin	dp/dzf	FP
100491AM16	0,09	3,21	–	2,76E-2	34,7	–	–	3,31E-2	–	–	–	84753	–	FSA
100491AM17	0,09	4,58	–	3,83E-2	34,9	–	–	4,60E-2	–	–	–	83616	–	FSA
100491AM18	0,09	6,46	–	2,86E-2	34,4	–	–	6,46E-2	–	–	–	84912	–	FSA
100491AM19	0,09	8,81	–	3,32E-2	35,2	–	–	8,77E-2	–	–	–	86829	–	FSA
100491AM20	0,09	12,37	–	3,20E-2	34,3	–	–	1,21E-1	–	–	–	88146	–	FSA
100491AM21	0,09	13,39	–	7,08E-3	39,9	59,9	4,29	1,50E-1	18108,3	634,1	1,45	98509	–	A
100491AM22	0,09	16,37	–	2,33E-2	39,0	–	–	1,60E-1	–	–	–	93682	–	A
100491AM23	0,09	18,95	–	2,71E-2	40,6	–	–	1,77E-1	–	–	–	92131	–	A
100491AM24	0,09	21,51	–	1,86E-2	39,7	–	–	1,95E-1	–	–	–	91055	–	A
100491AM25	0,09	23,31	–	2,04E-2	39,4	–	–	2,12E-1	–	–	–	93269	–	A
100491AM26	0,09	14,26	–	2,12E-2	44,5	54,0	3,90	1,49E-1	25656,9	622,6	5,76	91586	–	A
100491AM27	0,09	16,39	–	2,40E-2	40,1	52,7	4,28	1,70E-1	25765,4	625,4	3,71	92800	–	A
100491AM28	0,09	18,86	–	2,27E-2	38,8	51,7	4,39	1,92E-1	25799,4	620,2	3,63	92276	–	A
100491AM29	0,09	19,14	–	1,80E-2	37,6	45,1	4,51	2,01E-1	25923,5	634,0	6,72	102211	–	A
100491AM30	0,09	22,35	–	1,81E-2	33,6	–	–	2,15E-1	–	–	–	97095	–	A
100491AM31	0,14	13,80	–	1,36E-2	36,4	44,7	4,63	1,01E-1	28890,0	811,5	4,44	97723	–	FSA
100491AM32	0,13	16,74	–	1,48E-2	35,5	44,3	4,73	1,20E-1	28975,9	792,8	4,25	93151	–	A
100491AM33	0,13	19,31	–	3,04E-2	35,5	44,1	4,73	1,37E-1	28989,3	782,2	4,37	92193	–	A
100491AM34	0,14	19,90	–	7,21E-3	35,7	43,6	4,71	1,46E-1	28965,9	799,8	4,74	100757	–	A
100491AM35	0,13	23,02	–	1,61E-2	35,1	43,4	4,78	1,67E-1	29075,1	777,9	4,61	96620	–	A
100491AM36	0,20	15,02	–	1,24E-2	35,0	42,9	4,79	6,96E-2	29094,8	1372,0	2,75	89070	–	FSA
100491AM37	0,20	17,10	–	1,83E-2	35,6	43,8	4,72	8,08E-2	28987,9	1371,1	2,66	91373	–	A
100491AM38	0,17	19,66	–	1,88E-2	35,4	44,0	4,74	1,12E-1	28920,8	901,4	3,83	90807	–	A
100491AM39	0,19	19,72	–	1,70E-2	35,0	42,4	4,78	1,08E-1	29048,1	1322,5	3,06	101963	–	A
100491AM40	0,19	18,73	–	2,43E-2	33,8	41,1	4,91	1,16E-1	29135,7	1332,9	3,13	97730	–	A
100491PM11	0,45	0,37	–	8,08E-3	49,1	63,1	3,6	5,80E-4	46483,9	2068,5	1,64	67488	–	S
100491PM12	0,43	0,42	–	2,54E-3	48,5	59,9	3,6	8,95E-4	37999,0	1999,4	1,70	85250	–	S

100491PM13	0,41	0,52	–	4,25E-3	47.2	58,7	3,7	1,21E-3	38177,7	1959,5	1,72	88628	–	S
100491PM14	0,41	6,69	–	7,98E-3	45.6	57,6	3,8	1,65E-3	38219,2	1902,3	1,71	88366	–	S
100491PM15	0,40	1,06	–	7,33E-3	44.1	55,6	3,9	2,73E-3	38197,1	1856,6	1,83	87160	–	S
100491PM16	0,44	2,02	–	7,27E-3	44.4	55,4	3,9	4,34E-3	38354,2	2007,4	1,77	86567	–	S
100491PM17	0,40	3,94	–	-1,54E-3	44.3	54,9	3,9	9,39E-3	38365,5	1926,0	1,93	79934	–	FSA
100491PM18	0,42	7,51	–	1,43E-2	43.2	52,4	4,0	1,81E-2	38451,4	1938,8	2,23	91903	–	FSA
100491PM19	0,39	9,97	–	-2,79E-3	42.7	52,2	4,0	2,53E-2	38534,3	1894,2	2,20	90973	–	FSA
100491PM20	0,42	12,01	–	9,61E-3	42.2	50,7	4,1	3,02E-2	38580,9	1917,7	2,48	95130	–	FSA
100491PM21	0,50	2,58	–	3,83E-2	39.9	50,6	4,3	5,18E-4	37111,8	2091,2	1,69	89973	–	S
100491PM22	0,47	0,40	–	1,69E-2	39.4	52,5	4,3	8,29E-4	44063,6	2003,8	1,71	86077	–	S
100491PM23	0,46	0,51	–	2,57E-2	40.7	53,9	4,2	1,04E-3	44017,0	2032,3	1,67	69281	–	S
100491PM24	0,47	0,71	–	8,45E-3	42.0	55,3	4,1	1,32E-3	43882,9	2060,9	1,64	79361	–	S
100491PM25	0,47	1,04	–	1,65E-2	41.6	54,7	4,1	2,21E-3	43871,8	2031,9	1,68	88739	–	S
100491PM26	0,26	1,99	–	3,50E-2	42.6	–	–	7,32E-3	–	–	–	87180	–	S
100491PM27	0,38	3,81	–	1,08E-2	41.9	54,2	4,1	9,79E-3	46335,9	1883,1	2,05	90435	–	FSA
100491PM28	0,38	7,44	–	2,76E-3	41.7	53,0	4,1	1,96E-2	46505,4	1868,9	2,28	92228	–	FSA
100491PM29	0,36	9,74	–	1,29E-2	41.5	53,1	4,1	2,75E-2	46578,8	1822,9	2,29	93834	–	FSA
100491PM30	0,36	11,96	–	1,96E-2	40.6	51,5	4,2	3,42E-2	46752,6	1813,4	2,47	94820	–	FSA
100491PM31	0,20	17,37	–	1,76E-2	40.7	53,3	4,2	8,88E-2	46601,8	1388,7	2,76	104521	–	FSA
100491PM32	0,19	22,66	–	8,20E-3	40.4	49,3	4,2	1,16E-1	35763,1	1367,2	3,07	100067	–	A
100491PM33	0,14	27,86	–	2,70E-2	40.6	–	–	1,84E-1	–	–	–	101998	–	A
100491PM34	0,11	32,17	–	1,95E-2	40.4	–	–	2,55E-1	–	–	–	105238	–	A

October 1992, Flight Data
Water-Air, 9.525 mm i.d.

Code	V_{sl}	V_{sg}	α	$gz/g0$	Tf,avg	Ts,avg	Pr_L	x	q″	h_L	$\psi 2$	Pin	dp/dzf	FP
100692AM1	0,02	0,63	–	9,30E-3	35,8	–	–	1,75E-2	–	–	–	37681	–	FSA
100692AM2	0,01	0,88	–	3,74E-2	39,6	–	–	2,69E-2	–	–	–	39384	–	FSA
100692AM3	0,02	1,41	–	1,91E-2	39,4	–	–	2,61E-2	–	–	–	37502	–	FSA
100692AM4	0,02	2,67	–	1,94E-2	37,5	–	–	4,96E-2	–	–	–	38295	–	FSA
100692AM5	0,03	4,41	–	1,97E-2	34,9	–	–	6,49E-2	–	–	–	38929	–	FSA
100692AM6	0,04	5,34	–	1,75E-2	33,4	–	–	9,22E-2	–	–	–	63558	–	FSA
100692AM7	0,03	9,72	–	1,25E-2	31,6	–	–	1,77E-1	–	–	–	58077	–	FSA
100692AM8	0,03	13,77	–	1,58E-2	29,9	–	–	2,44E-1	–	–	–	65185	–	A
100692AM9	0,03	17,96	–	1,70E-2	28,1	–	–	3,19E-1	–	–	–	74535	–	A
100692AM11	0,02	0,44	–	1,73E-2	36,8	–	–	1,23E-2	–	–	–	52871	–	S
100692AM12	0,13	0,67	–	1,40E-2	34,1	–	–	2,92E-3	–	–	–	50030	–	S
100692AM14	0,16	1,96	–	8,00E-3	38,2	–	–	6,97E-3	–	–	–	52119	–	FSA
100692AM15	0,12	3,70	–	1,43E-2	37,3	46,1	4,54	1,68E-2	18083,3	569,6	3,54	46596	–	FSA
100692AM16	0,10	5,71	–	1,28E-2	35,6	–	–	3,64E-2	–	–	–	59683	–	FSA
100692AM18	0,06	14,96	–	7,00E-4	32,4	–	–	1,47E-1	–	–	–	60469	–	A
100692AM19	0,11	17,48	–	1,37E-2	32,0	–	–	1,25E-1	–	–	–	77564	–	A
100692AM20	0,16	21,06	–	2,33E-2	30,3	37,2	5,34	1,19E-1	18364,1	656,2	4,02	85388	–	A
100692AM21	0,23	0,61	–	2,53E-2	33,1	44,6	5,00	1,10E-3	24005,2	971,6	2,12	38750	–	S
100692AM22	0,23	0,82	–	2,40E-2	32,9	44,7	5,02	1,79E-3	23955,0	975,1	2,07	46500	–	S
100692AM23	0,23	1,66	–	1,63E-2	33,1	45,5	4,99	4,10E-3	23917,2	943,2	2,05	48010	–	S
100692AM24	0,25	3,23	–	2,48E-2	33,1	42,7	5,00	6,99E-3	24024,1	1008,9	2,49	48568	–	FSA
100692AM26	0,24	8,86	–	2,58E-2	32,8	41,4	5,03	2,65E-2	24049,3	972,4	2,89	63799	–	FSA
100692AM27	0,22	13,75	–	1,38E-2	32,5	40,1	5,07	4,91E-2	24099,5	944,5	3,42	73990	–	A
100692AM28	0,26	18,14	–	1,70E-2	31,0	39,0	5,25	6,03E-2	24049,3	1012,2	3,02	80858	–	A
100692AM29	0,09	22,85	–	1,11E-2	30,2	–	–	1,96E-1	–	–	–	78692	–	A
100692AM30	0,10	23,75	–	1,30E-2	29,4	–	–	1,90E-1	–	–	–	84906	–	A
100692AM31	0,47	0,60	–	4,20E-3	31,4	–	–	5,79E-4	–	–	–	39666	–	S
100692AM32	0,54	0,63	–	2,70E-2	31,9	42,8	5,14	8,38E-4	28703,0	1444,7	1,84	62117	–	S

100692AM33	0,56	0,93	–	2,55E-2	33,5	43,1	4,96	1,14E-3	27420,1	1490,6	2,00	60979	–	S
100692AM34	0,56	1,05	–	1,53E-2	38,4	48,4	4,43	1,27E-3	28086,7	1511,4	1,87	60131	–	S
100692AM35	0,57	1,08	–	1,40E-2	38,5	48,8	4,42	1,34E-3	28042,7	1534,4	1,81	63372	–	S
100692AM36	0,58	2,10	–	2,05E-2	38,8	47,7	4,40	2,61E-3	28055,3	1555,2	2,15	64944	–	S
100692AM37	0,59	4,43	–	1,95E-2	39,0	47,6	4,38	5,37E-3	28149,6	1573,5	2,13	64930	–	FSA
100692AM38	0,60	7,88	–	1,77E-2	38,8	–	–	1,06E-2	–	–	–	73326	–	FSA
100692AM39	0,60	12,77	–	-1,33E-3	37,7	45,9	4,50	2,09E-2	28233,5	1575,7	2,21	90035	–	FSA
100692AM40	0,57	16,72	–	8,33E-3	36,0	43,8	4,68	3,03E-2	28426,3	1529,0	2,48	95730	–	FSA
100892AM1	0,03	0,24	–	8,50E-3	32,1	–	–	4,23E-3	–	–	–	46517	–	S
100892AM2	0,04	0,25	–	4,00E-4	31,9	–	–	3,19E-3	–	–	–	44789	–	S
100892AM3	0,15	0,30	–	3,90E-3	34,8	–	–	8,59E-4	–	–	–	37981	–	S
100892AM4	0,16	0,23	–	1,20E-2	37,1	46,8	4,56	8,07E-4	18738,7	675,8	2,81	49506	–	S
100892AM5	0,20	0,28	–	1,12E-2	36,0	45,7	4,68	6,34E-4	18720,5	914,4	2,09	40956	–	S
100892AM6	0,25	0,28	–	1,40E-2	36,1	47,5	4,67	5,17E-4	24443,4	1016,6	2,10	40943	–	S
100892AM7	0,29	0,29	–	2,06E-2	36,0	46,5	4,68	4,41E-4	23445,5	1096,8	2,03	38722	–	S
100892AM8	0,36	0,24	–	1,63E-2	36,5	46,8	4,63	3,48E-4	23095,9	1231,2	1,82	46879	–	S
100892AM9	0,45	0,28	–	1,56E-2	37,9	48,8	4,48	2,84E-4	27685,6	1378,0	1,86	40025	–	S
100892AM10	0,50	0,23	–	2,88E-2	38,4	49,9	4,43	2,47E-4	28781,3	1476,8	1,72	48961	–	BS
100892AM11	0,03	0,47	–	7,90E-3	42,6	–	–	7,90E-3	–	–	–	49109	–	S
100892AM12	0,03	0,43	–	8,90E-3	37,8	–	–	7,60E-3	–	–	–	53974	–	S
100892AM13	0,12	0,59	–	9,00E-4	36,9	47,8	4,58	2,22E-3	19103,7	578,5	2,96	38874	–	S
100892AM14	0,15	0,57	–	5,56E-3	38,3	48,3	4,44	1,68E-3	19079,6	664,6	2,81	40101	–	S
100892AM15	0,20	0,55	–	6,70E-3	38,4	48,0	4,43	1,31E-3	19076,0	912,8	2,16	41866	–	S
100892AM16	0,25	0,54	–	1,07E-2	38,7	49,1	4,40	1,06E-3	22732,8	1013,0	2,14	42453	–	S
100892AM17	0,28	0,56	–	7,00E-4	38,4	49,4	4,43	9,16E-4	24200,2	1091,3	2,01	41046	–	S
100892AM18	0,40	0,56	–	1,82E-2	38,7	48,1	4,41	6,49E-4	22716,0	1291,5	1,87	41060	–	S
100892AM19	0,45	0,57	–	1,88E-2	39,1	49,7	4,37	5,67E-4	28272,6	1385,9	1,93	40370	–	S
100892AM20	0,57	0,36	–	6,70E-3	37,4	47,7	4,53	4,50E-4	28089,2	1541,8	1,79	63655	–	BS
100892AM21	0,04	0,60	–	1,27E-2	41,1	–	–	1,05E-2	–	–	–	57223	–	S
100892AM22	0,04	0,57	–	9,30E-3	35,5	–	–	1,01E-2	–	–	–	59975	–	S
100892AM23	0,12	0,89	–	1,17E-2	36,9	47,9	4,58	3,21E-3	18540,2	588,2	2,80	38612	–	S
100892AM24	0,15	0,88	–	1,83E-2	37,3	47,5	4,55	2,62E-3	18565,4	652,6	2,72	39419	–	S

(Continued)

October 1992, Flight Data
Water-Air, 9.525 mm i.d. (*Continued*)

Code	V_{sl}	V_{sg}	α	gz/g0	Tf,avg	Ts,avg	Pr_L	x	q″	h_L	$\psi 2$	Pin	dp/dzf	FP
100892AM25	0,19	0,74	–	1,74E-2	37,2	46,8	4,56	2,16E-3	18554,2	866,8	2,19	46334	–	S
100892AM26	0,29	0,88	–	1,39E-2	37,1	47,7	4,56	1,35E-3	24063,3	1094,0	2,07	39108	–	S
100892AM27	0,33	0,90	–	3,26E-2	37,9	48,2	4,48	1,18E-3	23878,8	1172,8	1,99	38433	–	S
100892AM28	0,39	0,86	–	7,44E-3	38,5	48,9	4,42	1,01E-3	24121,9	1273,8	1,82	40301	–	S
100892AM29	0,46	0,86	–	2,29E-2	38,2	48,4	4,45	8,29E-4	27344,6	1399,0	1,95	40577	–	S
100892AM30	0,58	0,53	–	1,51E-2	37,4	47,4	4,53	6,67E-4	27410,0	1552,3	1,78	65385	–	S
100892AM31	0,03	0,92	–	8,60E-3	40,1	–	–	1,70E-2	–	–	–	55826	–	S
100892AM32	0,04	0,81	–	9,60E-3	37,0	–	–	1,42E-2	–	–	–	63333	–	S
100892AM33	0,11	1,30	–	2,71E-2	36,2	46,9	4,66	5,03E-3	19893,0	573,4	3,16	39364	–	S
100892AM34	0,15	1,28	–	1,60E-2	36,9	47,1	4,58	3,77E-3	18503,9	662,5	2,70	40025	–	S
100892AM35	0,20	1,14	–	1,44E-2	36,7	45,9	4,61	2,89E-3	18533,0	908,5	2,19	44735	–	S
100892AM36	0,36	0,83	–	1,25E-2	37,0	46,8	4,57	1,66E-3	22934,9	1201,5	1,95	61827	–	S
100892AM37	0,40	1,21	–	2,32E-2	37,0	46,4	4,58	1,43E-3	23286,2	1290,6	1,91	42246	–	S
100892AM38	0,45	1,21	–	6,33E-3	37,1	46,1	4,56	1,26E-3	23241,5	1373,0	1,89	42584	–	S
100892AM39	0,46	1,25	–	1,36E-2	37,6	48,1	4,51	1,22E-3	28848,4	1404,3	1,99	41273	–	S
100892AM40	0,57	0,83	–	8,78E-3	37,1	47,6	4,56	1,04E-3	28644,3	1517,6	1,83	61600	–	S
100892AM41	0,04	0,25	–	2,70E-3	40,0	–	–	3,49E-3	–	–	–	46599	–	S
100892AM42	0,04	0,20	–	3,30E-3	37,5	–	–	3,29E-3	–	–	–	58772	–	S
100892AM43	0,11	0,27	–	1,78E-2	36,2	46,7	4,65	1,18E-3	18545,2	557,1	3,10	49506	–	S
100892AM44	0,16	0,31	–	1,34E-2	36,9	46,6	4,58	8,10E-4	18537,1	675,2	2,76	40956	–	S
100892AM45	0,22	0,31	–	4,22E-3	37,0	45,9	4,58	5,83E-4	18526,2	957,2	2,15	40943	–	S
101492AM1	1,11	0,55	–	2,09E-2	39,9	54,4	4,29	2,21E-4	46658,9	2186,0	1,51	40205	–	S
101492AM2	1,16	0,55	–	2,71E-2	40,5	54,9	4,24	2,10E-4	46479,5	2242,6	1,47	40267	–	S
101492AM3	1,17	1,12	–	2,38E-2	41,3	55,3	4,17	4,37E-4	46412,4	2245,1	1,50	41163	–	S
101492AM4	1,19	1,13	–	2,02E-2	41,9	55,8	4,11	4,27E-4	46256,5	2276,6	1,49	40722	–	S
101492AM5	1,19	1,59	–	1,82E-2	42,8	56,7	4,04	6,34E-4	47709,3	2289,3	1,53	43383	–	S
101492AM6	1,20	1,52	–	2,20E-2	43,7	57,4	3,97	6,34E-4	47626,2	2298,3	1,54	45631	–	S

101492AM7	1,20	1,98	–	2,39E-2	44,0	57,7	3,94	8,51E-4	47651,4	2304,2	1,55	47169	–	S
101492AM8	1,20	2,04	–	2,16E-2	44,5	58,2	3,90	8,53E-4	47603,1	2305,9	1,54	45859	–	S
101492AM9	1,21	2,33	–	2,04E-2	44,9	58,4	3,87	1,05E-3	47745,6	2320,3	1,57	50030	–	S
101492AM10	1,22	2,29	–	2,28E-2	45,2	58,7	3,84	1,04E-3	47625,5	2333,2	1,55	50885	–	S
101492AM11	1,19	3,20	–	2,35E-2	46,1	60.5	3,78	1,63E-3	51680,0	2305,5	1,59	55153	–	FSA
101492AM12	1,15	3,22	–	2,14E-2	46,0	60.4	3,79	1,66E-3	51664,9	2280,3	1,60	54857	–	FSA
101492AM13	1,17	5,50	–	2,60E-2	45,7	59.8	3,80	3,21E-3	51695,1	2316,9	1,62	64199	–	FSA
101492AM14	1,06	5,94	–	2,23E-2	45,4	59,9	3,83	4,04E-3	51879,5	2065,3	1,77	59435	–	FSA
101492AM15	1,23	7,40	–	2,21E-2	43,0	57.0	4,02	5,19E-3	52097,6	2323,8	1,63	79024	–	FSA
101492AM16	1,19	8,03	–	1,92E-2	42,9	–	–	5,38E-3	–	–	–	72723	–	FSA
101492AM18	1,21	10,50	–	2,28E-2	39,6	53,4	4,32	8,48E-3	52500,0	2267,4	1,71	88284	–	FSA
101492AM19	1,13	13,43	–	1,88E-2	38,5	52,5	4,42	1,33E-2	52575,5	2159,7	1,78	98895	–	FSA
101492AM20	1,19	12,66	–	2,08E-2	37,2	51,1	4,56	1,22E-2	52770,5	2240,3	1,73	104494	–	FSA
101492AM21	1,52	0,70	–	2,26E-2	41,2	57,8	4,17	2,10E-4	56363,9	2576,0	1,35	41439	–	S
101492AM22	1,55	0,68	–	2,23E-2	42,0	58,5	4,10	2,05E-4	56133,6	2608,0	1,34	42604	–	S
101492AM23	1,57	1,26	–	2,81E-2	42,8	58,7	4,04	4,07E-4	56082,2	2633,3	1,37	46321	–	S
101492AM24	1,57	1,37	–	2,02E-2	43,3	59,2	4,00	4,08E-4	56061,0	2633,0	1,37	46328	–	S
101492AM25	1,58	1,81	–	2,63E-2	43,7	59,3	3,96	6,49E-4	56015,5	2642,7	1,39	51740	–	S
101492AM26	1,57	1,86	–	2,54E-2	44,1	59,7	3,93	6,54E-4	55977,8	2636,4	1,40	50340	–	S
101492AM27	1,61	2,12	–	2,65E-2	44,5	59,9	3,90	7,91E-4	55933,7	2676,0	1,39	54967	–	S
101492AM28	1,57	2,21	–	2,40E-2	44,7	60,2	3,88	8,11E-4	55949,1	2646,0	1,41	52740	–	S
101492AM29	1,57	3,01	–	2,25E-2	44,8	60,1	3,88	1,23E-3	56117,4	2639,3	1,42	58449	–	S
101492AM30	1,71	2,90	–	2,42E-2	43,2	58,3	4,01	1,10E-3	56139,2	2766,9	1,37	60345	–	S
101492AM31	1,64	4,17	–	2,31E-2	44,2	60,9	3,92	1,96E-3	62002,5	2703,5	1,41	70384	–	FSA
101492AM32	1,71	3,88	–	1,75E-2	42,5	59,4	4,06	1,90E-3	62179,9	2728,2	1,39	75280	–	FSA
101492AM33	1,65	6,58	–	2,87E-2	42,9	59,3	4,03	4,18E-3	62031,2	2599,2	1,49	88794	–	FSA
101492AM34	1,73	6,14	–	2,59E-2	40,9	57,3	4,20	3,75E-3	62232,4	2724,8	1,43	94661	–	FSA
101492AM35	1,71	8,62	–	2,41E-2	40,4	56,6	4,24	6,03E-3	62216,3	2708,0	1,46	107824	–	FSA
101492AM36	1,74	8,34	–	2,06E-2	39,6	55,7	4,32	5,90E-3	62215,6	2728,9	1,45	111216	–	FSA
101492AM37	1,78	9,49	–	2,49E-2	39,5	55,5	4,33	7,24E-3	62103,8	2750,2	1,45	122117	–	FSA
101492AM38	1,73	9,71	–	2,16E-2	39,3	55,2	4,35	7,45E-3	61920,7	2709,4	1,48	119249	–	FSA

(Continued)

October 1992, Flight Data
Water-Air, 9.525 mm i.d. *(Continued)*

Code	V_{sl}	V_{sg}	α	$gz/g0$	Tf,avg	Ts,avg	Pr_L	x	q″	h_L	$\psi 2$	Pin	dp/dzf	FP
101492AM39	1,67	10,68	–	2,21E-2	39,0	54,9	4,38	8,81E-3	62047,9	2667,1	1,50	124544	–	FSA
101492AM40	1,83	10,11	–	1,95E-2	38,2	–	–	8,05E-3	–	–	–	131239	–	FSA
101492PM1	0,89	0,67	–	1,20E-2	36,4	51,4	4,64	2,87E-4	46051,9	1955,9	1,60	36261	–	S
101492PM2	0,87	0,78	–	1,42E-2	36,8	52,9	4,59	4,35E-4	47534,4	1906,2	1,57	43638	–	S
101492PM3	0,92	1,35	–	3,46E-2	37,6	52,5	4,52	6,13E-4	47425,0	2001,3	1,62	39095	–	S
101492PM4	0,92	2,10	–	1,71E-2	36,8	51,5	4,60	1,26E-3	47692,6	1947,9	1,69	48506	–	S
101492PM5	0,90	3,45	–	2,10E-2	37,3	51,8	4,54	2,12E-3	47619,9	1957,1	1,71	50003	–	S
101492PM6	0,89	5,75	–	1,51E-2	37,3	51,2	4,54	4,23E-3	47569,6	1951,3	1,79	59780	–	FSA
101492PM7	0,86	8,66	–	1,93E-2	37,7	51,3	4,50	7,70E-3	47454,6	1870,5	1,91	66399	–	FSA
101492PM8	0,94	11,29	–	9,89E-3	36,8	49,9	4,59	1,09E-2	47424,3	1971,8	1,88	81320	–	FSA
101492PM9	0,91	14,74	–	2,36E-2	36,1	49,1	4,67	1,74E-2	47513,7	1901,9	1,96	93234	–	FSA
101492PM10	0,85	20,26	–	2,19E-2	36,0	48,4	4,68	3,11E-2	47167,1	1823,1	2,12	113140	–	FSA
101492PM11	1,97	0,49	–	3,10E-2	40,1	56,0	4,27	1,35E-4	55281,5	2893,7	1,23	49527	–	S
101492PM12	1,98	0,63	–	1,66E-2	40,3	55,8	4,25	1,96E-4	55292,3	2895,7	1,26	54932	–	S
101492PM13	2,05	0,85	–	1,59E-2	39,7	55,4	4,31	2,91E-4	56474,6	2935,5	1,25	62607	–	S
101492PM14	2,04	1,61	–	1,40E-2	39,9	55,2	4,29	5,50E-4	56381,2	2971,5	1,27	63848	–	S
101492PM15	2,02	2,42	–	2,38E-2	40,3	55,2	4,26	9,21E-4	56262,2	2992,0	1,29	71915	–	S
101492PM16	2,27	3,81	–	7,56E-3	40,5	55,1	4,24	1,65E-3	55993,8	3157,8	1,24	91055	–	FSA
101492PM17	2,12	5,72	–	1,28E-2	40,8	55,3	4,21	2,99E-3	55877,2	3032,4	1,31	101660	–	FSA
101492PM18	2,00	7,81	–	1,20E-2	40,4	54,8	4,25	5,15E-3	55864,6	2916,9	1,37	118877	–	FSA
101492PM19	1,72	10,94	–	1,38E-2	39,6	54,1	4,32	9,36E-3	55996,6	2640,3	1,50	126985	–	FSA
101492PM20	1,13	18,05	–	2,62E-2	38,3	52,9	4,45	2,29E-2	56441,1	2157,3	1,84	127916	–	FSA
101492PM21	2,42	0,43	–	3,09E-2	40,4	57,5	4,25	1,11E-4	61901,8	3203,3	1,16	58056	–	S
101492PM22	2,54	0,59	–	2,74E-2	41,4	58,5	4,15	1,46E-4	62153,5	3351,0	1,11	59504	–	S
101492PM23	2,47	0,85	–	2,29E-2	41,3	58,2	4,16	2,33E-4	62098,2	3288,0	1,15	63131	–	S
101492PM24	2,57	1,42	–	2,34E-2	41,6	58,1	4,14	4,44E-4	62016,0	3337,5	1,16	72970	–	S
101492PM25	2,61	2,11	–	2,69E-2	41,8	58,0	4,13	7,54E-4	61889,3	3337,8	1,18	83064	–	S
101492PM26	2,53	3,54	–	2,50E-2	41,6	57,6	4,14	1,56E-3	61685,4	3268,0	1,22	98419	–	S
101492PM27	2,55	4,72	–	2,39E-2	41,0	56,8	4,19	2,53E-3	61543,4	3306,6	1,22	123414	–	FSA

101492PM28	2,24	7,25	–	2,04E-2	40,7	56,5	4,21	4,66E-3	61662,9	3079,5	1,31	128392	–	FSA
101492PM29	2,32	8,14	–	2,94E-2	40,2	55,8	4,27	5,61E-3	61701,3	3123,3	1,30	142327	–	FSA
101492PM30	1,52	12,08	–	2,30E-2	39,7	52,0	4,31	1,05E-2	47340,4	2485,4	1,58	115043	–	FSA
101492PM31	3,05	0,38	–	1,98E-2	40,2	57,8	4,26	8,34E-5	65606,1	3677,1	1,05	63255	–	S
101492PM32	3,10	0,53	–	1,38E-2	41,4	58,7	4,15	1,20E-4	65272,7	3673,6	1,05	65847	–	S
101492PM33	3,08	0,74	–	1,93E-2	41,6	58,9	4,14	1,88E-4	65297,9	3667,6	1,06	72384	–	S
101492PM34	3,17	1,28	–	2,81E-2	42,3	59,2	4,08	3,54E-4	65010,7	3745,3	1,06	80878	–	S
101492PM35	3,07	2,02	–	2,04E-2	43,6	60,3	3,97	6,21E-4	64843,0	3698,1	1,08	86918	–	S
101492PM36	3,04	2,43	–	1,87E-2	44,5	61,1	3,90	8,63E-4	64685,4	3617,0	1,11	95806	–	S
101492PM37	2,67	3,62	–	1,40E-2	45,5	60,9	3,82	1,41E-3	60715,1	3478,7	1,17	97482	–	S
101492PM38	2,11	5,04	–	1,66E-2	45,3	60,2	3,84	2,54E-3	57318,7	2998,9	1,32	93655	–	FSA
101492PM39	0,76	9,30	–	4,00E-3	44,7	–	–	8,44E-3	–	–	–	63441	–	FSA
101492PM40	0,35	25,72	–	7,60E-3	35,5	–	–	3,60E-2	–	–	–	44361	–	A

February 1993, Flight Data
50% Glycol/Water-Air, 9.525 mm i.d., L (pressure taps): dz1 = 16 cm, dz2 = 39 cm

Code	V_{sl}	V_{sg}	α	$gz/g0$	Tf,avg	Ts,avg	Pr_L	x	q''	h_L	$\psi 2$	Pin	dp/dzf	FP
022693AM1	0,70	0,97	–	1,83E-2	29,3	42,8	36,59	5,83E-4	27976,1	653,0	3,19	37012	–	S
022693AM2	0,75	1,04	–	1,05E-2	29,5	42,9	36,32	6,66E-4	27870,5	678,2	3,09	45879	–	S
022693AM3	0,95	1,26	–	2,64E-3	28,7	41,8	37,41	7,13E-4	28042,2	726,2	2,95	52885	8868	S
022693AM4	0,92	2,41	–	1,67E-2	29,4	–	–	1,51E-3	–	–	–	50023	–	S
022693AM5	0,83	3,57	–	1,48E-2	30,2	41,7	35,52	2,43E-3	28053,7	698,3	3,52	56263	13245	S
022693AM6	1,00	5,20	–	9,90E-3	29,5	40,7	36,41	3,95E-3	28039,0	914,6	2,76	74887	–	FSA
022693AM7	0,83	8,03	–	2,07E-2	29,9	41,2	35,90	8,04E-3	27939,3	693,3	3,58	80996	–	FSA
022693AM8	1,03	9,13	–	2,47E-2	30,3	40,6	35,38	7,95E-3	27891,3	927,9	2,97	87091	–	FSA
022693AM9	0,88	11,88	–	1,55E-2	30,9	41,8	34,74	2,10E-5	27786,1	653,0	3,67	86070	–	FSA
022693AM10	0,17	16,11	–	-9,09E-4	31,4	42,9	34,13	6,19E-2	27717,5	482,3	4,99	81154	–	FSA
022693AM11	1,12	0,79	–	1,58E-2	32,0	45,3	32,22	5,22E-4	27694,4	1067,9	2,19	44631	–	S
022693AM12	1,12	1,29	–	2,85E-2	33,1	45,1	31,52	6,44E-4	27631,5	1060,3	2,29	47438	11465	S
022693AM13	1,17	1,59	–	2,11E-2	33,8	45,2	31,52	7,35E-4	27631,5	1108,1	2,19	54119	11973	S
022693AM14	1,09	2,12	–	2,47E-2	33,8	45,5	31,51	1,05E-3	27619,2	1039,8	2,29	53464	13545	S
022693AM15	1,07	2,57	–	1,23E-2	34,3	45,7	30,92	1,30E-3	27513,3	1047,7	2,34	54774	13655	S
022693AM16	1,13	3,28	–	2,42E-2	34,8	45,6	30,47	1,84E-3	27525,9	1092,7	2,36	63675	15563	S
022693AM17	1,10	5,82	–	1,83E-2	35,0	45,5	30,23	3,76E-3	27486,5	1058,6	2,52	70198	–	FSA
022693AM18	1,10	7,69	–	1,44E-2	33,3	43,8	32,01	5,99E-3	27508,3	1053,0	2,52	86829	–	FSA
022693AM19	1,06	10,65	–	1,05E-2	34,0	44,3	31,32	9,02E-3	27272,4	1096,9	2,43	99336	–	FSA
022693AM21	1,61	0,80	–	1,41E-2	34,3	46,5	30,97	2,57E-4	28273,2	1445,1	1,62	52395	12533	S
022693AM22	1,63	1,29	–	1,63E-2	35,4	46,7	29,87	4,63E-4	28266,4	1467,5	1,71	58911	14654	S
022693AM23	1,65	1,78	–	1,52E-2	35,8	47,0	29,43	6,88E-4	28242,8	1487,9	1,72	63813	16978	S
022693AM24	1,64	2,16	–	7,67E-3	36,7	47,5	28,61	8,50E-4	28247,6	1508,2	1,76	65668	18438	S
022693AM25	1,61	2,78	–	1,14E-2	36,0	46,7	29,32	1,28E-3	28178,0	1471,3	1,81	74935	–	S
022693AM26	1,64	4,10	–	1,61E-2	36,8	47,2	28,50	2,08E-3	28169,9	1495,6	1,85	83760	–	S
022693AM27	1,67	6,46	–	2,61E-2	37,9	47,4	27,56	3,96E-3	28021,4	1543,2	1,95	104004	–	FSA

February 1993, Flight Data
59% Glycol/Water-Air, 9.525 mm i.d., L (pressure taps): dz1 = 16 cm, dz2 = 39 cm

Code	V_{sl}	V_{sg}	α	gz/g0	Tf,avg	Ts,avg	Pr_L	x	q″	h_L	ψ2	Pin	dp/dzf	FP
022693PM1	0,89	0,88	–	9,70E-3	29,9	48,5	60,38	4,03E-4	28645,3	689,7	2,22	40839	–	S
022693PM2	0,94	0,97	–	1,12E-2	30,2	48,1	59,65	4,94E-4	28635,2	701,8	2,27	49023	–	S
022693PM3	0,91	1,43	–	1,77E-2	30,4	47,6	59,12	7,49E-4	28659,7	686,2	2,42	46624	–	S
022693PM4	0,89	2,61	–	1,39E-2	31,0	46,9	57,78	1,38E-3	28646,0	679,4	2,66	46396	–	S
022693PM5	0,90	3,85	–	2,97E-2	31,6	46,3	56,50	2,33E-3	28617,6	673,8	2,89	52326	–	FSA
022693PM6	0,90	5,87	–	6,40E-3	31,3	45,5	57,18	4,35E-3	28542,1	679,4	2,97	66813	–	FSA
022693PM7	0,91	8,16	–	1,78E-2	31,2	44,4	57,50	7,10E-3	28557,2	680,0	3,18	79803	–	FSA
022693PM8	0,92	9,84	–	1,34E-2	31,3	44,7	57,25	8,39E-3	28497,3	686,5	3,11	81258	–	FSA
022693PM9	0,24	17,53	–	8,50E-3	30,4	–	–	4,16E-2	–	–	–	59246	–	FSA
022693PM10	0,23	16,97	–	1,61E-2	30,6	–	–	5,40E-2	–	–	–	77236	–	FSA
022693PM11	1,35	0,69	–	1,86E-2	32,9	48,9	53,59	4,35E-4	28243,5	748,5	2,43	51402	–	S
022693PM12	1,27	1,09	–	1,73E-2	33,1	48,6	52,59	6,38E-4	28252,9	938,1	2,04	55967	–	S
022693PM13	1,26	1,49	–	2,96E-2	33,6	48,4	52,59	6,56E-4	28252,9	938,1	2,04	57532	–	S
022693PM14	1,24	1,98	–	2,65E-2	34,1	48,2	51,50	8,81E-4	28212,6	931,8	2,15	57394	–	S
022693PM15	1,26	2,09	–	1,61E-2	34,0	48,1	51,69	1,09E-3	28182,4	929,8	2,17	67275	14508	S
022693PM16	1,29	3,04	–	1,41E-2	34,3	47,8	51,12	1,58E-3	28197,7	950,2	2,22	68633	–	S
022693PM17	1,39	4,59	–	8,90E-3	35,0	47,0	49,97	2,87E-3	28149,7	1010,8	2,34	88966	–	FSA
022693PM18	1,33	7,04	–	7,40E-3	35,0	47,3	49,84	5,08E-3	28129,6	953,5	2,43	95358	–	FSA
022693PM19	1,23	9,51	–	1,54E-2	35,2	47,4	49,59	8,64E-3	28112,0	725,6	3,19	111678	–	FSA
022693PM20	0,23	18,46	–	1,36E-2	34,6	–	–	5,42E-2	–	–	–	72060	–	FSA
022693PM21	1,54	0,73	–	1,09E-2	34,6	42,3	50,61	2,63E-4	14080,3	1071,1	1,71	56608	12342	S
022693PM22	1,52	1,26	–	1,09E-2	34,5	49,1	50,78	4,85E-4	28974,8	1083,4	1,84	59931	14688	S
022693PM23	1,50	1,78	–	3,28E-2	35,1	49,0	49,69	7,38E-4	28957,2	1082,5	1,93	63903	16441	S
022693PM24	1,50	2,15	–	4,10E-3	35,4	–	–	9,32E-4	–	–	–	65827	–	S
022693PM25	1,51	2,69	–	1,19E-2	36,8	49,5	46,71	1,34E-3	28854,1	1112,3	2,07	77431	–	S
022693PM26	1,50	3,96	–	5,50E-3	37,1	49,4	46,24	2,21E-3	28856,6	1112,0	2,13	86712	–	S
022693PM27	1,50	6,42	–	6,27E-3	37,7	49,2	45,34	4,25E-3	28719,1	1132,7	2,22	104611	–	FSA
022693PM28	1,39	7,88	–	9,82E-3	37,7	49,9	45,21	5,76E-3	28748,9	1042,2	2,29	103522	–	FSA

(Continued)

(Continued)

Code	V_{sl}	V_{sg}	α	gz/g0	Tf,avg	Ts,avg	Pr_L	x	q''	h_L	$\psi 2$	Pin	dp/dzf	FP
022693PM29	0,35	13,24	–	1,66E-2	36,7	–	–	2,90E-2	–	–	–	80196	7067	FSA
022693PM30	0,23	19,60	–	1,80E-2	34,8	–	–	5,40E-2	–	–	–	67668	–	A
022693PM31	1,92	0,57	–	1,25E-2	35,1	50,4	49,62	1,87E-4	28833,5	1316,0	1,44	64868	13798	S
022693PM32	1,90	0,74	–	1,25E-2	35,5	53,0	48,90	2,50E-4	34381,4	1315,3	1,50	65971	15307	S
022693PM33	1,93	1,03	–	1,65E-2	37,0	51,5	46,43	3,39E-4	31365,0	1353,2	1,61	65089	17525	S
022693PM34	1,90	1,70	–	1,65E-2	37,7	51,3	45,25	6,28E-4	31294,2	1357,1	1,71	72460	20438	S
022693PM35	1,90	2,38	–	1,45E-2	37,9	51,4	44,95	1,05E-3	31203,0	1353,1	1,72	86574	–	S
022693PM36	1,92	4,02	–	2,24E-2	39,3	51,4	42,78	1,99E-3	31089,3	1404,9	1,86	100177	–	FSA
022693PM37	1,92	5,50	–	2,88E-2	39,6	51,6	42,38	3,33E-3	31033,2	1411,0	1,86	122883	–	FSA
022693PM38	2,05	5,96	–	1,05E-2	38,3	–	–	3,89E-3	–	–	–	136969	–	FSA
022693PM39	2,00	7,29	–	1,76E-2	39,2	50,7	42,96	5,19E-3	30935,7	1422,7	1,92	147601	–	FSA
022693PM40	1,98	8,33	–	1,71E-2	38,6	50,3	43,93	6,55E-3	30921,8	1399,2	1,91	160502	–	FSA

February 1993, Flight Data
59% Glycol/Water-Air, 9.525 mm i.d., L (pressure taps): dz1 = 16 cm, dz2 = 39 cm

Code	V_{sl}	V_{sg}	α	$gz/g0$	Tf,avg	Ts,avg	Pr_L	x	q''	h_L	$\psi 2$	Pin	dp/dzf	FP
022793AM1	0,82	0,78	–	1,04E-2	28,8	50,7	92,27	4,29E-4	27525,8	665,2	1,88	46362	–	S
022793AM2	0,84	0,87	–	1,00E-2	28,7	50,8	92,84	5,52E-4	27603,1	670,3	1,85	54505	–	S
022793AM3	0,79	1,37	–	3,30E-3	29,5	49,7	89,94	8,18E-4	27659,2	655,7	2,08	48479	–	S
022793AM4	0,77	2,32	–	1,09E-2	31,7	48,9	82,10	1,53E-3	27677,4	641,9	2,51	52237	–	S
022793AM5	0,79	3,44	–	1,39E-2	32,1	48,2	81,05	2,46E-3	27696,9	645,4	2,65	58725	13616	S
022793AM6	0,73	5,90	–	1,59E-2	32,6	47,8	79,41	4,91E-3	27683,0	638,2	2,85	66647	–	FSA
022793AM7	0,82	7,55	–	1,43E-2	33,1	47,4	77,74	7,78E-3	27666,8	643,6	3,00	86925	–	FSA
022793AM8	0,83	9,48	–	1,29E-2	32,9	46,9	78,29	9,13E-3	27641,6	649,7	3,06	84560	–	FSA
022793AM9	0,16	15,20	–	1,05E-2	32,3	–	–	6,02E-2	–	–	–	68516	7526	FSA
022793AM10	0,85	11,11	–	5,67E-3	31,9	45,7	81,68	1,50E-2	27593,5	648,6	3,09	117808	–	FSA
022793AM11	1,21	0,69	–	1,14E-2	33,4	52,3	75,63	4,68E-4	28328,1	698,1	2,28	50754	9955	S
022793AM12	1,09	1,03	–	8,82E-3	33,8	51,6	76,94	8,69E-4	28378,9	688,4	2,39	58897	10119	S
022793AM13	1,04	1,56	–	4,78E-3	33,3	50,6	76,94	8,08E-4	28378,9	688,4	2,39	54788	11879	S
022793AM14	1,07	1,97	–	4,56E-3	33,7	50,0	75,92	1,04E-3	28328,6	690,4	2,52	57222	13244	S
022793AM15	1,13	2,22	–	1,25E-2	33,0	48,9	77,90	1,21E-3	28259,8	702,6	2,55	62731	15077	S
022793AM16	1,11	2,85	–	9,90E-3	33,3	48,7	77,01	1,76E-3	28290,6	705,2	2,62	72756	15321	S
022793AM17	1,10	4,92	–	9,64E-3	33,8	48,0	75,63	3,63E-3	28419,6	692,7	2,89	82595	18443	FSA
022793AM18	0,76	6,68	–	1,30E-2	32,9	–	–	8,55E-3	–	–	–	99922	14188	FSA
022793AM19	1,13	8,76	–	1,64E-2	33,1	46,4	77,67	9,12E-3	28345,4	696,0	3,10	120132	–	FSA
022793AM20	1,10	10,88	–	1,44E-2	34,0	47,0	74,98	1,24E-2	28335,9	677,9	3,23	121200	–	FSA
022793AM21	1,63	0,71	–	7,90E-3	34,4	54,0	73,80	2,48E-4	32612,3	776,4	2,15	58959	13565	S
022793AM22	1,60	1,10	–	5,60E-3	35,1	52,2	71,73	4,54E-4	30685,4	770,2	2,33	68812	15458	S
022793AM23	1,59	1,64	–	7,91E-3	35,5	51,5	70,70	6,91E-4	30665,3	765,7	2,50	69274	17652	S
022793AM24	1,57	2,07	–	1,08E-2	36,0	51,6	69,13	8,65E-4	30656,1	761,6	2,60	71805	–	S
022793AM25	1,66	2,57	–	1,33E-2	36,4	51,3	68,00	1,21E-3	30622,5	1000,1	2,08	81320	21956	S
022793AM26	1,65	3,68	–	2,07E-2	37,0	51,3	66,56	1,99E-3	30603,5	1007,1	2,14	93289	–	S
022793AM27	1,50	6,07	–	7,50E-3	37,1	51,0	66,21	4,23E-3	30627,5	750,0	2,98	110389	–	FSA
022793AM28	1,56	6,39	–	1,52E-2	35,2	49,2	71,48	4,84E-3	30619,5	764,0	2,88	126703	–	FSA
022793AM29	1,57	8,01	–	8,82E-3	36,7	51,2	67,41	6,65E-3	33478,1	754,0	3,08	132963	–	FSA

(Continued)

(Continued)

Code	V_{sl}	V_{sg}	α	gz/g0	Tf,avg	Ts,avg	Pr_L	x	q″	h_L	$\psi 2$	Pin	dp/dzf	FP
022793AM30	1,67	8,45	–	4,36E-3	36,2	50,4	68,55	7,61E-3	33469,0	1004,3	2,38	157027	–	FSA
022793AM31	1,92	0,60	–	3,90E-3	35,8	55,4	69,72	1,84E-4	33294,0	1123,5	1,52	62276	–	S
022793AM32	1,92	0,70	–	7,00E-3	36,6	55,0	67,59	2,44E-4	33289,8	1133,8	1,60	69543	–	S
022793AM33	1,89	1,00	–	1,03E-2	37,3	54,4	65,85	3,40E-4	33304,1	1130,0	1,73	67371	–	S
022793AM34	1,92	1,62	–	1,37E-2	36,6	53,6	67,50	6,18E-4	33150,6	1131,8	1,74	76507	21050	S
022793AM35	1,91	2,39	–	1,65E-2	38,1	60,1	63,82	1,03E-3	48253,5	1143,3	1,93	86153	23732	S
022793AM36	1,93	3,64	–	1,16E-2	39,2	59,8	61,17	1,96E-3	48223,7	1179,8	2,01	110810	26945	S
022793AM37	1,88	5,35	–	1,14E-2	40,6	60,3	58,24	3,49E-3	48153,3	1145,3	2,15	126765	–	FSA
022793AM38	1,87	5,91	–	1,57E-2	39,8	59,6	59,79	4,16E-3	48134,5	1150,2	2,14	138755	–	FSA
022793AM39	1,94	7,15	–	1,38E-2	40,7	59,9	57,96	5,36E-3	47972,2	1179,2	2,14	151049	–	FSA
022793AM40	1,94	8,12	–	6,09E-3	40,0	59,1	59,33	6,56E-3	48043,1	1188,8	2,15	165618	–	FSA

February 1993, Flight Data
Water-Air, 9.525 mm i.d., L (pressure taps): dz1 = 16 cm, dz2 = 39 cm

Code	V_{sl}	V_{sg}	α	$gz/g0$	Tf,avg	Ts,avg	Pr_L	x	q''	h_L	$\psi 2$	Pin	dp/dzf	FP
022393AM1	0,57	0,31	–	4,91E-3	35,5	46,2	4,73	2,57E-4	24268,9	1465,8	1,55	39419	–	S
022393AM2	0,50	0,54	–	8,91E-3	37,0	46,9	4,57	4,80E-4	24150,0	1441,3	1,70	39708	2143	S
022393AM3	0,50	0,67	–	1,09E-2	38,0	47,9	4,48	5,94E-4	24175,2	1461,1	1,67	41115	2206	S
022393AM4	0,49	0,84	–	7,50E-3	39,6	49,0	4,32	7,50E-4	24222,9	1444,3	1,80	39543	–	S
022393AM5	0,52	0,80	–	1,62E-2	39,6	49,0	4,31	8,27E-4	24091,3	1491,6	1,74	51657	–	S
022393AM6	0,50	1,15	–	1,78E-2	40,7	50,0	4,22	1,11E-3	24058,5	1464,7	1,77	43832	3193	S
022393AM7	0,52	1,22	–	-1,29E-3	41,7	50,5	4,13	1,31E-3	24006,3	1499,0	1,81	50961	–	S
022393AM8	0,50	1,81	–	1,50E-2	42,3	51,0	4,08	1,59E-3	24016,7	1488,3	1,88	41163	4158	S
022393AM9	0,50	1,93	–	1,62E-2	42,9	51,5	4,03	1,91E-3	24092,9	1465,7	1,93	44707	4405	S
022393AM10	0,48	2,08	–	1,19E-2	41,7	–	–	2,25E-3	–	–	–	47139	–	S
022393AM12	0,49	3,73	–	1,81E-2	36,3	–	–	4,04E-3	–	–	–	47344	–	FSA
022393AM13	0,20	4,97	–	8,60E-3	37,2	47,0	4,56	1,25E-2	24421,7	934,1	2,65	47134	–	FSA
022393AM14	0,54	5,58	–	1,46E-2	35,0	42,8	4,79	5,90E-3	24379,0	1489,2	2,12	52009	5472	FSA
022393AM15	0,52	6,99	–	8,36E-3	35,7	43,3	4,71	7,38E-3	24424,4	1460,2	2,21	49734	7277	FSA
022393AM16	0,45	0,40	–	1,25E-2	36,5	45,8	4,62	3,99E-4	24371,4	1356,5	1,96	39832	–	S
022393AM17	0,47	0,53	–	1,16E-2	35,4	44,3	4,74	5,01E-4	24305,5	1403,6	1,95	40667	1987	S
022393AM18	0,46	0,66	–	1,02E-2	36,1	45,1	4,67	6,60E-4	24278,1	1382,0	1,97	41294	2206	S
022393AM19	0,04	0,09	–	1,72E-2	36,3	–	–	1,25E-3	–	–	–	43114	–	S
022493PM1	0,34	0,34	–	2,45E-3	35,4	47,1	4,74	3,78E-4	23681,2	1197,7	1,67	34289	–	S
022493PM2	0,36	0,56	–	-1,10E-3	34,8	45,9	4,80	6,57E-4	23694,7	1214,0	1,77	37219	–	S
022493PM3	0,36	0,81	–	9,27E-3	35,3	45,8	4,75	8,29E-4	23694,9	1220,5	1,85	33013	–	S
022493PM4	0,36	1,03	–	-3,00E-4	35,2	45,7	4,76	1,19E-3	23672,1	1217,5	1,85	37150	–	S
022493PM5	0,36	1,66	–	6,40E-3	35,7	44,9	4,71	1,72E-3	23646,9	1215,4	2,10	33158	–	S
022493PM6	0,36	2,38	–	6,60E-3	35,5	44,7	4,73	2,61E-3	23697,2	1214,8	2,11	35268	–	S
022493PM7	0,37	3,41	–	5,56E-4	34,8	44,3	4,80	3,64E-3	23730,8	1219,4	2,04	34709	–	FSA
022493PM8	0,36	4,42	–	4,56E-3	35,5	44,0	4,72	5,54E-3	23711,2	1203,0	2,34	39674	3579	FSA
022493PM9	0,36	6,50	–	1,91E-3	35,4	43,3	4,74	8,94E-3	23740,7	1212,8	2,48	44638	4178	FSA
022493PM10	0,36	8,40	–	1,75E-2	33,5	41,4	4,95	1,42E-2	23787,8	1199,1	2,52	54602	4111	FSA
022493PM11	0,24	0,31	–	1,27E-2	35,2	46,0	4,76	5,45E-4	23674,6	1003,4	2,17	38019	–	S

(Continued)

Code	V_{sl}	V_{sg}	α	gz/g0	Tf,avg	Ts,avg	Pr_L	x	q″	h_L	ψ2	Pin	dp/dzf	FP
022493PM12	0,22	0,64	–	1,29E-2	34,5	45,7	4,84	1,07E-3	23678,9	955,7	2,20	33199	–	S
022493PM13	0,22	0,93	–	1,27E-2	34,7	45,5	4,82	1,51E-3	23672,1	944,1	2,28	31352	–	S
022493PM14	0,22	1,08	–	1,13E-2	34,8	45,8	4,81	2,04E-3	23669,3	937,3	2,27	35433	–	S
022493PM15	0,22	1,93	–	1,29E-2	34,4	44,7	4,85	3,07E-3	23680,4	955,3	2,39	31476	–	S
022493PM16	0,23	2,48	–	1,09E-2	34,4	44,2	4,85	4,42E-3	23709,8	964,9	2,48	35971	–	S
022493PM17	0,22	3,69	–	2,07E-2	33,7	43,9	4,93	6,08E-3	23714,0	943,1	2,45	31958	–	FSA
022493PM18	0,21	4,58	–	1,02E-2	34,0	43,8	4,89	9,50E-3	23724,9	917,4	2,64	38171	–	FSA
022493PM19	0,23	6,37	–	4,30E-3	32,4	41,9	5,08	1,43E-2	23800,3	949,2	2,63	45100	–	FSA
022493PM20	0,23	9,22	–	3,00E-3	32,0	41,2	5,12	2,78E-2	23845,6	721,1	3,61	49534	–	FSA
022493PM21	0,19	0,39	–	5,82E-3	35,5	45,0	4,73	9,99E-4	17005,8	617,5	2,85	30497	–	S
022493PM22	0,12	0,70	–	8,64E-3	33,1	43,2	5,00	1,98E-3	16962,4	587,7	2,78	30366	–	S
022493PM23	0,13	0,81	–	-8,00E-4	32,6	42,2	5,06	2,34E-3	17015,4	608,2	2,85	32972	–	S
022493PM24	0,14	1,28	–	-1,65E-3	32,7	42,2	5,05	3,17E-3	17016,5	624,6	2,78	29752	–	S
022493PM25	0,14	1,61	–	-7,78E-4	33,2	43,1	4,99	5,13E-3	17000,3	587,9	2,83	33889	–	S
022493PM26	0,13	2,74	–	6,52E-3	32,2	41,2	5,11	7,37E-3	17061,4	602,7	3,03	30379	–	FSA
022493PM27	0,13	3,41	–	1,27E-2	32,6	42,4	5,06	1,15E-2	16995,0	573,8	2,92	34385	–	FSA
022493PM28	0,13	4,19	–	9,72E-3	31,4	40,2	5,21	1,53E-2	17112,1	601,6	3,14	41322	–	FSA
022493PM29	0,13	7,33	–	6,31E-3	29,5	38,5	5,44	2,55E-2	17162,1	595,2	3,13	38784	–	FSA
022493PM30	0,13	9,75	–	6,47E-3	29,0	38,7	5,51	3,93E-2	17157,6	602,1	2,88	46362	–	FSA
022493PM31	0,48	0,80	–	4,10E-3	34,5	43,9	4,84	5,92E-4	25692,0	1439,9	1,92	33668	–	S
022493PM32	0,44	1,14	–	5,30E-3	29,0	35,7	5,51	9,97E-4	17013,4	1284,7	2,01	33020	–	S
022493PM33	0,50	1,01	–	1,01E-2	28,4	35,1	5,59	1,11E-3	16965,3	1372,7	1,88	43818	2832	S
022493PM34	0,49	1,98	–	1,09E-3	31,2	41,6	5,23	1,66E-3	31911,6	1392,6	2,22	36137	4890	S
022493PM35	0,53	2,02	–	8,45E-3	32,3	42,6	5,09	2,05E-3	31856,7	1442,8	2,18	46900	4631	S
022493PM36	0,52	2,61	–	1,39E-2	33,7	44,1	4,93	2,58E-3	31727,4	1442,5	2,14	45155	–	S
022493PM37	0,53	3,48	–	1,11E-2	34,7	44,7	4,81	3,73E-3	31783,5	1465,5	2,21	50327	6524	S
022493PM38	0,53	4,76	–	9,18E-3	35,8	45,4	4,70	4,94E-3	31756,1	1469,8	2,27	48886	7050	S
022493PM39	0,54	5,50	–	9,10E-3	35,6	–	–	6,03E-3	–	–	–	52726	7450	S
022493PM40	0,55	8,43	–	1,09E-2	37,0	46,2	4,57	9,30E-3	31759,5	1511,9	2,32	55057	8817	FSA

February 1994, Flight Data
Water-Air, 9.525 mm i.d., L (pressure taps): dz1 = 69.0 cm, dz2 = 38.5 cm, and dz3 = 30.5 cm

Code	V_{sl}	V_{sg}	α	$gz/g0$	Tf,avg	Ts,avg	Pr_L	x	q''	h_L	$\psi 2$	Pin	dp/dzf	FP
020294PM3	0,04	0,19	–	9,99E-4	30,1	43,2	5,36	3,82E-3	14838,2	525,9	2,18	78327	–	S
020294PM4	0,04	0,39	–	9,29E-4	31,3	44,0	5,22	9,17E-3	14683,3	497,7	2,35	78224	–	S
020294PM6	0,10	0,14	0,562	5,18E-3	30,7	39,1	5,30	1,34E-3	17839,8	670,7	3,19	79076	–	S
020294PM7	0,12	0,19	0,555	-4,08E-3	29,8	37,0	5,40	1,53E-3	17880,2	704,4	3,57	79083	–	S
020294PM8	0,10	0,39	0,727	1,13E-3	30,4	39,7	5,33	3,65E-3	17882,4	672,6	2,88	79235	–	S
020294PM9	0,11	0,59	0,773	4,47E-3	30,7	38,5	5,29	5,08E-3	17861,3	688,0	3,38	79430	–	S
020294PM10	0,10	0,79	0,713	1,14E-3	31,1	38,9	5,24	7,40E-3	17799,6	668,1	3,45	79116	–	S
020294PM12	0,26	0,09	0,386	-2,62E-2	29,4	36,7	5,45	3,47E-4	19587,6	1552,5	1,77	79553	–	S
020294PM13	0,24	0,19	0,437	-9,63E-3	29,8	36,7	5,41	7,44E-4	22934,9	1542,5	2,18	79718	–	S
020294PM14	0,23	0,34	0,620	-3,71E-3	30,7	37,9	5,29	1,33E-3	27964,9	1544,0	2,58	79984	–	S
020294PM15	0,24	0,64	0,690	3,16E-3	31,0	38,2	5,25	2,42E-3	27887,4	1584,3	2,49	80191	–	S
020294PM16	0,12	0,09	–	-2,17E-3	32,6	43,2	5,06	6,68E-4	27563,2	736,8	3,56	79099	–	S
020294PM17	0,38	0,19	0,390	-8,22E-3	30,2	38,4	5,35	4,70E-4	31919,4	2241,5	1,78	79910	–	S
020294PM18	0,41	0,34	0,415	-4,31E-3	30,5	38,3	5,32	8,04E-4	36756,9	2322,4	2,05	80678	–	S
020294PM19	0,39	0,49	0,662	-1,01E-2	31,8	40,6	5,15	1,59E-3	36657,9	1822,1	2,33	80695	–	S
020294PM20	0,29	0,69	0,736	1,60E-2	32,6	–	–	2,22E-3	–	–	–	81499	–	S
020294PM22	0,85	0,09	0,139	4,58E-3	32,7	38,0	5,05	1,00E-4	25335,5	4346,0	1,11	81438	–	B
020294PM23	0,69	0,14	0,152	1,10E-2	33,0	–	–	1,94E-4	–	–	–	81851	–	S
020294PM24	0,92	0,19	0,165	1,16E-2	32,6	38,3	5,06	2,53E-4	27735,5	3775,1	1,32	81322	–	S
020294PM25	0,88	0,39	0,218	-4,89E-3	33,0	38,4	5,02	4,66E-4	30732,6	4179,1	1,41	83306	–	S
020294PM26	1,47	0,09	0,165	8,61E-3	33,6	39,6	4,94	8,92E-5	30511,9	4900,1	1,08	81412	2320	B
020294PM27	1,47	0,14	0,120	-2,68E-3	33,1	37,5	4,99	9,79E-5	30477,0	6713,2	1,08	86725	2996	B
020294PM28	1,38	0,19	0,126	8,97E-3	34,1	39,6	4,89	1,59E-4	36272,8	5717,2	1,19	84645	2564	S
020294PM29	1,53	0,29	0,140	-6,23E-3	33,8	38,4	4,92	1,96E-4	36364,0	6928,9	1,18	88425	3958	S
020294PM30	1,25	0,39	0,178	-1,19E-2	34,8	40,1	4,81	3,29E-4	37485,5	5751,2	1,27	85836	2735	S
020294PM32	2,48	0,09	0,118	-1,12E-3	34,5	39,5	4,84	4,16E-5	47799,9	10429,4	0,97	96847	7095	B
020294PM33	2,52	0,15	0,108	-2,08E-3	35,6	40,4	4,72	6,42E-5	47689,7	10720,8	0,99	98313	7505	B
020294PM34	2,51	0,19	0,106	7,03E-3	36,8	42,5	4,59	8,62E-5	59231,1	10791,3	1,02	98707	7647	B
020294PM35	2,50	0,24	0,117	2,25E-3	36,6	42,3	4,61	1,10E-4	59273,2	10740,9	1,03	99184	7747	B

(Continued)

February 1994, Flight Data
Water-Air, 9.525 mm i.d., L (pressure taps): dz1 = 69.0 cm, dz2 = 38.5 cm, and dz3 = 30.5 cm *(Continued)*

Code	V_{sl}	V_{sg}	α	$gz/g0$	Tf,avg	Ts,avg	Pr_L	x	q''	h_L	$\psi 2$	Pin	dp/dzf	FP
020294PM36	2,51	0,29	0,127	-1,69E-2	38,2	44,5	4,45	1,32E-4	70843,9	11042,4	1,09	100094	8018	B
020294PM37	2,47	0,39	0,126	-1,09E-2	39,5	45,7	4,33	1,81E-4	70457,1	10972,8	1,11	100557	8289	S
020394AM12	0,13	0,96	0,657	1,05E-2	26,7	29,7	5,62	7,23E-3	12782,9	712,0	6,04	81804	–	S
020394AM13	0,12	1,96	0,571	-2,85E-4	28,5	37,4	4,98	1,60E-2	45066,0	700,0	7,34	82954	–	S
020394AM14	0,11	3,96	0,707	6,00E-3	31,3	–	–	3,22E-2	–	–	–	81933,29	–	FSA
020394AM15	0,04	6,96	–	5,28E-3	34,6	53,2	3,87	1,48E-1	43578,6	487,1	4,91	80484	–	A
020394AM16	0,10	9,96	–	4,26E-3	31,8	40,5	4,58	7,63E-2	43620,5	715,0	7,20	88625	–	A
020394AM18	0,11	13,94	0,872	-5,63E-4	29,8	39,0	4,74	1,18E-1	48493,9	678,1	7,94	86985	2447	A
020394AM19	0,07	15,89	–	2,00E-3	29,7	–	–	1,95E-1	–	–	–	86945,95	2017	A
020394AM22	0,19	0,96	0,724	-1,44E-3	29,1	42,9	4,66	4,65E-3	51722,3	1371,2	2,77	81479	–	S
020394AM23	0,20	1,46	0,600	-1,60E-2	30,1	44,0	4,53	6,77E-3	51579,4	1429,2	2,63	81250	–	S
020394AM24	0,21	2,46	0,633	5,69E-3	30,9	42,8	4,56	1,28E-2	51360,4	1297,6	3,39	81945	–	FSA
020394AM25	0,22	3,46	0,633	1,60E-2	27,8	–	–	1,54E-2	–	–	–	84119	1786	FSA
020394AM27	0,18	9,89	0,813	1,00E-2	27,0	–	–	5,45E-2	–	–	–	90607	–	A
020394AM29	0,15	13,99	0,931	1,70E-2	26,0	–	–	8,97E-2	–	–	–	87567	2023	A
020494AM2	0,09	0,56	0,740	1,30E-2	28,7	41,7	5,56	4,97E-3	27162,6	684,8	3,07	78391	–	S
020494AM3	0,07	0,96	0,608	1,13E-3	30,2	44,9	5,35	1,24E-2	26135,3	611,1	2,96	79518	–	S
020494AM5	0,07	1,96	0,642	2,02E-2	30,7	40,1	5,29	2,44E-2	28010,5	608,7	4,93	79736	–	S
020494AM9	0,07	13,93	0,875	-3,76E-3	29,6	37,8	5,43	1,62E-1	33733,5	605,9	7,03	86875	1756	A
020494AM10	0,07	17,95	–	1,53E-2	28,1	36,6	5,63	2,19E-1	36251,7	597,9	7,47	94161	2480	A
020494AM12	0,41	0,97	0,671	8,25E-3	29,9	35,6	5,39	2,22E-3	27430,2	2374,2	2,07	82217	–	S
020494AM13	0,34	1,46	–	7,95E-3	31,3	41,2	5,22	4,33E-3	46707,8	1993,6	2,41	82721	1737	S
020494AM14	0,32	2,46	0,590	2,00E-2	32,4	–	–	7,38E-3	–	–	–	83429,5	2174	S
020494AM15	0,52	3,45	0,562	-1,31E-2	28,5	29,8	5,58	7,60E-3	5788,9	2516,0	1,83	87409	4372	FSA
020494AM16	0,41	6,96	0,682	4,29E-3	29,6	35,1	5,43	1,77E-2	42374,2	2338,8	3,48	90796	4278	FSA
020494AM17	0,33	9,96	–	1,00E-2	30,8	–	–	3,20E-2	–	–	–	93082,5	4770	FSA
020494AM18	0,25	11,95	–	1,90E-2	31,3	–	–	4,81E-2	–	–	–	91703,5	4732	FSA
020494AM19	0,26	13,98	0,885	1,80E-2	34,4	–	–	5,61E-2	–	–	–	96530	5444	A
020494AM22	0,88	0,96	0,495	6,97E-3	30,2	37,1	5,35	1,08E-3	49089,2	4383,7	1,70	85622	3642	FSA

020494AM23	0,84	1,46	0,578	1,31E-2	31,6	37,8	5,18	1,64E-3	49053,3	4547,7	1,84	88573	4423	FSA
020494AM24	0,88	1,96	0,590	5,25E-3	33,1	40,5	5,00	2,26E-3	59347,4	4509,1	1,81	88496	5301	FSA
020494AM25	0,74	3,96	0,637	3,70E-2	34,6	–	–	5,75E-3	–	–	–	94461,5	6877	FSA
020494AM26	0,49	5,96	–	1,30E-2	34,2	–	–	1,26E-2	–	–	–	92393	5505	FSA
020494AM29	0,71	11,93	0,824	3,10E-2	32,0	–	–	2,15E-2	–	–	–	113078	11108	FSA
020494AM30	0,60	8,83	0,756	3,50E-2	30,8	–	–	1,72E-2	–	–	–	102956	–	FSA
020494AM32	1,54	0,77	0,312	7,00E-3	34,9	–	–	5,19E-4	–	–	–	91704	5905	S
020494AM33	1,41	1,47	0,467	2,90E-2	35,6	–	–	1,13E-3	–	–	–	95841	8127	S
020494AM34	1,60	1,96	0,515	-2,95E-3	33,1	36,3	5,00	1,45E-3	34441,8	7066,7	1,61	100878	10937	S
020494AM35	1,52	3,95	0,638	7,64E-3	34,3	38,6	4,86	3,45E-3	53714,3	6858,9	1,98	112711	17048	FSA
020494AM36	0,21	6,95	0,714	1,60E-2	37,0	–	–	3,03E-2	–	–	–	83788,04	–	FSA
020494AM38	0,41	5,98	0,697	5,00E-3	33,5	–	–	1,42E-2	–	–	–	85519	–	FSA
020494AM40	1,55	8,15	0,733	3,04E-2	30,3	32,7	5,34	8,40E-3	32181,0	6747,6	2,19	138678	26384	FSA

October 1991, Ground Data
Water-Air, 9.525 mm i.d.

Code	V_{sl}	V_{sg}	α	gz/g0	Tf,avg	Ts,avg	Pr_L	x	q″	h_L	$\psi 2$	Pin	dp/dzf	FP
021292A1	0,09	0,24	–	1,0	35,8	45,7	4,70	4,04E-3	19198,7	643,3	3,06	135890	–	S
021292A2	0,09	0,32	–	1,0	35,3	46,3	4,75	5,40E-3	18809,4	634,2	2,76	135558	–	S
021292A3	0,09	0,66	–	1,0	35,3	45,9	4,75	1,07E-2	18871,0	648,8	2,78	135558	–	S
021292A4	0,09	0,84	–	1,0	35,6	46,7	4,71	1,38E-2	18789,6	641,7	2,69	135558	–	S
021292A5	0,09	1,67	–	1,0	35,4	45,1	4,74	2,67E-2	18945,1	644,7	3,11	135558	–	S
021292A6	0,09	2,52	–	1,0	35,6	41,1	4,72	4,08E-2	19119,5	628,3	6,62	135558	–	SC
021292A7	0,09	1,24	–	1,0	30,9	40,8	5,27	2,04E-2	14479,5	634,5	2,34	135890	–	SC
021292A8	0,09	1,25	–	1,0	33,0	43,5	5,01	2,06E-2	16164,5	636,8	2,45	135558	–	SC
021292A9	0,09	9,91	–	1,0	29,6	35,0	5,43	1,46E-1	18554,0	621,3	6,04	135890	–	A
021292A13	0,09	14,97	–	1,0	31,8	39,6	5,15	2,10E-1	25841,9	613,2	6,01	136222	–	A
021292A14	0,13	0,18	–	1,0	32,9	45,6	5,02	2,13E-3	25400,8	792,1	2,56	135890	–	S
021292A15	0,15	0,32	–	1,0	33,3	45,6	4,97	3,20E-3	27592,1	858,0	2,67	135558	–	S
021292A16	0,13	0,66	–	1,0	33,2	45,8	4,99	7,54E-3	27353,8	798,7	2,78	135558	–	S
021292A17	0,13	0,83	–	1,0	33,2	45,8	4,99	9,65E-3	27279,5	787,6	2,81	135558	–	S
021292A18	0,13	1,65	–	1,0	32,8	44,6	5,04	1,90E-2	27385,9	783,1	3,00	135558	–	SC
021292A19	0,13	2,50	–	1,0	33,0	42,7	5,01	2,83E-2	27886,7	783,9	3,73	135558	–	C
021292A20	0,13	4,17	–	1,0	32,6	42,0	5,06	4,62E-2	28721,3	784,2	3,99	135558	–	C
021292A21	0,13	5,77	–	1,0	30,1	39,1	5,37	6,41E-2	27913,0	773,2	4,12	135890	–	CA
021292A22	0,14	8,26	–	1,0	30,7	39,3	5,29	8,59E-2	28807,6	797,3	4,32	135890	–	A
021292A28	0,46	0,17	–	1,0	37,8	53,4	4,49	5,55E-4	45824,2	1862,5	1,61	135226	–	BS
021292A29	0,45	0,24	–	1,0	37,8	50,0	4,49	8,24E-4	38058,0	1863,7	1,71	135226	–	S
021292A30	0,44	0,30	–	1,0	35,5	47,4	4,73	1,11E-3	36756,8	1825,7	1,74	143856	–	S
021292A31	0,43	0,42	–	1,0	37,1	49,3	4,57	1,46E-3	38353,0	1830,8	1,76	134562	–	S
021292A32	0,42	0,66	–	1,0	37,3	49,4	4,55	2,39E-3	39045,1	1810,2	1,81	135890	–	S
021292A33	0,44	1,25	–	1,0	37,2	48,5	4,55	4,39E-3	37826,2	1838,4	1,86	136222	–	S
021292A34	0,42	2,49	–	1,0	36,6	47,2	4,61	9,08E-3	37856,6	1803,8	2,04	136521	–	S
021292A35	0,42	4,97	–	1,0	35,9	45,4	4,68	1,79E-2	38127,5	1808,2	2,29	136886	–	CA
021292A36	0,42	6,56	–	1,0	34,1	43,2	4,89	2,38E-2	37223,8	1794,7	2,35	137550	–	A
021292A39	0,45	0,24	–	1,0	38,7	52,1	4,41	8,16E-4	42808,2	1873,2	1,73	135558	–	S

021292A40	0,45	0,32	–	1,0	39,7	53,0	4,31	1,08E-3	43902,7	1876,7	1,80	135558	–	S
020792B1	1,41	0,17	–	1,0	38,1	53,7	4,47	1,87E-4	60604,3	3341,6	1,19	135226	–	B
020792B2	1,42	0,32	–	1,0	40,7	56,1	4,22	3,45E-4	63857,1	3371,9	1,27	135226	–	BS
020792B3	1,37	0,65	–	1,0	40,8	55,4	4,21	7,16E-4	62026,1	3319,4	1,32	135890	–	S
020792B4	1,39	1,66	–	1,0	41,4	54,9	4,16	1,82E-3	61818,5	3336,4	1,43	137217	–	SC
020792B5	1,44	3,32	–	1,0	40,1	53,2	4,27	3,56E-3	64194,5	3370,7	1,51	138545	–	C
020792B6	1,40	4,90	–	1,0	41,0	53,8	4,19	5,46E-3	63374,3	3336,6	1,55	140537	–	C
020792B7	1,35	6,43	–	1,0	40,9	53,4	4,20	7,53E-3	63401,1	3283,5	1,62	143192	–	C
020792B9	2,30	0,07	–	1,0	46,5	65,9	3,74	4,66E-5	89307,9	4283,9	1,12	134562	–	B
020792B10	2,30	0,17	–	1,0	47,7	67,1	3,66	1,11E-4	91775,6	4297,9	1,14	135558	–	BS
020792B11	2,37	0,32	–	1,0	44,1	63,3	3,93	2,07E-4	90239,6	4312,2	1,13	139209	–	BS
020792B12	2,34	0,47	–	1,0	46,4	65,2	3,76	3,11E-4	91285,7	4310,5	1,17	140205	–	BS
020792B13	2,27	0,67	–	1,0	47,9	66,5	3,65	4,51E-4	91044,1	4271,5	1,20	140205	–	S
020792B14	2,42	1,64	–	1,0	50,0	67,9	3,51	1,05E-3	92968,1	4413,3	1,23	143192	–	SC
020792B15	2,18	3,24	–	1,0	51,1	68,6	3,44	2,36E-3	92044,1	4221,0	1,31	146511	–	C
020792B16	2,19	6,25	–	1,0	46,7	63,7	3,73	4,68E-3	91100,1	4179,9	1,35	150163	–	C
020792B18	3,57	0,08	–	1,0	50,3	67,6	3,49	3,18E-5	88681,5	5250,8	1,02	138877	–	B
020792B19	3,52	0,18	–	1,0	49,7	67,4	3,53	7,94E-5	92482,9	5215,6	1,04	139873	–	B
020792B20	3,45	0,32	–	1,0	50,5	67,7	3,47	1,42E-4	90180,3	5169,2	1,06	140537	–	BS
020792B21	3,38	0,49	–	1,0	51,2	68,6	3,43	2,23E-4	91345,5	5134,3	1,07	142196	–	BS
020792B22	3,36	0,64	–	1,0	51,5	68,9	3,41	2,95E-4	92865,0	5125,7	1,09	142860	–	BS

October 1992, Ground Data
Water-Air, 9.525 mm i.d.

Code	V_{sl}	V_{sg}	α	$gz/g0$	Tf,avg	Ts,avg	Pr_L	x	q″	h_L	$\psi 2$	Pin	dp/dzf	FP
091692A11	0,09	0,18	–	1,0	36,2	45,7	4,65	2,89E-3	19188,0	504,5	3,93	99633	–	S
091692A12	0,09	0,32	–	1,0	35,8	45,8	4,69	4,21E-3	19118,8	505,5	3,72	100598	–	S
091692A13	0,09	0,66	–	1,0	36,2	46,4	4,65	6,46E-3	19116,7	508,0	3,60	100115	–	SC
091692A14	0,09	0,83	–	1,0	36,9	47,5	4,58	1,26E-2	19093,6	507,7	3,49	98667	–	SC
091692A15	0,09	1,65	–	1,0	37,4	47,6	4,53	2,07E-2	19096,1	512,9	3,58	99633	–	SC
091692A16	0,09	2,50	–	1,0	37,2	45,4	4,55	4,03E-2	19211,0	509,6	4,46	100529	–	C
091692A17	0,09	4,17	–	1,0	36,7	44,9	4,61	6,70E-2	19219,4	503,9	4,51	100115	–	CA
091692A18	0,09	5,77	–	1,0	35,6	44,1	4,71	1,02E-1	19355,7	507,1	4,42	101357	–	A
091692A19	0,09	8,26	–	1,0	34,8	43,2	4,81	1,52E-1	19380,4	493,8	4,60	103563	–	A
091692A21	0,26	0,23	–	1,0	38,5	48,5	4,42	1,00E-3	23692,9	1041,5	2,27	99426	–	S
091692A22	0,26	0,39	–	1,0	37,8	47,6	4,49	1,70E-3	23747,4	1042,3	2,34	99357	–	S
091692A23	0,26	0,82	–	1,0	37,4	47,5	4,53	3,51E-3	23768,4	1044,1	2,26	99012	–	SC
091692A24	0,26	1,59	–	1,0	36,8	46,9	4,59	6,92E-3	23927,6	1036,0	2,29	99288	–	SC
091692A25	0,25	2,81	–	1,0	36,6	45,6	4,62	1,25E-2	23611,6	1022,3	2,56	100253	–	C
091692A26	0,25	5,55	–	1,0	36,0	44,7	4,68	2,53E-2	23701,3	1011,0	2,70	101770	–	CA
091692A27	0,24	9,73	–	1,0	35,1	43,6	4,78	4,58E-2	23734,8	992,3	2,80	104183	–	A
091692A28	0,22	13,55	–	1,0	34,3	42,5	4,86	7,07E-2	23793,5	944,3	3,11	107907	–	A
091692A29	0,24	15,66	–	1,0	33,3	41,5	4,97	8,01E-2	23927,7	974,1	3,02	114526	–	A
091692A30	0,16	16,97	–	1,0	31,7	40,2	5,16	1,23E-1	23789,3	683,3	4,07	118180	–	A
091692A31	0,65	0,22	–	1,0	36,2	47,0	4,65	3,96E-4	28208,3	1643,0	1,60	103425	–	S
091692A32	0,65	0,37	–	1,0	40,4	50,8	4,24	6,58E-4	27969,3	1671,6	1,63	103770	–	S
091692A33	0,65	0,55	–	1,0	40,8	51,1	4,21	9,83E-4	27990,3	1671,8	1,65	103839	–	S
091692A34	0,65	0,60	–	1,0	41,2	51,4	4,18	1,08E-3	27818,4	1674,0	1,65	104252	–	S
091692A35	0,65	0,66	–	1,0	41,6	51,8	4,14	1,16E-3	27726,2	1678,0	1,63	103563	–	S
091692A36	0,65	1,31	–	1,0	41,7	51,7	4,13	2,33E-3	27940,0	1676,9	1,69	104183	–	S
091692A37	0,64	2,65	–	1,0	41,7	51,2	4,13	4,94E-3	27780,6	1665,1	1,77	108045	–	S
091692A38	0,62	5,26	–	1,0	40,7	49,9	4,21	1,02E-2	28044,8	1627,4	1,92	109079	–	CA
091692A39	0,61	9,75	–	1,0	39,5	47,9	4,33	2,05E-2	28124,4	1607,4	2,11	117008	–	A
091692A40	0,61	12,99	–	1,0	38,5	46,9	4,42	2,87E-2	28308,9	1594,8	2,16	122731	–	A

020293A1	0,24	0,24	–	1,0	30,6	40,5	5,30	1,11E-3	23190,3	985,9	2,38	100184	–	S
020293A2	0,25	0,39	–	1,0	30,8	40,8	5,28	1,78E-3	23048,6	1000,9	2,28	99357	–	S
020293A3	0,24	0,88	–	1,0	30,9	41,3	5,27	4,12E-3	23038,0	977,4	2,26	97978	–	S
020293A4	0,24	1,75	–	1,0	31,1	41,4	5,24	8,30E-3	23131,5	974,3	2,29	98461	–	SC
020293A5	0,25	6,28	–	1,0	31,1	39,7	5,24	2,82E-2	23310,1	989,2	2,76	99219	–	CA
020293A6	0,23	11,09	–	1,0	30,7	39,0	5,29	5,32E-2	23260,4	952,9	2,96	101150	–	A
020293A7	0,22	15,57	–	1,0	30,2	38,5	5,36	7,91E-2	24107,6	921,3	3,14	103287	–	A
020293A8	0,56	0,40	–	1,0	29,8	41,1	5,40	8,48E-4	28674,1	1488,9	1,71	102942	–	S
020293A9	0,56	0,59	–	1,0	30,8	42,1	5,28	1,24E-3	28549,1	1496,3	1,70	102253	–	S
020293A10	0,56	0,66	–	1,0	31,8	43,2	5,16	1,36E-3	28424,3	1502,9	1,68	102322	–	S
020293A11	0,57	0,74	–	1,0	32,6	44,1	4,93	3,01E-3	28366,9	1514,0	1,82	102115	–	S
020293A12	0,57	1,46	–	1,0	33,7	44,7	4,79	6,42E-3	28481,9	1507,7	1,96	102598	–	S
020293A13	0,56	3,08	–	1,0	34,9	44,8	4,79	6,48E-3	28481,9	1508,4	1,92	103563	–	SC
020293A14	0,56	11,22	–	1,0	35,4	44,2	4,74	2,53E-2	28332,0	1516,4	2,16	114664	–	CA
020293A15	0,55	15,18	–	1,0	35,2	43,6	4,76	3,57E-2	28385,4	1490,3	2,32	117353	–	A
091692B4	0,15	0,11	–	1,0	36,0	44,3	4,67	8,54E-4	18414,5	652,8	3,37	101288	–	S
091692B5	0,21	0,11	–	1,0	35,5	44,0	4,73	6,14E-4	18796,8	923,5	2,39	101701	–	S
091692B6	0,25	0,11	–	1,0	35,8	45,8	4,70	5,10E-4	23249,3	1019,9	2,27	100874	–	S
091692B7	0,31	0,11	–	1,0	35,8	46,0	4,70	4,17E-4	23189,0	1127,7	2,02	99909	–	S
091692B8	0,40	0,11	–	1,0	36,3	46,4	4,65	3,20E-4	23853,0	1287,6	1,85	100115	–	BS
091692B9	0,51	0,11	–	1,0	37,6	49,6	4,52	2,51E-4	28833,8	1464,3	1,64	101494	–	BS
091692B10	0,60	0,11	–	1,0	38,3	50,1	4,44	2,14E-4	28557,1	1591,0	1,54	101425	–	BS
091692B14	0,15	0,23	–	1,0	38,4	46,4	4,43	1,66E-3	18565,4	665,5	3,47	100115	–	S
091692B15	0,20	0,23	–	1,0	38,8	46,8	4,40	1,29E-3	18323,9	917,2	2,46	100253	–	S
091692B16	0,25	0,23	–	1,0	38,9	48,8	4,38	1,03E-3	23928,5	1030,8	2,36	100529	–	S
091692B17	0,30	0,23	–	1,0	38,7	48,5	4,41	8,53E-4	23978,8	1130,3	2,17	100874	–	S
091692B18	0,40	0,23	–	1,0	38,9	48,5	4,39	6,38E-4	23515,9	1305,9	1,88	101012	–	S
091692B19	0,49	0,23	–	1,0	39,9	51,3	4,29	5,23E-4	28944,5	1452,4	1,76	101425	–	S
091692B20	0,61	0,23	–	1,0	40,1	51,5	4,27	4,23E-4	28909,3	1614,6	1,59	100667	–	BS
091692B24	0,15	0,35	–	1,0	39,2	47,5	4,35	2,53E-3	18879,8	661,5	3,44	99288	–	S
091692B25	0,20	0,34	–	1,0	39,3	47,3	4,34	1,91E-3	18731,4	924,7	2,53	99633	–	S
091692B26	0,31	0,34	–	1,0	39,4	49,0	4,33	1,26E-3	23485,8	1140,7	2,17	100529	–	S
091692B27	0,35	0,34	–	1,0	39,1	48,3	4,37	1,11E-3	23349,9	1212,6	2,08	100598	–	S

(Continued)

October 1992, Ground Data
Water-Air, 9.525 mm i.d. ***(Continued)***

Code	V_{sl}	V_{sg}	α	gz/g0	Tf,avg	Ts,avg	Pr_L	x	q″	h_L	ψ2	Pin	dp/dzf	FP
091692B28	0,40	0,34	–	1,0	39,1	48,3	4,36	9,55E-4	23143,6	1307,7	1,95	101012	–	S
091692B29	0,50	0,34	–	1,0	39,9	50,9	4,29	7,70E-4	28114,4	1465,7	1,76	101357	–	S
091692B30	0,60	0,34	–	1,0	40,2	50,9	4,27	6,39E-4	27918,2	1608,0	1,64	100736	–	S
091692B34	0,15	0,52	–	1,0	38,1	46,8	4,47	3,84E-3	18769,2	654,7	3,25	98736	–	SC
091692B35	0,20	0,51	–	1,0	39,3	47,6	4,35	2,85E-3	18643,3	924,3	2,41	99150	–	SC
091692B36	0,35	0,51	–	1,0	39,5	49,2	4,32	1,64E-3	23158,7	1220,4	1,98	100184	–	S
091692B37	0,40	0,51	–	1,0	39,6	49,3	4,31	1,43E-3	24074,4	1305,5	1,92	100322	–	S
091692B38	0,45	0,51	–	1,0	40,2	49,8	4,26	1,27E-3	24245,5	1388,8	1,85	100529	–	S
091692B39	0,51	0,51	–	1,0	40,1	50,9	4,27	1,14E-3	28174,7	1472,4	1,79	100874	–	S
091692B40	0,60	0,51	–	1,0	40,3	50,9	4,25	9,57E-4	27963,4	1602,8	1,68	101288	–	S
091792A1	1,37	0,20	–	1,0	37,9	53,1	4,48	1,78E-4	45599,1	2411,8	1,26	106859	–	B
091792A2	1,35	0,20	–	1,0	38,7	54,0	4,40	1,80E-4	45688,0	2403,0	1,27	107803	–	BS
091792A3	1,35	0,43	–	1,0	39,5	54,4	4,33	3,88E-4	45738,3	2403,3	1,31	109017	–	S
091792A4	1,35	0,42	–	1,0	40,0	54,5	4,28	3,74E-4	45607,4	2410,5	1,31	108472	–	BS
091792A5	1,34	0,63	–	1,0	40,4	54,7	4,24	5,72E-4	45325,7	2406,0	1,33	109286	–	S
091792A6	1,32	0,61	–	1,0	41,0	55,6	4,19	5,71E-4	45305,6	2395,7	1,34	110561	–	S
091792A7	1,32	0,83	–	1,0	41,2	55,3	4,17	7,77E-4	45297,2	2389,8	1,36	110092	–	S
091792A8	1,37	0,82	–	1,0	40,4	54,3	4,25	7,46E-4	45295,5	2425,5	1,34	111637	–	S
091792A9	1,31	1,02	–	1,0	40,9	54,8	4,20	9,76E-4	45033,9	2375,2	1,38	111368	–	S
091792A10	1,32	1,01	–	1,0	40,8	54,6	4,21	9,65E-4	45084,2	2382,5	1,38	111906	–	S
091792A11	1,39	1,50	–	1,0	42,2	58,0	4,09	1,38E-3	52268,6	2467,4	1,38	114471	–	S
091792A12	1,42	1,51	–	1,0	42,6	58,2	4,06	1,35E-3	52429,6	2497,0	1,36	114202	–	S
091792A13	1,35	2,86	–	1,0	43,1	58,3	4,01	2,85E-3	52329,0	2440,2	1,44	121131	–	FSA
091792A14	1,37	2,90	–	1,0	43,5	58,6	3,98	2,81E-3	52077,4	2458,3	1,43	119380	–	FSA
091792A15	1,34	4,53	–	1,0	43,7	58,4	3,96	4,78E-3	52238,4	2431,7	1,48	127599	–	FSA
091792A16	1,33	4,49	–	1,0	43,9	58,9	3,94	4,80E-3	52308,9	2430,8	1,49	128943	–	FSA
091792A17	1,30	6,81	–	1,0	43,9	58,5	3,95	7,86E-3	52389,4	2398,2	1,52	136018	–	A
091792A18	1,31	6,82	–	1,0	44,1	58,7	3,93	7,82E-3	52399,4	2406,7	1,53	136018	–	A
091792A19	1,40	9,04	–	1,0	44,1	58,0	3,93	1,05E-2	52057,3	2484,6	1,51	147463	–	A

091792A20	1,35	9,24	–	1,0	44,3	58,6	3,92	1,09E-2	52429,6	2445,3	1,54	144430	–	A
091792A21	3,01	0,25	–	1,0	45,3	61,4	3,84	1,11E-4	56444,4	3659,1	0,99	120194	–	B
091792A22	3,01	0,24	–	1,0	45,4	61,4	3,83	1,06E-4	56273,3	3659,5	0,98	119856	–	B
091792A23	2,98	0,48	–	1,0	45,8	61,5	3,80	2,19E-4	55904,4	3649,0	1,00	123896	–	BS
091792A24	3,08	0,47	–	1,0	46,1	61,6	3,78	2,09E-4	55870,9	3707,1	0,99	124234	–	BS
091792A25	2,99	0,73	–	1,0	46,1	61,6	3,78	3,42E-4	55810,5	3657,3	1,01	126254	–	S
091792A26	3,00	0,73	–	1,0	46,0	61,6	3,79	3,42E-4	55911,1	3658,2	1,01	126923	–	S
091792A27	2,95	0,92	–	1,0	45,7	61,0	3,81	4,38E-4	55770,2	3622,9	1,02	126585	–	S
091792A28	3,01	0,87	–	1,0	45,5	60,9	3,82	4,24E-4	55921,2	3658,1	1,02	132308	–	S
091792A29	2,88	1,29	–	1,0	45,4	60,7	3,83	6,68E-4	56122,4	3576,6	1,04	134328	–	S
091792A30	3,03	1,38	–	1,0	45,9	61,0	3,80	6,35E-4	55921,2	3672,5	1,03	126585	–	S
091792A31	2,73	2,30	–	1,0	46,2	62,9	3,77	1,18E-3	61495,6	3505,7	1,08	126923	–	FSA
091792A32	3,02	2,11	–	1,0	46,6	62,8	3,74	1,06E-3	61103,2	3687,5	1,04	138707	–	FSA
091792A33	2,60	3,81	–	1,0	45,7	61,9	3,81	2,47E-3	61173,6	3410,8	1,13	152511	–	FSA
091792A34	2,95	3,65	–	1,0	46,0	62,4	3,78	2,17E-3	62391,2	3637,4	1,07	159585	–	FSA
091792A35	2,73	4,96	–	1,0	47,1	63,3	3,71	3,29E-3	61815,9	3511,3	1,11	164970	–	A
091792A36	2,73	4,95	–	1,0	47,2	63,6	3,70	3,29E-3	61777,4	3513,8	1,11	165308	–	A
091792A37	2,52	6,33	–	1,0	47,8	64,0	3,65	4,58E-3	61505,7	3382,1	1,15	166652	–	A
091792A38	2,49	6,16	–	1,0	48,2	64,4	3,63	4,63E-3	61646,6	3370,2	1,16	171368	–	A
091792A39	2,33	7,78	–	1,0	47,4	63,6	3,68	6,31E-3	61888,1	3252,7	1,21	173051	–	A
091792A40	2,11	8,05	–	1,0	47,6	63,8	3,67	7,00E-3	61867,9	3095,3	1,26	167997	–	A
020293B1	1,21	0,23	–	1,0	34,2	50,5	4,87	2,26E-4	46961,7	2238,3	1,31	104301	–	S
020293B2	1,21	0,47	–	1,0	34,8	50,6	4,81	4,66E-4	46928,1	2244,7	1,35	105583	–	S
020293B3	1,19	0,71	–	1,0	35,3	50,9	4,75	7,21E-4	46915,5	2229,1	1,38	106183	–	S
020293B4	1,20	0,96	–	1,0	35,7	51,0	4,70	9,70E-4	46842,3	2242,8	1,39	106790	–	S
020293B5	1,20	1,21	–	1,0	35,9	51,2	4,69	1,23E-3	46877,2	2238,7	1,39	107534	–	S
020293B6	1,19	1,80	–	1,0	36,4	51,1	4,64	1,87E-3	46454,7	2236,3	1,43	109617	–	S
020293B7	1,22	3,46	–	1,0	37,1	51,3	4,57	3,67E-3	47858,9	2260,0	1,52	114533	–	S
020293B8	1,20	5,43	–	1,0	37,3	51,1	4,54	6,17E-3	47683,9	2242,7	1,57	121131	–	FSA
020293B9	1,16	8,26	–	1,0	37,5	51,1	4,52	1,01E-2	47658,8	2211,3	1,63	126585	–	A
020293B10	1,12	11,18	–	1,0	37,1	50,6	4,57	1,47E-2	47584,0	2170,6	1,66	132984	–	A
020293B11	1,64	0,60	–	1,0	35,4	53,0	4,67	7,02E-4	56349,9	2631,0	1,29	109755	–	S
020293B12	1,65	0,93	–	1,0	36,0	53,4	4,60	8,72E-4	56410,6	2651,2	1,31	111099	–	S

(Continued)

(Continued)

Code	V_{sl}	V_{sg}	α	gz/g0	Tf,avg	Ts,avg	Pr_L	x	q″	h_L	ψ2	Pin	dp/dzf	FP
020293B13	1,67	1,16	–	1,0	36,7	53,9	4,60	8,85E-4	56410,6	2654,4	1,26	112782	–	S
020293B14	1,67	1,70	–	1,0	37,4	54,1	4,54	1,33E-3	56394,6	2658,9	1,30	116153	–	S
020293B15	1,65	2,74	–	1,0	37,5	54,1	4,52	2,26E-3	57009,2	2644,0	1,33	120525	–	S
020293B16	1,65	4,99	–	1,0	38,1	53,9	4,46	4,50E-3	56487,2	2644,9	1,38	131977	–	FSA
020293B17	1,64	7,37	–	1,0	38,7	54,1	4,40	7,14E-3	56249,4	2646,9	1,41	142409	–	A
020293B18	1,64	8,59	–	1,0	39,1	54,6	4,37	8,85E-3	56216,5	2647,4	1,41	151504	–	A
020293B19	1,62	9,62	–	1,0	39,6	54,9	4,32	1,04E-2	56180,5	2636,2	1,42	156889	–	A
020293C1	0,89	0,25	–	1,0	38,2	54,8	4,45	3,25E-4	46637,9	1954,2	1,45	102549	–	S
020293C2	0,90	0,36	–	1,0	38,3	54,6	4,45	4,68E-4	46557,3	1966,3	1,47	102818	–	S
020293C3	0,91	0,56	–	1,0	38,3	54,3	4,44	7,21E-4	46699,0	1973,7	1,50	103425	–	S
020293C4	0,90	1,08	–	1,0	38,3	53,9	4,44	1,40E-3	46696,2	1968,8	1,54	104301	–	S
020293C5	0,88	1,84	–	1,0	38,3	53,6	4,44	2,48E-3	46778,3	1939,5	1,61	105307	–	S
020293C6	0,91	3,50	–	1,0	38,6	52,9	4,42	4,78E-3	46774,7	1966,8	1,69	110430	–	SC
020293C7	0,90	5,66	–	1,0	38,7	52,7	4,40	8,05E-3	46688,1	1961,8	1,73	114740	–	FSA
020293C8	0,93	8,88	–	1,0	38,5	52,1	4,43	1,24E-2	47032,9	1987,3	1,77	116691	–	FSA
020293C9	0,86	12,12	–	1,0	38,3	51,7	4,44	1,99E-2	47034,8	1913,8	1,87	127937	–	A
020293C10	2,02	0,23	–	1,0	36,9	54,2	4,58	1,43E-4	55639,0	2924,3	1,12	109755	–	S
020293C11	1,98	0,35	–	1,0	37,7	54,8	4,46	3,20E-4	55680,8	2934,2	1,16	110761	–	S
020293C12	2,02	0,52	–	1,0	38,2	55,1	4,42	6,09E-4	55428,6	2934,7	1,16	112113	–	S
020293C13	2,02	0,97	–	1,0	38,5	55,1	4,42	6,30E-4	55428,6	2934,7	1,16	115815	–	S
020293C14	1,99	1,62	–	1,0	39,2	55,2	4,36	1,10E-3	55569,0	2915,4	1,21	119856	–	S
020293C15	2,01	3,00	–	1,0	39,7	55,2	4,31	2,17E-3	55528,8	2933,2	1,25	129281	–	FSA
020293C16	1,99	4,64	–	1,0	40,1	55,4	4,27	3,67E-3	55267,1	2923,3	1,27	140727	–	FSA
020293C17	2,02	6,85	–	1,0	41,0	56,0	4,19	5,82E-3	55300,7	2953,8	1,28	154193	–	FSA
020293C18	2,53	0,33	–	1,0	40,2	58,3	4,26	1,73E-4	61718,5	3310,0	1,05	117836	–	S
020293C19	2,50	0,49	–	1,0	40,9	58,8	4,20	2,62E-4	61197,8	3299,6	1,06	118842	–	S
020293C20	2,52	0,93	–	1,0	41,5	59,1	4,15	5,06E-4	60976,5	3316,1	1,07	122883	–	S
020293C21	2,51	1,52	–	1,0	42,1	59,1	4,10	8,70E-4	61011,9	3315,7	1,10	128606	–	S
020293C22	2,50	2,81	–	1,0	42,7	59,4	4,05	1,74E-3	61320,5	3314,8	1,13	139382	–	S
020293C23	2,50	4,41	–	1,0	43,4	59,9	3,99	2,98E-3	61110,7	3318,6	1,15	151835	–	FSA

February 1993, Ground Data
50%Glycol/Water-Air, 9.525 mm i.d., L (pressure taps): dz1 = 16 cm, dz2 = 39 cm

Code	V_{sl}	V_{sg}	α	$gz/g0$	Tf,avg	Ts,avg	Pr_L	x	q″	h_L	$\psi 2$	Pin	dp/dzf	FP
020693A1	0,91	0,24	–	1,0	30,8	48,8	35,33	2,79E-4	27246,0	729,1	2,08	104707	3618	BS
020693A2	0,92	0,36	–	1,0	30,8	47,3	35,37	4,17E-4	27080,0	888,8	1,85	104370	4054	S
020693A3	0,90	0,55	–	1,0	30,8	46,5	35,32	6,56E-4	27164,5	722,3	2,41	105376	4281	S
020693A4	0,93	1,03	–	1,0	30,9	45,2	35,19	1,22E-3	27388,6	891,1	2,17	107397	5579	S
020693A5	0,89	1,72	–	1,0	31,1	44,7	35,03	2,15E-3	27282,2	715,3	2,82	108741	6913	S
020693A6	0,91	3,33	–	1,0	31,3	43,9	34,71	4,21E-3	27389,9	877,7	2,51	113795	11025	SC
020693A7	0,93	5,26	–	1,0	31,5	43,5	34,45	6,89E-3	27534,8	893,0	2,61	120525	–	CA
020693A8	0,92	7,47	–	1,0	31,6	43,2	34,34	1,11E-2	27620,3	883,5	2,73	135680	–	CA
020693A9	0,91	11,05	–	1,0	32,0	43,0	33,91	1,67E-2	27449,3	879,2	2,87	137362	–	CA
020693A10	0,85	14,32	–	1,0	32,0	42,9	33,91	2,24E-2	27542,6	698,8	3,66	132646	–	A
020693A11	1,12	0,23	–	1,0	30,7	50,8	34,79	4,62E-4	32354,6	1040,2	1,77	107059	4578	BS
020693A12	1,11	0,47	–	1,0	31,2	48,9	34,27	6,78E-4	32551,8	1050,2	1,87	107059	5547	S
020693A13	1,12	0,70	–	1,0	31,7	48,4	34,27	6,89E-4	32551,8	1050,2	1,87	108741	6412	S
020693A14	1,10	0,93	–	1,0	32,0	48,2	33,94	9,34E-4	32628,3	1039,1	1,95	109417	6849	S
020693A15	1,11	1,15	–	1,0	32,4	48,1	33,45	1,16E-3	32563,9	1047,6	2,00	110761	7588	S
020693A16	1,10	1,74	–	1,0	32,7	48,0	33,15	1,81E-3	32580,0	1043,0	2,06	112451	8906	S
020693A17	1,12	3,26	–	1,0	33,6	47,6	32,12	3,53E-3	32484,4	1070,2	2,19	120194	–	SC
020693A18	1,12	4,99	–	1,0	34,2	47,3	31,57	5,84E-3	32463,8	1078,3	2,31	130626	–	CA
020693A19	1,11	7,60	–	1,0	34,2	47,1	31,52	9,29E-3	32430,2	1070,3	2,38	135680	–	CA
020693A20	1,15	10,24	–	1,0	34,4	46,8	31,29	1,28E-2	32249,9	1099,9	2,41	143761	–	CA
020693A21	1,66	0,30	–	1,0	33,0	49,9	32,82	1,99E-4	32049,7	1453,5	1,31	110430	7927	B
020693A22	1,65	0,56	–	1,0	33,1	48,7	32,65	3,88E-4	32018,5	1444,1	1,44	113457	9195	BS
020693A23	1,67	0,87	–	1,0	33,3	47,8	32,46	6,17E-4	31899,7	1456,2	1,53	116491	10776	S
020693A24	1,69	1,08	–	1,0	33,4	47,7	32,37	7,65E-4	32041,6	1468,0	1,54	117836	11879	S
020693A25	1,65	1,61	–	1,0	33,7	47,6	32,04	1,19E-3	32357,9	1449,6	1,62	120862	13392	S
020693A26	1,67	2,52	–	1,0	34,2	47,5	31,52	1,97E-3	32601,1	1468,8	1,69	129281	–	S
020693A27	1,65	4,53	–	1,0	34,8	47,3	30,91	3,94E-3	32635,0	1464,0	1,80	142409	–	SC
020693A28	1,67	6,81	–	1,0	35,4	47,5	30,34	6,22E-3	32776,1	1486,7	1,84	152173	–	CA

(Continued)

(Continued)

Code	V_{sl}	V_{sg}	α	gz/g0	Tf,avg	Ts,avg	Pr_L	x	q″	h_L	$\psi 2$	Pin	dp/dzf	FP
020693A29	1,66	7,81	–	1,0	36,0	47,9	29,70	7,83E-3	32593,7	1492,0	1,87	166314	–	A
020693A30	1,62	8,69	–	1,0	36,4	48,1	29,29	9,14E-3	32291,9	1471,5	1,91	170693	–	A
020693A31	2,12	0,22	–	1,0	34,6	52,7	31,14	1,21E-4	37413,8	1796,1	1,16	114133	9915	B
020693A32	2,11	0,32	–	1,0	34,8	52,6	30,88	1,80E-4	37833,4	1795,3	1,19	115815	10747	B
020693A33	2,12	0,47	–	1,0	35,2	52,2	30,48	2,67E-4	37435,0	1807,2	1,24	118511	11868	BS
020693A34	2,12	0,91	–	1,0	35,5	51,5	30,25	5,35E-4	37794,2	1811,8	1,32	123896	14678	S
020693A35	2,14	1,47	–	1,0	35,9	51,1	29,77	8,93E-4	37479,2	1835,6	1,36	129619	17728	S
020693A36	2,11	2,73	–	1,0	36,5	51,0	29,20	1,83E-3	37538,6	1826,9	1,44	141065	–	S
020693A37	2,12	4,20	–	1,0	37,7	51,3	28,15	3,07E-3	37535,7	1855,4	1,51	155206	–	SC
020693A38	2,11	6,03	–	1,0	38,4	51,5	27,51	4,93E-3	37446,0	1867,7	1,56	173382	–	CA
020693A39	2,09	8,22	–	1,0	38,8	48,8	27,19	7,43E-3	28609,4	1844,3	1,58	189881	–	CA
020693A40	2,06	9,10	–	1,0	36,5	45,2	29,23	8,75E-3	24752,4	1763,6	1,64	198969	–	A

February 1993, Ground Data
59% Glycol/Water-Air, 9.525 mm i.d., L (pressure taps): dz1 = 16 cm, dz2 = 39 cm

Code	V_{sl}	V_{sg}	α	$gz/g0$	Tf,avg	Ts,avg	Pr_L	x	q″	h_L	$\psi 2$	Pin	dp/dzf	FP
020693B1	0,93	0,23	–	1,0	29,3	48,3	59,23	2,64E-4	22663,7	703,1	1,70	106728	–	S
020693B2	0,93	0,35	–	1,0	30,0	46,7	57,81	4,00E-4	22775,4	698,3	1,95	105714	–	S
020693B3	0,91	0,54	–	1,0	30,5	45,7	56,67	6,34E-4	22825,7	690,3	2,19	107059	–	S
020693B4	0,93	1,03	–	1,0	31,0	47,6	55,60	1,19E-3	28628,5	696,1	2,50	107734	–	S
020693B5	0,92	1,71	–	1,0	31,5	47,3	54,64	2,02E-3	28558,1	693,4	2,61	109417	6402	S
020693B6	0,92	3,29	–	1,0	32,4	47,0	52,86	4,12E-3	28578,2	688,6	2,85	115477	10648	SC
020693B7	0,89	5,30	–	1,0	33,0	46,9	51,69	7,09E-3	28613,9	680,9	3,04	120194	–	SC
020693B8	0,90	8,25	–	1,0	33,5	46,9	50,68	1,11E-2	28585,3	682,6	3,16	123558	–	CA
020693B9	0,90	11,07	–	1,0	34,2	47,0	49,39	1,66E-2	28603,4	679,7	3,32	138038	–	A
020693B10	0,79	17,68	–	1,0	33,9	46,2	49,95	3,02E-2	28536,4	655,7	3,60	139720	–	A
020693B11	1,29	0,23	–	1,0	32,6	52,5	51,43	3,93E-4	28197,8	951,0	1,72	109079	5034	BS
020693B12	1,28	0,46	–	1,0	33,1	50,5	50,90	5,90E-4	28319,4	951,3	1,86	109417	6165	S
020693B13	1,27	0,69	–	1,0	33,4	49,5	50,90	5,95E-4	28319,4	951,3	1,86	110430	7106	S
020693B14	1,29	0,92	–	1,0	33,8	49,1	50,29	7,81E-4	28231,4	966,4	1,91	111099	7944	S
020693B15	1,27	1,13	–	1,0	33,9	49,1	49,95	9,94E-4	28295,3	956,4	1,97	112451	8495	S
020693B16	1,29	1,71	–	1,0	34,2	48,8	49,45	1,52E-3	28437,8	967,4	2,02	115477	10116	S
020693B17	1,28	3,23	–	1,0	34,8	48,3	48,33	3,03E-3	28203,1	969,7	2,18	121876	14386	SC
020693B18	1,27	5,11	–	1,0	35,5	48,2	47,20	5,06E-3	28367,6	973,5	2,31	128275	–	SC
020693B19	1,29	7,57	–	1,0	36,0	48,1	46,39	7,86E-3	28134,9	990,3	2,36	137024	–	SC
020693B20	1,31	9,89	–	1,0	36,4	48,2	45,75	1,10E-2	28155,9	1007,5	2,38	149821	–	CA
020693B21	1,50	0,30	–	1,0	36,5	53,9	45,56	2,17E-4	27834,1	1132,6	1,41	109755	6523	BS
020693B22	1,51	0,58	–	1,0	36,5	52,1	45,49	4,19E-4	27913,8	1133,6	1,59	110761	–	S
020693B23	1,51	0,91	–	1,0	36,7	51,1	45,27	6,73E-4	28015,4	1133,6	1,72	113119	–	S
020693B24	1,49	1,13	–	1,0	36,5	50,9	45,57	8,48E-4	28069,9	1123,0	1,75	114133	–	S
020693B25	1,50	1,66	–	1,0	36,6	50,4	45,32	1,29E-3	27953,6	1126,5	1,82	118173	12324	S
020693B26	1,50	2,65	–	1,0	36,9	50,0	44,92	2,16E-3	28052,1	1128,6	1,92	123896	14993	SC
020693B27	1,49	4,85	–	1,0	37,5	49,4	43,97	4,27E-3	27654,9	1132,7	2,07	134328	–	SC
020693B28	1,51	7,25	–	1,0	37,9	49,5	43,40	6,76E-3	27913,8	1146,6	2,12	144092	–	CA

(Continued)

(Continued)

Code	V_{sl}	V_{sg}	α	gz/g0	Tf,avg	Ts,avg	Pr_L	x	q″	h_L	ψ2	Pin	dp/dzf	FP
020693B29	1,51	8,49	–	1,0	38,3	49,5	42,78	8,45E-3	27631,8	1152,1	2,16	154193	–	CA
020693B30	1,50	9,30	–	1,0	38,6	49,7	42,33	9,67E-3	27546,9	1152,0	2,18	160591	–	CA
020693B31	1,91	0,23	–	1,0	36,7	51,8	45,14	1,33E-4	25398,2	1366,5	1,24	112451	–	B
020693B32	1,91	0,33	–	1,0	36,6	51,0	45,41	1,97E-4	25498,9	1360,3	1,30	114802	–	BS
020693B33	1,92	0,48	–	1,0	36,5	50,1	45,50	2,88E-4	25396,1	1363,7	1,38	117160	–	BS
020693B34	1,91	0,93	–	1,0	36,5	48,9	45,47	5,82E-4	25567,0	1355,0	1,54	121200	14010	S
020693B35	1,90	1,51	–	1,0	36,5	48,5	45,59	9,91E-4	25538,7	1342,2	1,60	125916	16358	S
020693B36	1,92	2,78	–	1,0	36,8	47,9	45,02	1,97E-3	25607,9	1360,1	1,71	138369	–	S
020693B37	1,93	4,32	–	1,0	37,3	48,0	44,26	3,30E-3	25790,2	1374,4	1,78	150828	–	SC
020693B38	1,93	6,27	–	1,0	37,6	48,0	43,72	5,29E-3	25684,4	1378,8	1,82	166652	–	CA
020693B39	1,90	8,84	–	1,0	38,1	48,1	42,97	8,00E-3	25820,6	1369,9	1,91	176415	–	CA
020693B40	1,80	10,15	–	1,0	38,3	48,3	42,71	9,82E-3	25614,1	1320,1	1,98	179442	–	A

APPENDIX B

Appendix B presents the flow image, void fraction time trace, and void fraction PDF for flows with varying flow rates. Flight data is compared to corresponding ground data. The images are presented by increasing values of, α, void fraction.

Bubble Flow Regime at μ-g, $\alpha_{ave} = 0.106$

Flow 1: (96F3P23) $V_{SL} = 1.73\,m/s$, $V_{SG} = 0.20\,m/s$

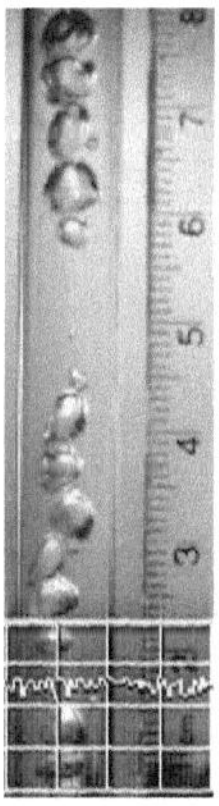

Figure 1. Flow image at μ-g for flow 1. The μ-g flow shows two separate trails of small rounded bubbles. The bubbles appear to flow near or at the tube center. The liquid slug separating them contains no bubbles

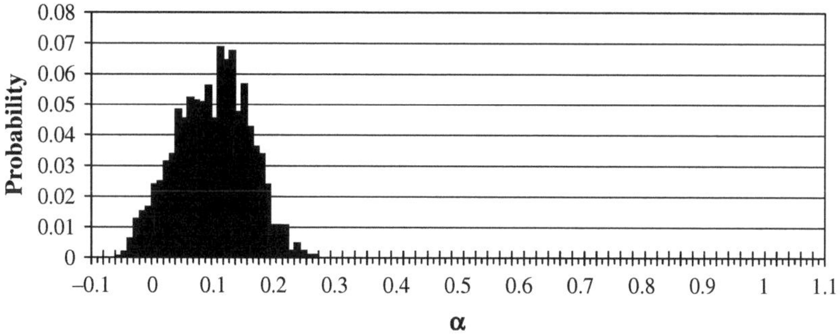

Figure 2. Void fraction PDF at μ-g for flow 1

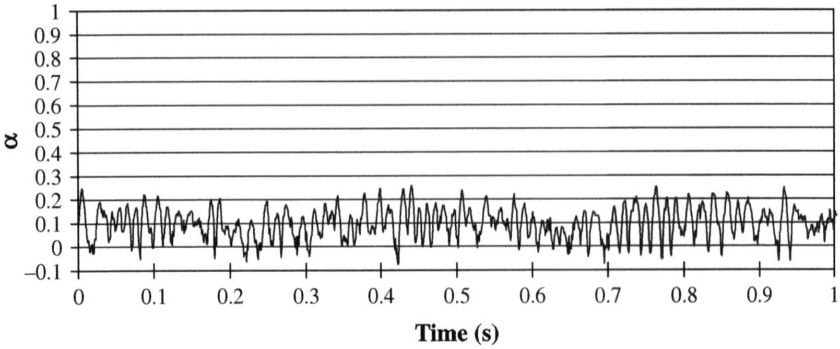

Figure 3. Void fraction time trace at μ-*g* for flow 1

Bubble Flow Regime at 1-*g*, $\alpha_{\text{ave}} = 0.961$
Flow 1: (96G3P20) $V_{\text{SL}} = 1.74\,\text{m/s}$, $V_{\text{SG}} = 0.20\,\text{m/s}$

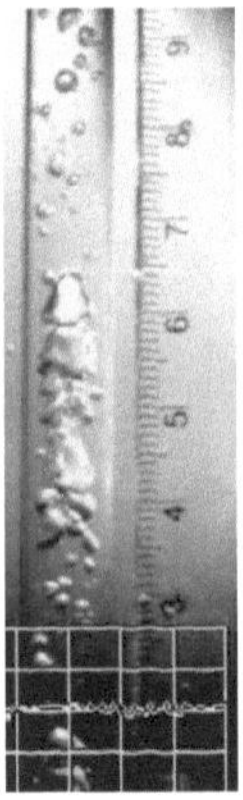

Figure 4. Flow image at 1-*g* for flow 1. The 1-*g* flow also shows a trail of medium-sized bubbles. In this case, the bubbles are packed more closely together causing the individual bubble shape to be flattened. The liquid slug is longer than in the μ-*g* case and contains small disbursed bubbles. The void fraction is significantly greater than that at μ-*g*

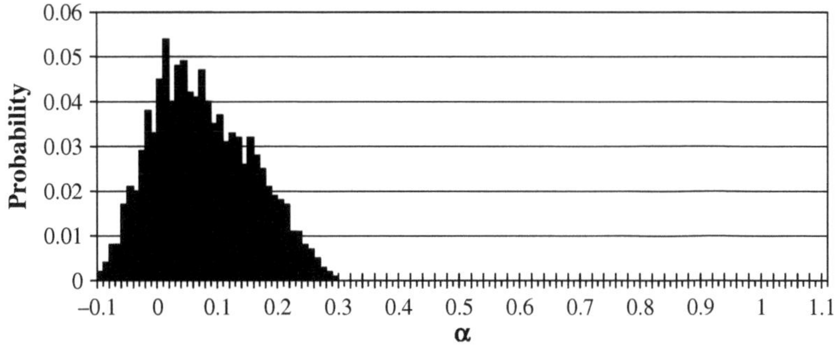

Figure 5. Void fraction PDF at 1-*g* for flow 1

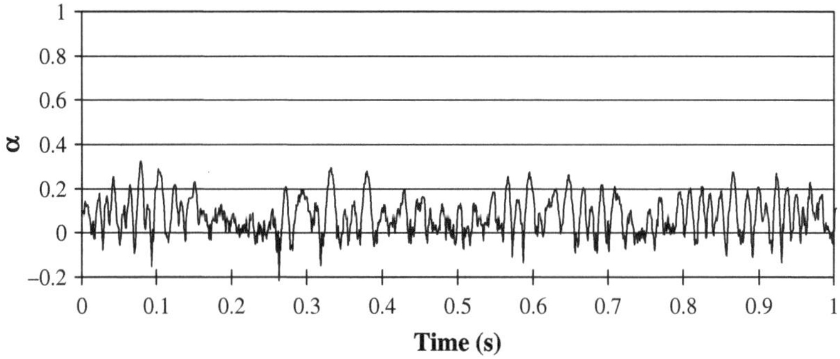

Figure 6. Void fraction time trace at 1-*g* for flow 1

Bubble Flow Regime at μ-*g*, $\alpha_{ave} = 0.145$
Flow 2: (96F3P38) $V_{SL} = 2.51\,m/s$, $V_{SG} = 0.59\,m/s$

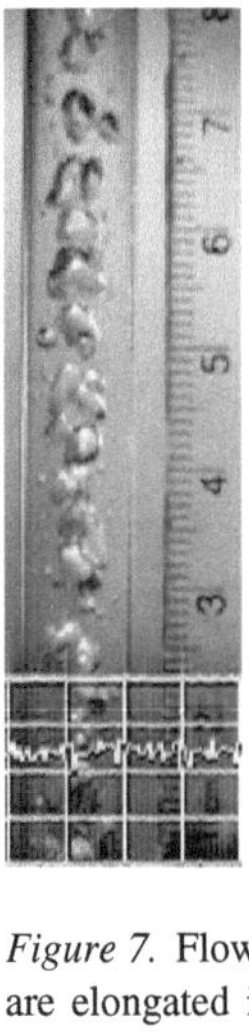

Figure 7. Flow image at μ-*g* for flow 2. The μ-*g* flow shows a loose trail of small bubbles. The bubbles are elongated in the direction of the stream flow. Very narrow liquid slugs separate small groups of bubbles

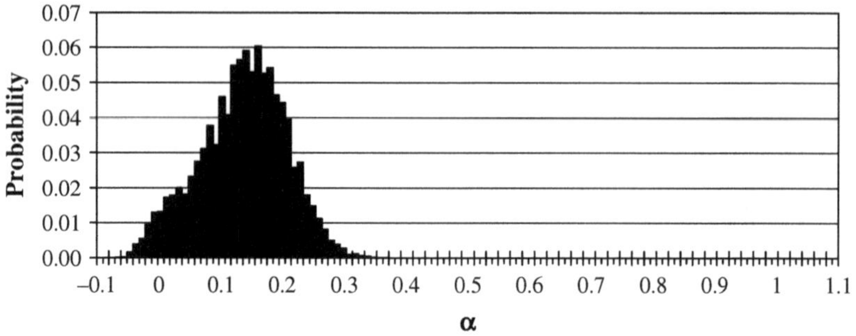

Figure 8. Void fraction PDF at μ-*g* for flow 2

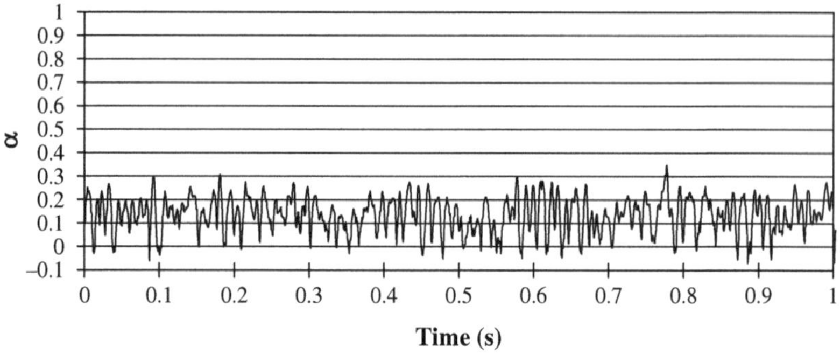

Figure 9. Void fraction time trace at μ-*g* for flow 2

Bubble Flow Regime at 1-g, $\alpha_{ave} = 0.201$
Flow 2: (96G3P33) $V_{SL} = 2.49\,\mathrm{m/s}$, $V_{SG} = 0.59\,\mathrm{m/s}$

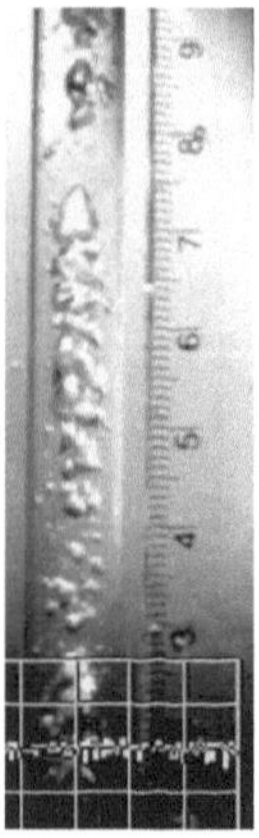

Figure 10. Flow image at 1-*g* for flow 2. The 1-*g* flow shows a more densely packed trail of irregular shaped bubbles. The liquid slugs preceding and following the trail contain small clusters of irregular shaped bubbles

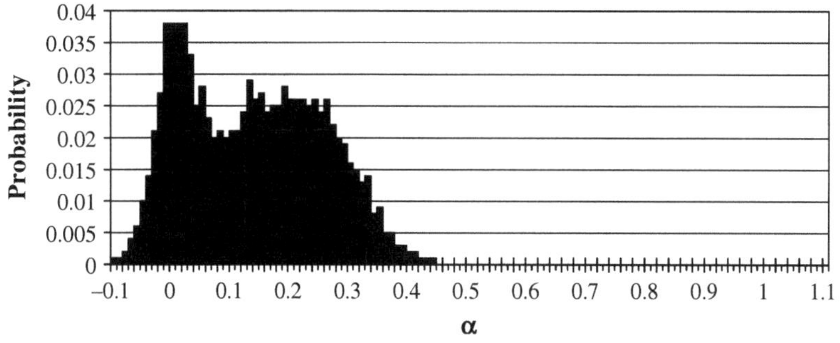

Figure 11. Void fraction PDF at 1-g for flow 2

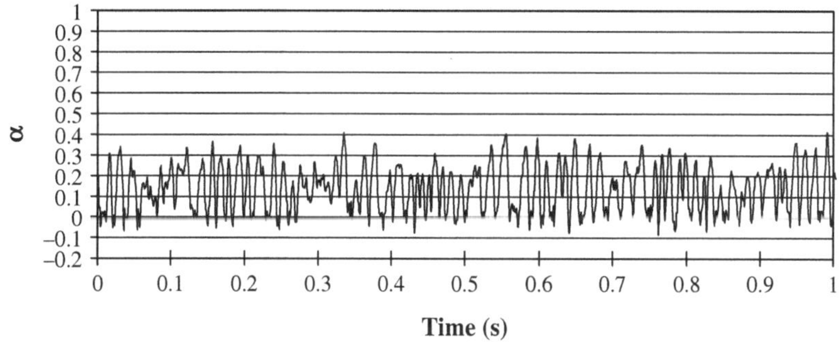

Figure 12. Void fraction time trace at 1-g for flow 2

Bubble Flow Regime at μ-g, $\alpha_{ave} = 0.161$
Flow 3: (96F3P5) $V_{SL} = 1.78\,m/s$, $V_{SG} - 0.40\,m/s$

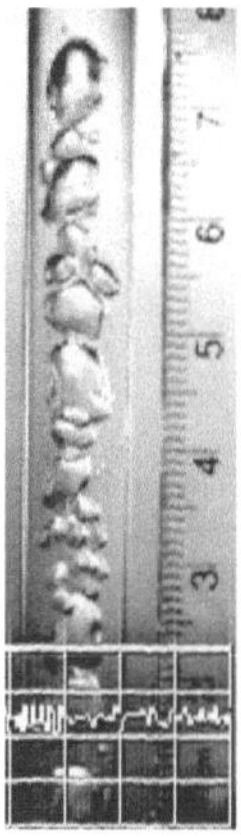

Figure 13. Flow image at μ-g for flow 3. The μ-g flow shows a packed trail of small bubbles with diameters less than one tube diameter. The bubbles flow at the center of the tube

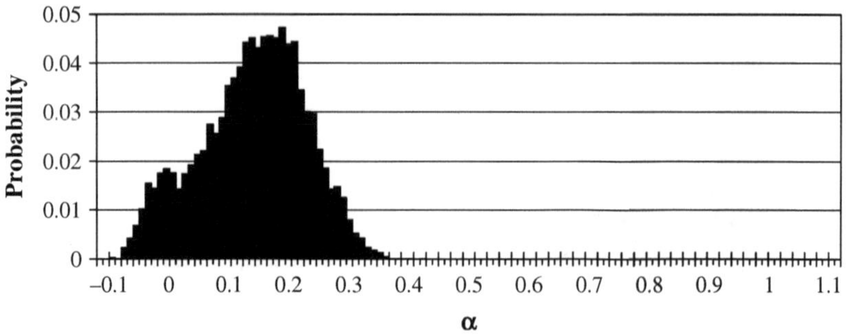

Figure 14. Void fraction PDF at μ-g for flow 3

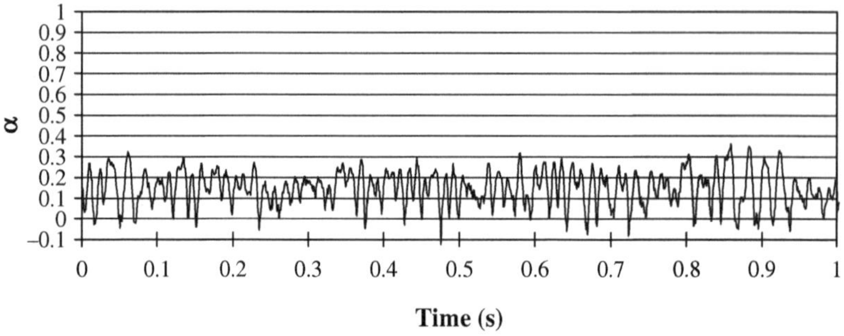

Figure 15. Void fraction time trace at μ-g for flow 3

Bubble Flow Regime at 1-g, $\alpha_{ave} = 0.192$
Flow 3: (96G3P4) $V_{SL} = 1.79\,\text{m/s}$, $V_{SG} = 0.40\,\text{m/s}$

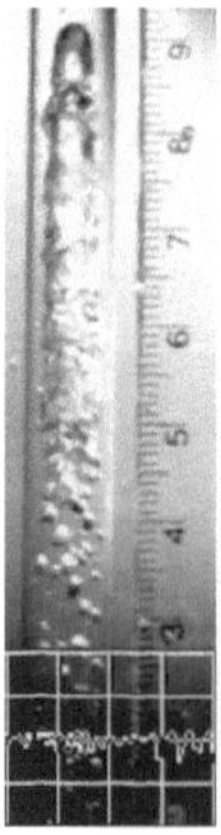

Figure 16. Flow image at 1-g for flow 3. The 1-g flow shows tightly packed bubbles that have merged to form one long bubble. The small disbursed bubbles flow in a zigzag motion from the tube center to the wall and vice versa.

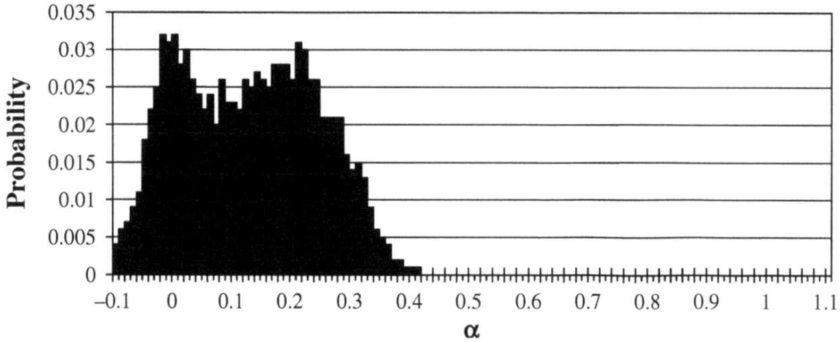

Figure 17. Void fraction PDF at 1-*g* for flow 3

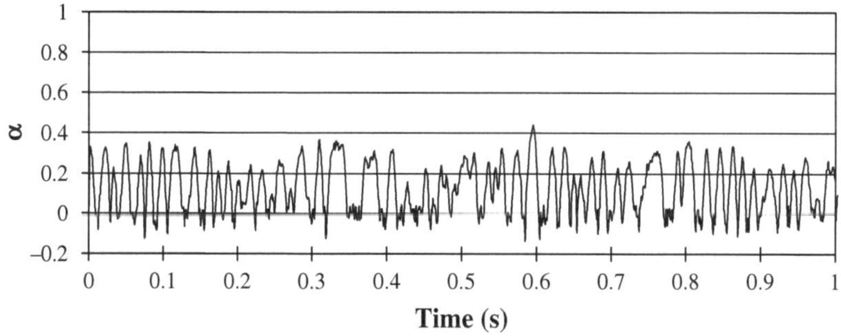

Figure 18. Void fraction time trace at 1-*g* for flow 3

Slug Flow Regime at μ-*g*, $\alpha_{ave} = 0.164$
Flow 4: (96F3P26) $V_{SL} = 1.72\,m/s$, $V_{SG} = 0.40\,m/s$

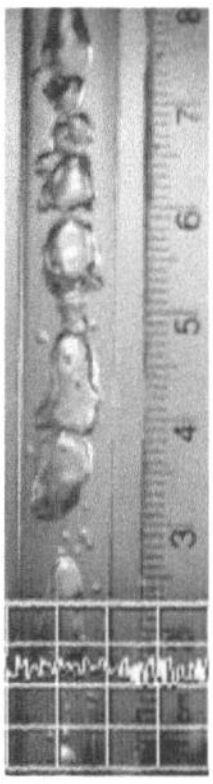

Figure 19. Flow image at μ-*g* for flow 4. The μ-*g* flow shows a trail of slightly distorted bubbles of approximately tube diameter length

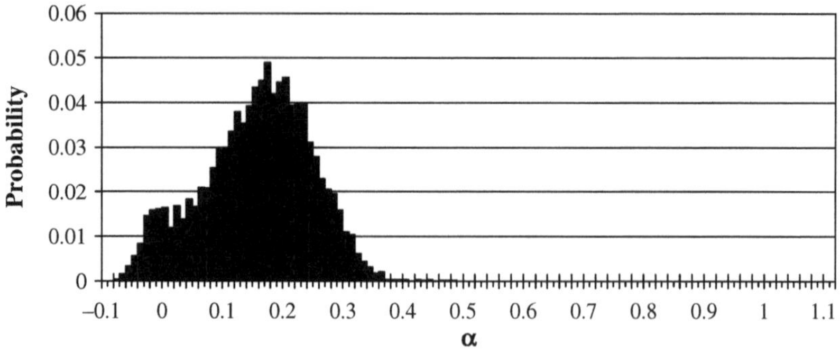

Figure 20. Void fraction PDF at μ-g for flow 4

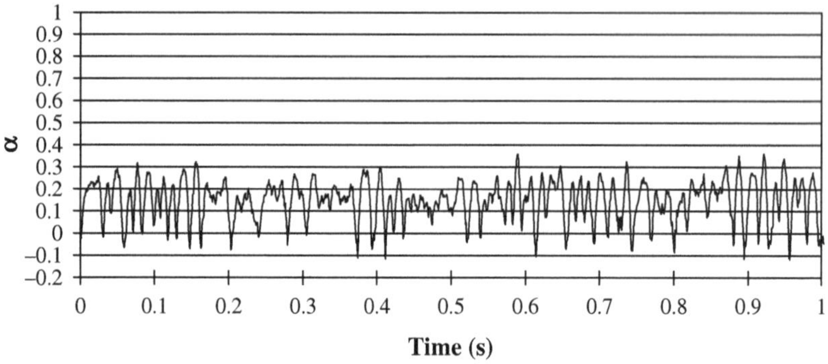

Figure 21. Void fraction time trace at μ-g for flow 4

Slug Flow Regime at 1-g, $\alpha_{ave} = 0.200$
Flow 4: (96G3P22) $V_{SL} = 1.73\,m/s$, $V_{SG} = 0.40\,m/s$

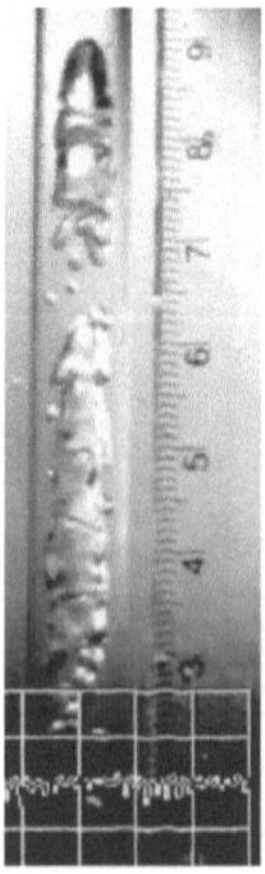

Figure 22. Flow image at 1-g for flow 4. The 1-g flow shows two separate trails of medium-sized bubbles. The bubbles are more closely packed and have diameters less than the tube diameter

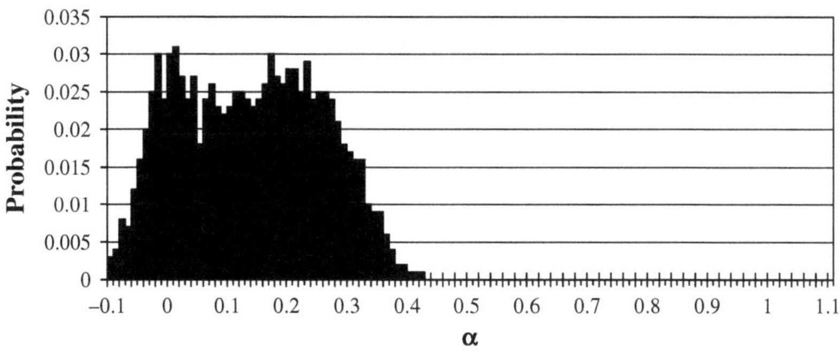

Figure 23. Void fraction PDF at 1-g for flow 4

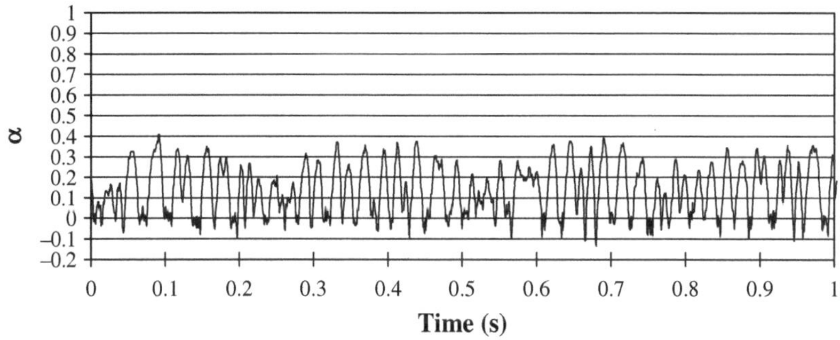

Figure 24. Void fraction time trace at 1-g for flow 4

Slug Flow Regime at μ-g, $\alpha_{ave} = 0.200$
Flow 5: (96F3P40) $V_{SL} = 2.50\,m/s$, $V_{SG} = 0.99\,m/s$

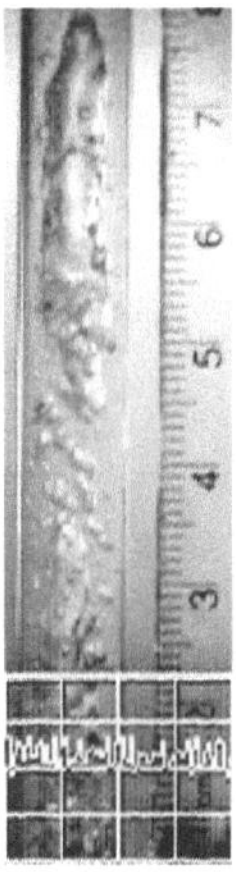

Figure 25. Flow image at μ-g for flow 5. The μ-g flow shows a short bubble with irregular edges and a distorted nose, followed by a liquid slug. The slug contains small bubbles

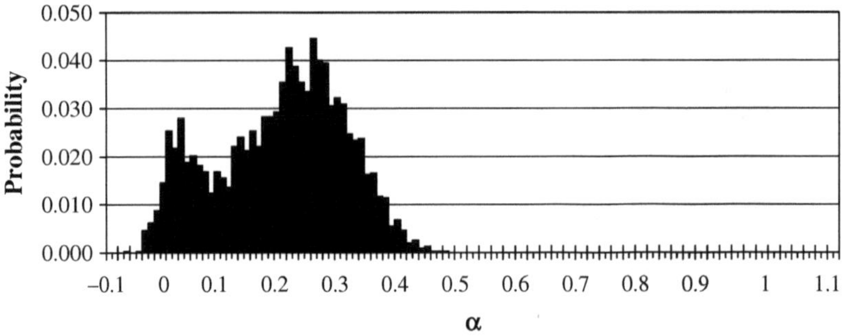

Figure 26. Void fraction PDF at μ-g for flow 5

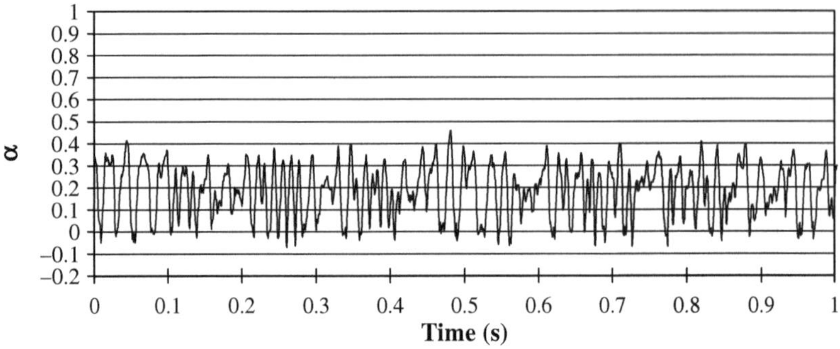

Figure 27. Void fraction time trace at μ-g for flow 5

Slug Flow Regime at μ-g, $\alpha_{ave} = 0.256$
Flow 5: (96F3P40) $V_{SL} = 2.49\,\mathrm{m/s}$, $V_{SG} = 0.98\,\mathrm{m/s}$

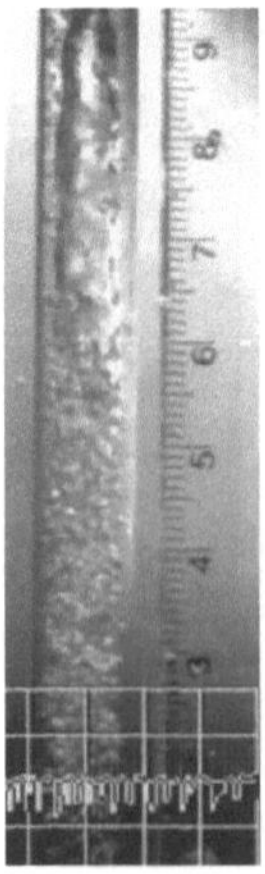

Figure 28. Flow image at 1-g for flow 5. The 1-g flow shows a similar bubble. The liquid slug is longer and contains more disbursed bubbles than the μ-g case

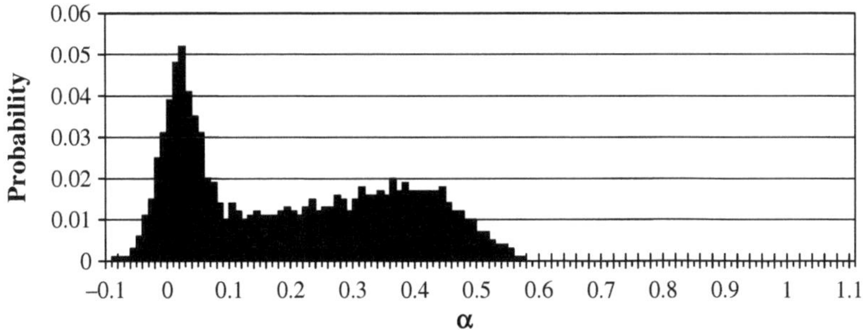

Figure 29. Void fraction PDF at 1-g for flow 5

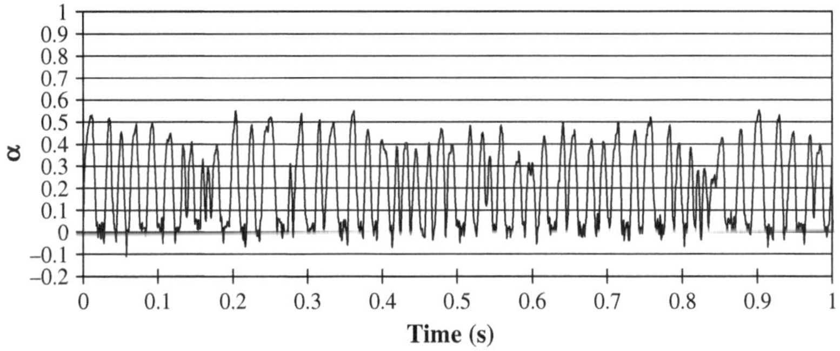

Figure 30. Void fraction time trace at 1-g for flow 5

Bubble Flow Regime at μ-g, $\alpha_{\text{ave}} = 0.209$
Flow 6: (96F3P44) $V_{\text{SL}} = 0.73\,\text{m/s}$, $V_{\text{SG}} = 0.20\,\text{m/s}$

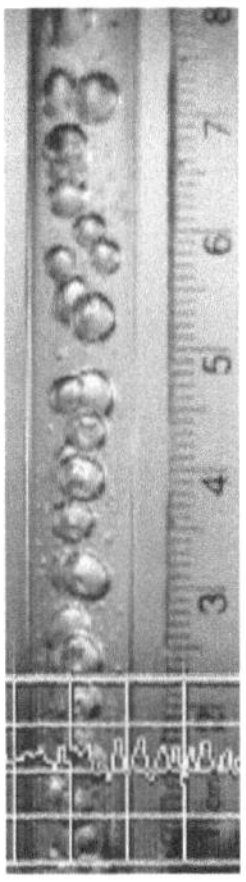

Figure 31. Flow image at μ-g for flow 6. The μ-g flow shows a group of spherical medium-sized bubbles together with some small disbursed bubbles. The bubbles tend to flow at or near the tube center

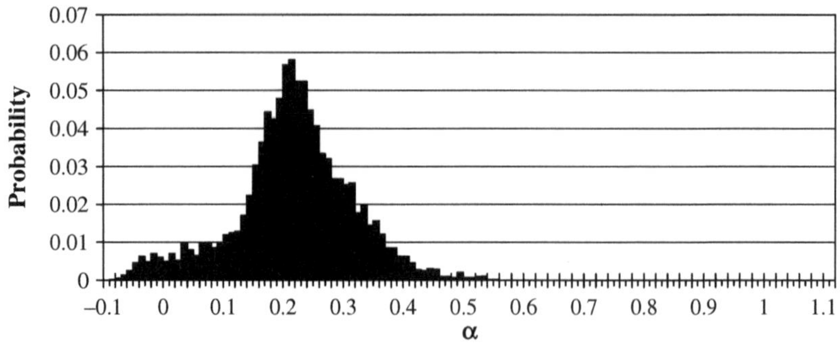

Figure 32. Void fraction PDF at μ-g for flow 6

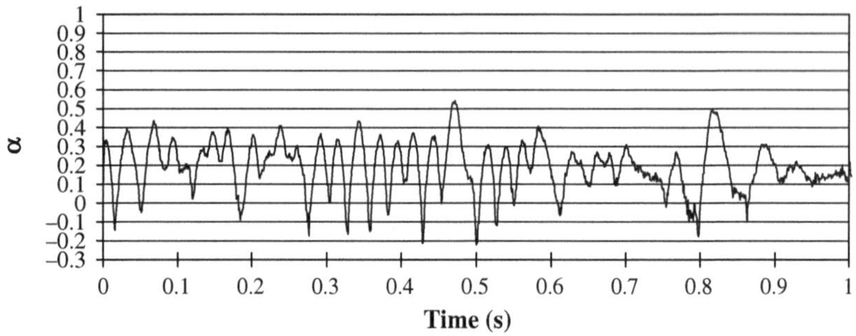

Figure 33. Void fraction time trace at μ-g for flow 6

Bubble Flow Regime at 1-g, $\alpha_{ave} = 0.234$
Flow 6: (96G3P39) $V_{SL} = 0.73\,m/s$, $V_{SG} = 0.20\,m/s$

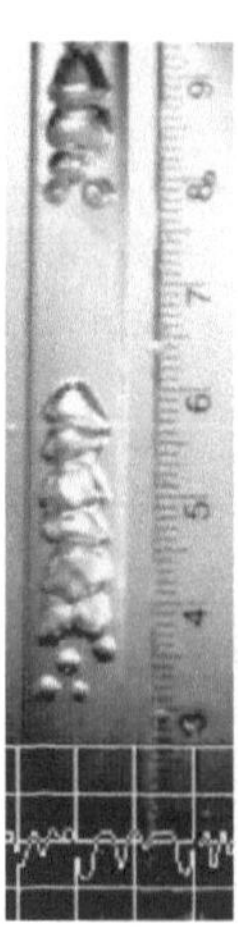

Figure 34. Flow image at 1-g for flow 6. The 1-g flow shows a trail of irregular shaped medium-sized bubbles. The small round bubbles at the end of the trail flow in a zigzag motion from the tube center to the wall and vice versa

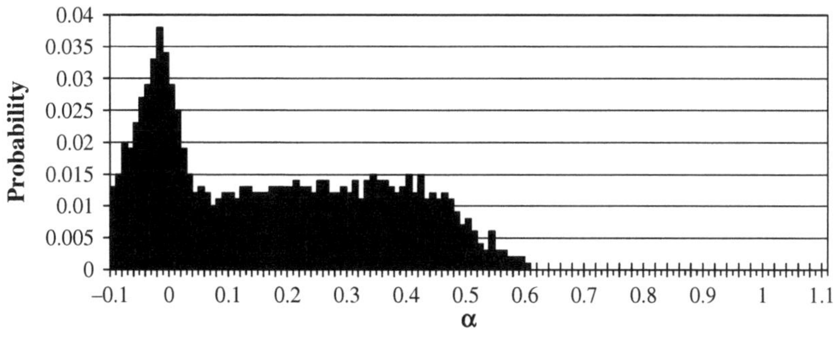

Figure 35. Void fraction PDF at 1-*g* for flow 6

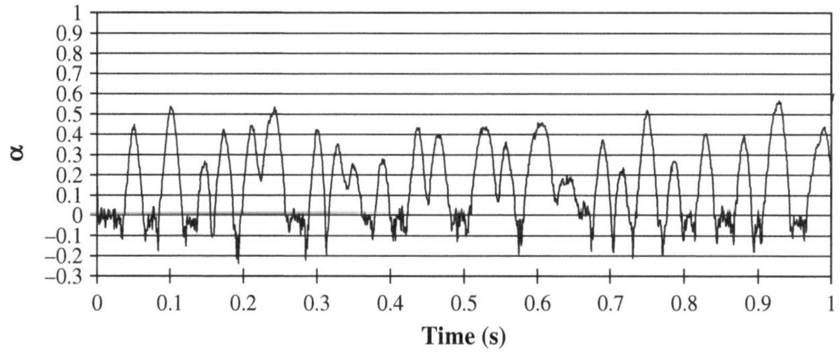

Figure 36. Void fraction time trace at 1-*g* for flow 6

Slug Flow Regime at μ-*g*, $\alpha_{ave} = 0.273$
Flow 7: (96F3P10) $V_{SL} = 1.77\,\text{m/s}$, $V_{SG} = 0.99\,\text{m/s}$

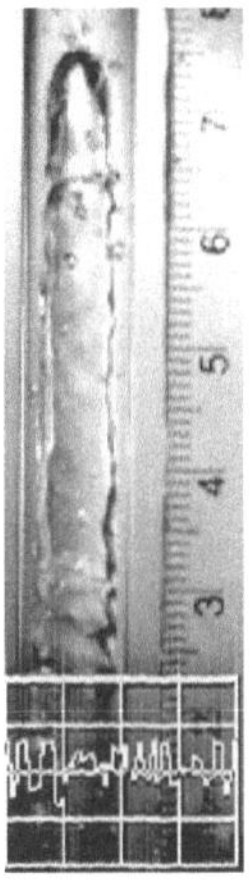

Figure 37. Flow image at μ-*g* for flow 7. The μ-*g* flow shows a Taylor bubble with a wavy surface, it appears to be capped by a shorter bubble with a rounded nose. Some tiny bubbles are disbursed in the flow

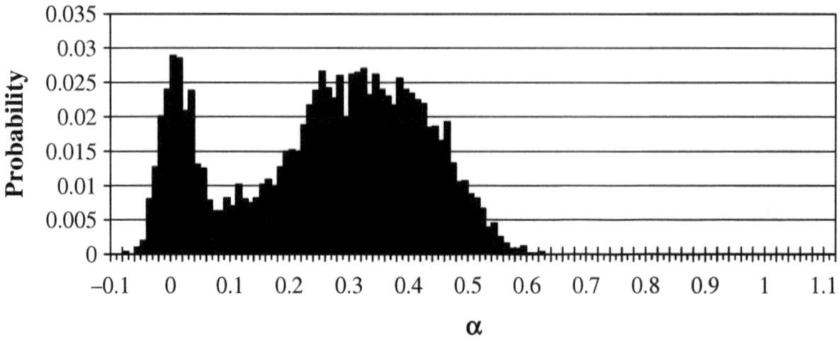

Figure 38. Void fraction PDF at μ-g for flow 7

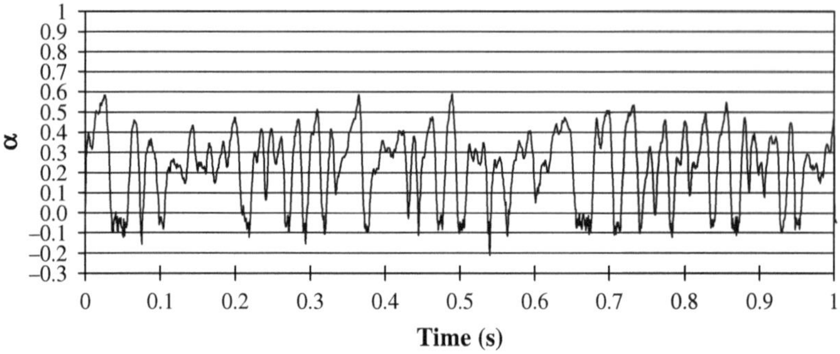

Figure 39. Void fraction time trace at μ-g for flow 7

Slug Flow Regime at μ-g, $\alpha_{ave} = 0.307$
Flow 7: (96G3P9) $V_{SL} = 1.77\,m/s$, $V_{SG} = 0.98\,m/s$

Figure 40. Flow image at 1-g for flow 7. The 1-g flow shows a distorted Taylor bubble. The bubble appears to break up near the bottom of the image into numerous small bubbles. Disbursed bubbles are seen throughout the viewing section

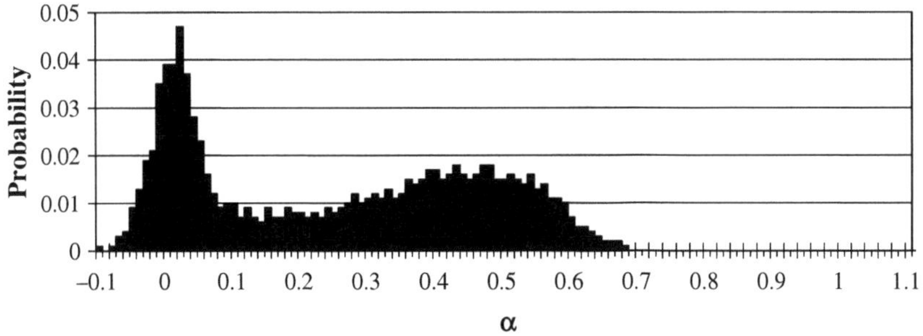

Figure 41. Void fraction PDF at 1-*g* for flow 7

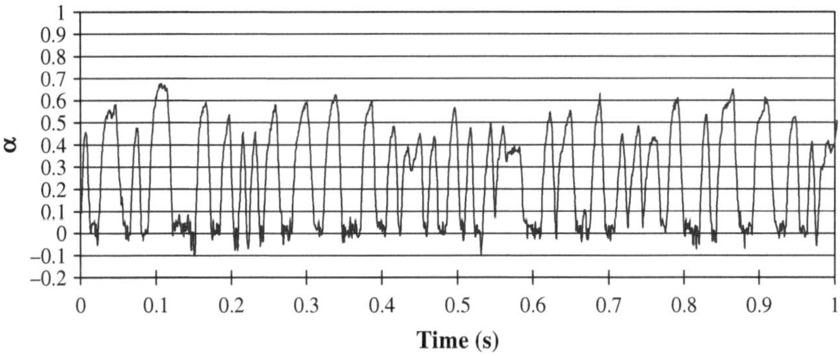

Figure 42. Void fraction time trace at 1-*g* for flow 7

Bubble Flow Regime at μ-*g*, $\alpha_{ave} = 0.278$
Flow 8: (96F4P42) $V_{SL} = 0.43\,m/s$, $V_{SG} = 0.19\,m/s$

Figure 43. Flow image at μ-*g* for flow 8. The μ-*g* flow shows a repeating pattern of medium bubble – narrow slug – small bubble – narrow slug. The medium bubbles tend to be bell-shaped and flow at the center of the tube. The small bubbles are spherical and move toward the tube center

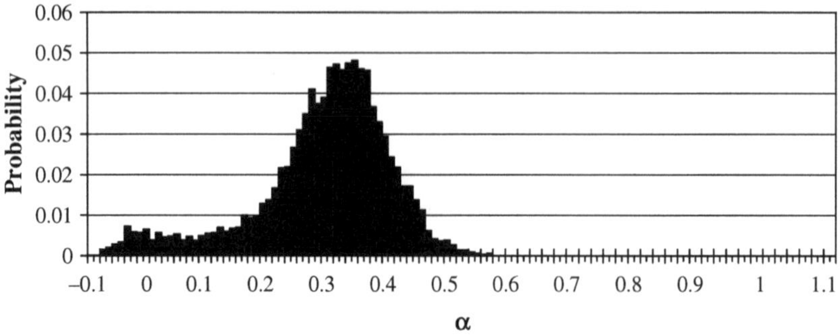

Figure 44. Void fraction PDF at μ-g for flow 8

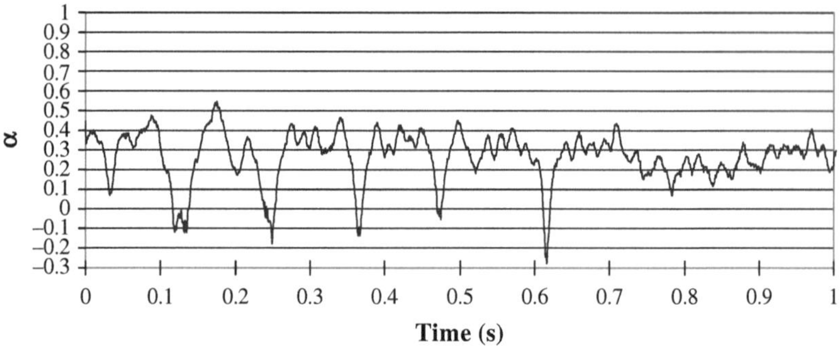

Figure 45. Void fraction time trace at μ-g for flow 8

Bubble Flow Regime at 1-g, $\alpha_{ave} = 0.247$
Flow 8: (96G4P41) $V_{SL} = 0.43\,m/s$, $V_{SG} = 0.19\,m/s$

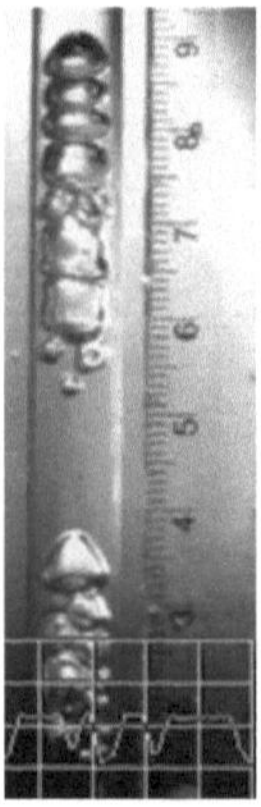

Figure 46. Flow image at 1-g for flow 8. The 1-g flow shows a trail of medium-sized bubbles. The leaders are compressed into ovals while the followers are stretched into irregular shapes. The tail has small spherical bubbles. A short liquid slug containing no bubbles separates the two groups

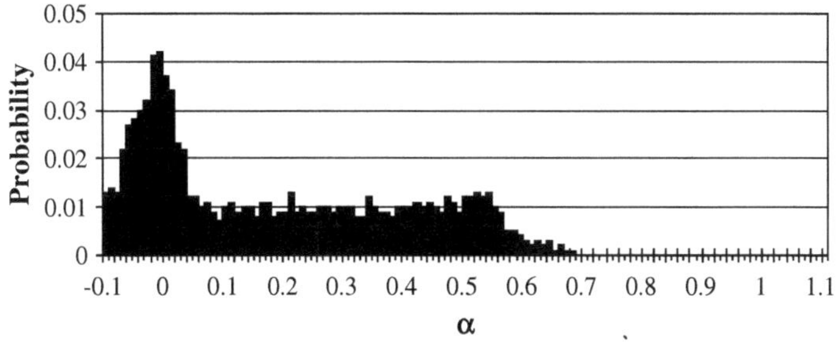

Figure 47. Void fraction PDF at 1-g for flow 8

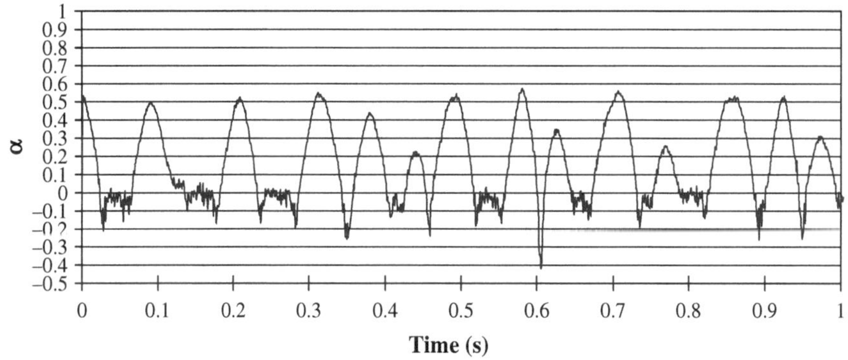

Figure 48. Void fraction time trace at 1 g for flow 8

Slug Flow Regime at μ-g, $\alpha_{\text{ave}} = 0.383$
Flow 9. (96F4P48) $V_{SL} = 0.24\,\text{m/s}$, $V_{SG} = 0.19\,\text{m/s}$

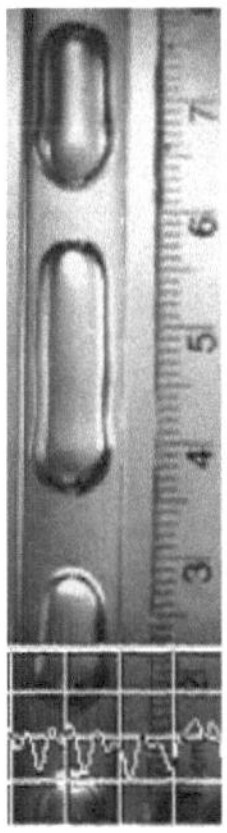

Figure 49. Flow image at μ-g for flow 9. The μ-g flow shows a series of rounded, smooth, uniform bubbles of approximately two to three tube diameters

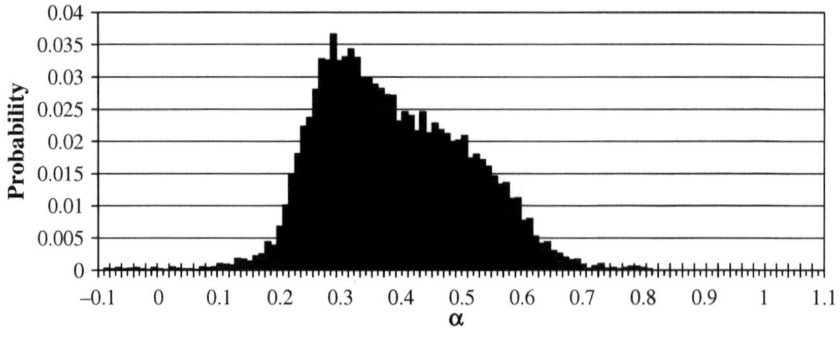

Figure 50. Void fraction PDF at μ-*g* for flow 9

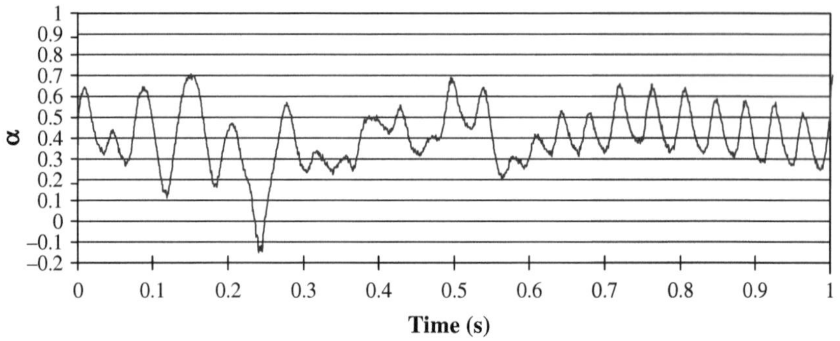

Figure 51. Void fraction time trace at μ-*g* for flow 9

Slug Flow Regime at 1-*g*, $\alpha_{ave} = 0.334$
Flow 9: (96G4P47) $V_{SL} = 0.24\,m/s$, $V_{SG} = 0.18\,m/s$

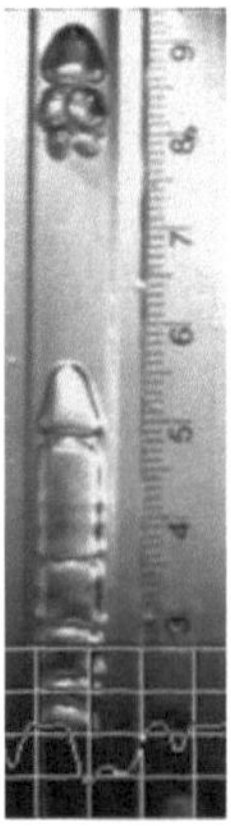

Figure 52. Flow image at 1-*g* for flow 9. The 1-*g* flow shows bubbles of one to two tube diameters forming clusters of varying lengths

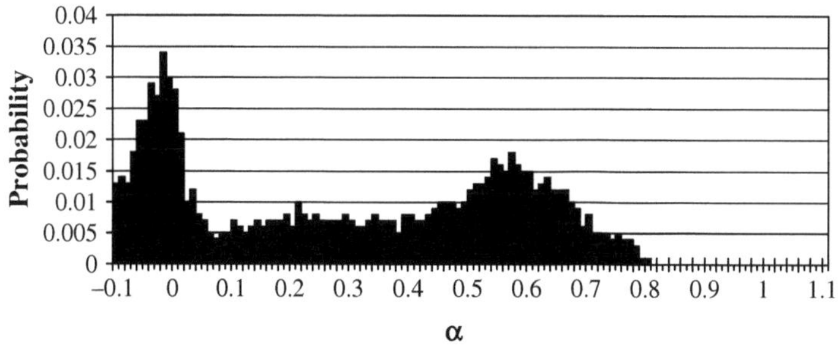

Figure 53. Void fraction PDF at 1-g for flow 9

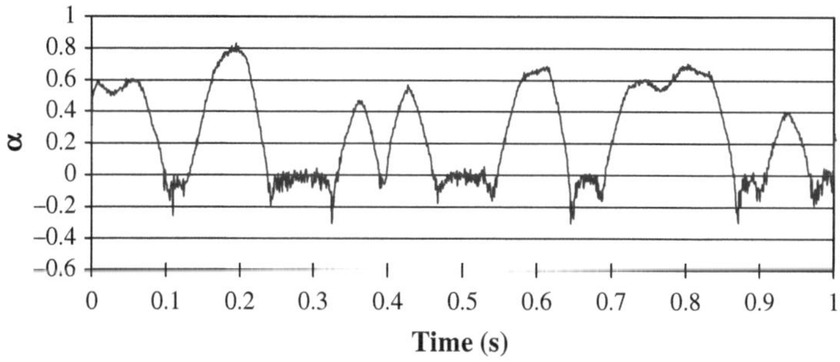

Figure 54. Void fraction time trace at 1-g for flow 9

Slug Flow Regime at μ-g, $\alpha_{ave} = 0.424$
Flow 10: (96F3P50) $V_{SL} = 0.73\,m/s$, $V_{SG} = 0.79\,m/s$

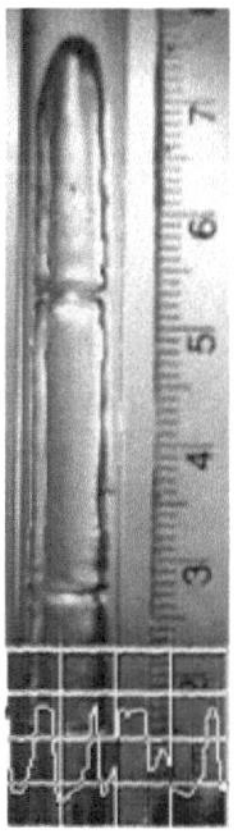

Figure 55. Flow image at μ-g for flow 10. The μ-g flow shows a long Taylor bubble with a rounded nose and a relatively smooth surface

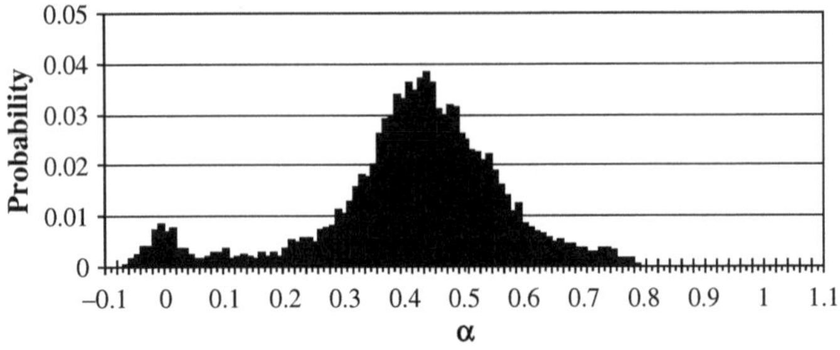

Figure 56. Void fraction PDF at μ-g for flow 10

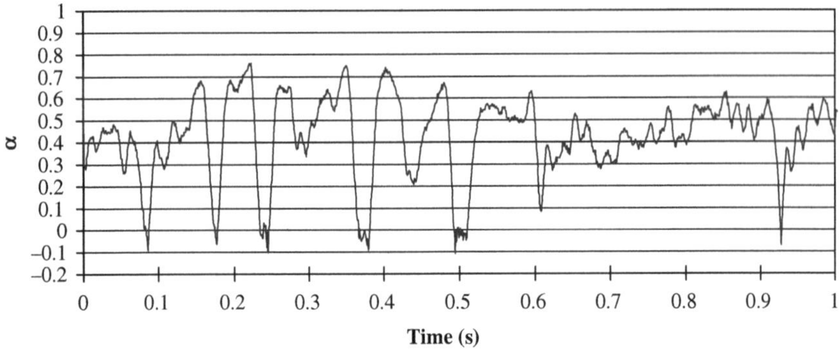

Figure 57. Void fraction time trace at μ-g for flow 10

Slug Flow Regime at 1-g, $\alpha_{\text{ave}} = 0.432$
Flow 10: (96G3P44) $V_{\text{SL}} = 0.74\,\text{m/s}$, $V_{\text{SG}} = 0.78\,\text{m/s}$

Figure 58. Flow image at 1-g for flow 10. The 1-g flow also shows a long Taylor bubble with many interfaces and rough edges. The nose is smaller in diameter

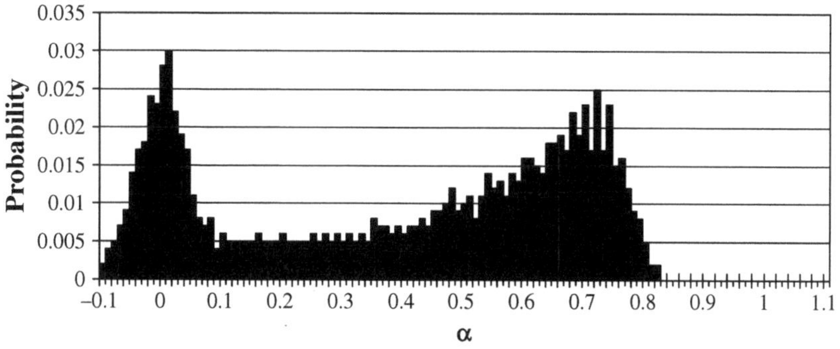

Figure 59. Void fraction PDF at 1-g for flow 10

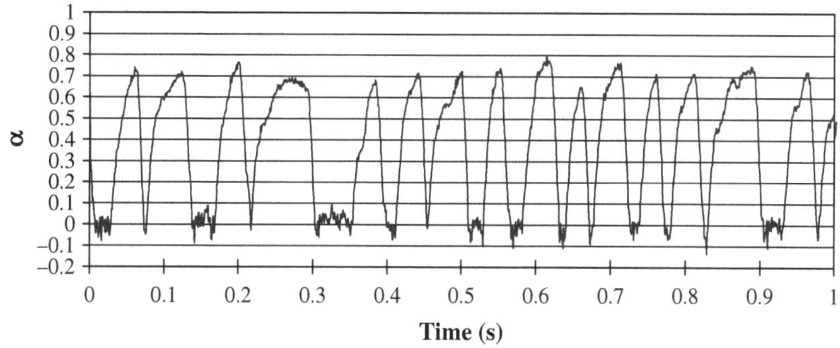

Figure 60. Void fraction time trace at 1-g for flow 10

Transitional Flow Regime at μ-g, $\alpha_{ave} = 0.724$
Flow 11: (96F4P5) $V_{SL} = 0.11\,\text{m/s}$, $V_{SG} = 1.59\,\text{m/s}$

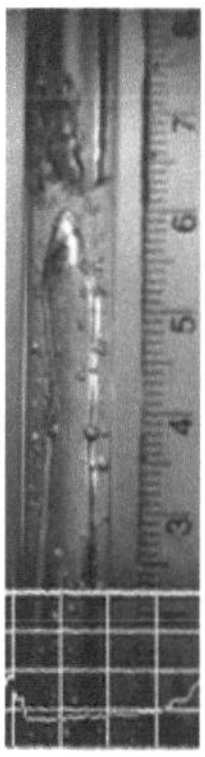

Figure 61. Flow image at μ-g for flow 11. The μ-g flow shows a liquid slug that separates two large bubbles. Small bubbles are disbursed in the thick liquid film. Few bubbles are present at the tail region of the slug

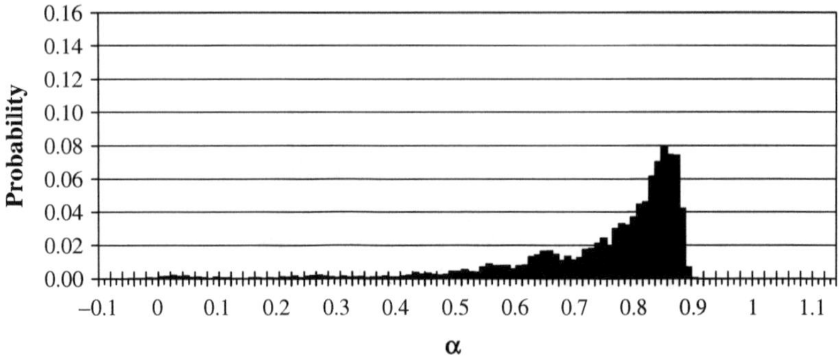

Figure 62. Void fraction PDF at μ-g for flow 11

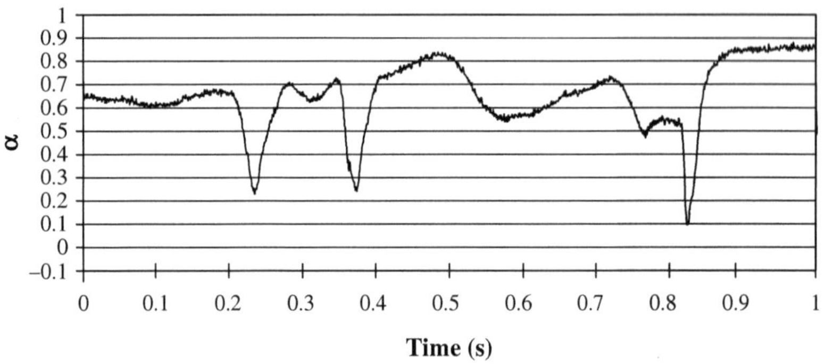

Figure 63. Void fraction time trace at μ-g for flow 11

Transitional Flow Regime at 1-g, $\alpha_{ave} = 0.727$
Flow 11: (96G4P5) $V_{SL} = 0.11\,m/s$, $V_{SG} = 1.58\,m/s$

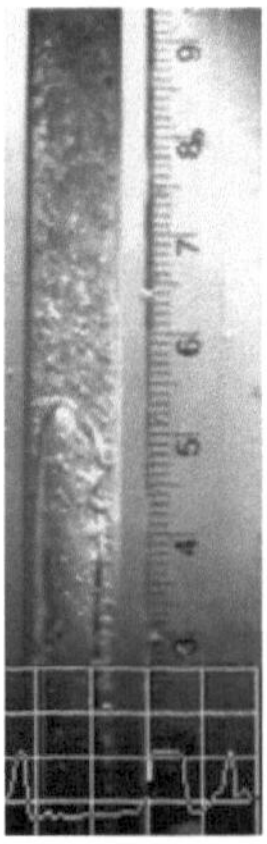

Figure 64. Flow image at 1-g for flow 11. The 1-g flow shows a similar situation where one long bubble follows the tail of another. However, this tail leaves a long trail of small disbursed bubbles in the long liquid slug

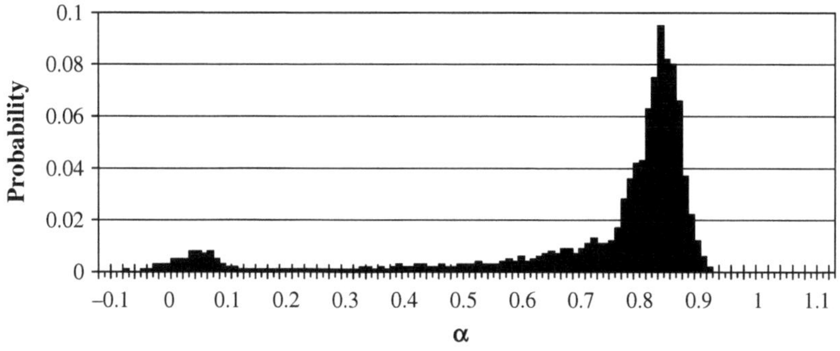

Figure 65. Void fraction PDF at 1-g for flow 11

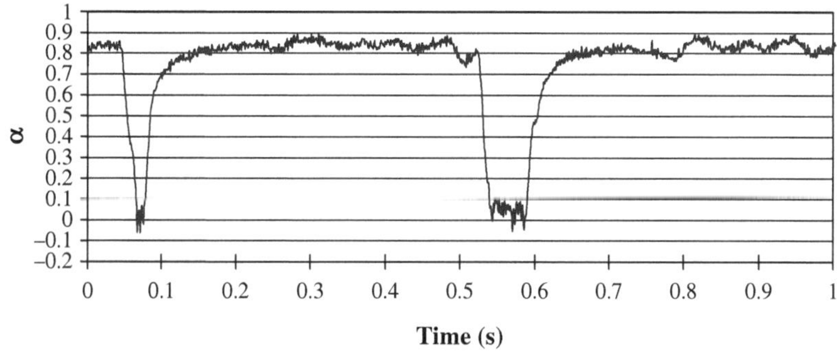

Figure 66. Void fraction time trace at 1-g for flow 11

Transitional Flow Regime at μ-g, $\alpha_{ave} = 0.740$
Flow 12: (96F4P15) $V_{SL} = 0.16\,m/s$, $V_{SG} = 1.89\,m/s$

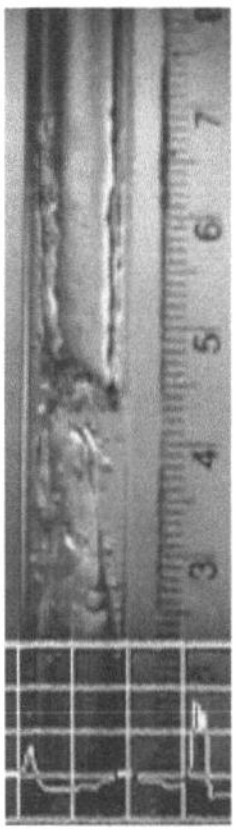

Figure 67. Flow image at μ-g for flow 12. The μ-g flow shows two Taylor bubbles touching nose to tail. The leading bubble has a slightly irregular film and the tail curves to mirror the rounded nose of the next bubble. A few small bubbles are disbursed in the narrow liquid slug that separates the two Taylor bubbles

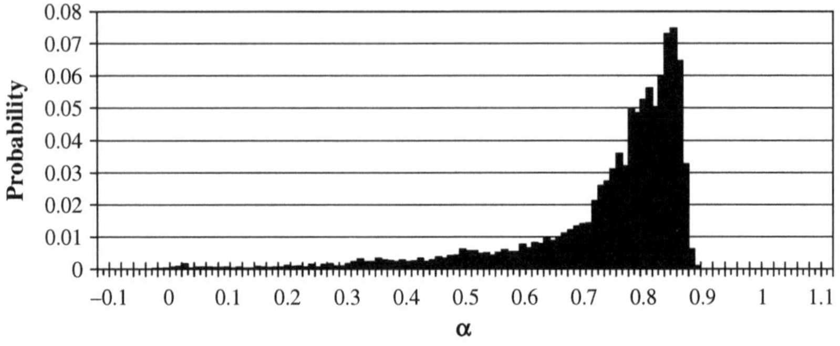

Figure 68. Void fraction PDF at μ-*g* for flow 12

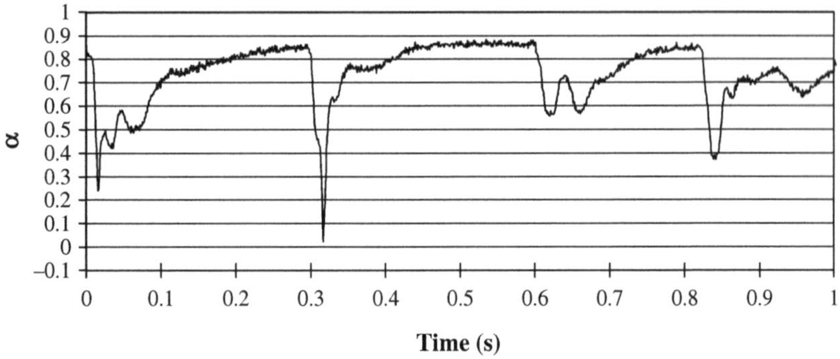

Figure 69. Void fraction time trace at μ-*g* for flow 12

Transitional Flow Regime at 1-*g*, $\alpha_{ave} = 0.712$
Flow 12: (96G4P14) $V_{SL} = 0.16\,m/s$, $V_{SG} = 1.88\,m/s$

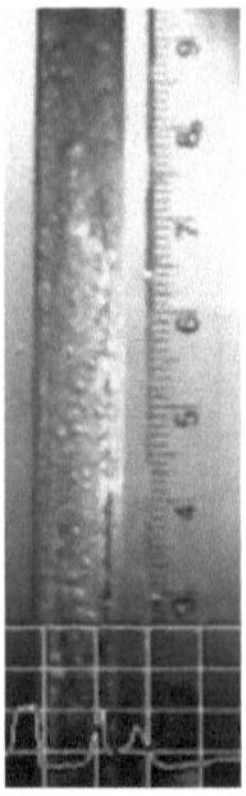

Figure 70. Flow image at 1-*g* for flow 12. The 1-*g* flow shows a distorted Taylor bubble. Many small bubbles are disbursed throughout the viewing section, particularly in the liquid slug in front of the Taylor bubble. The liquid slug is significantly longer than that for the μ-*g* flow

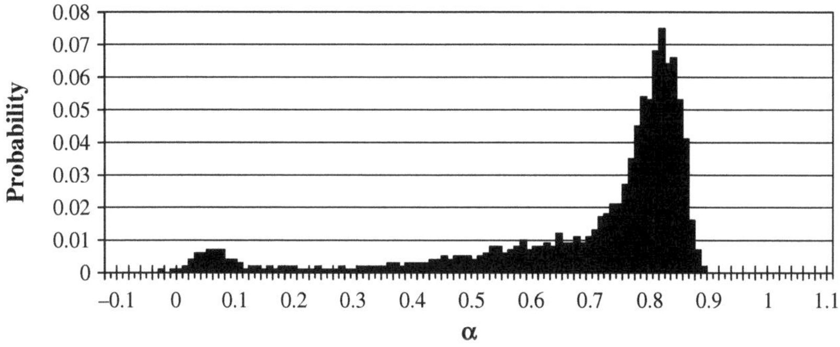

Figure 71. Void fraction PDF at 1-*g* for flow 12

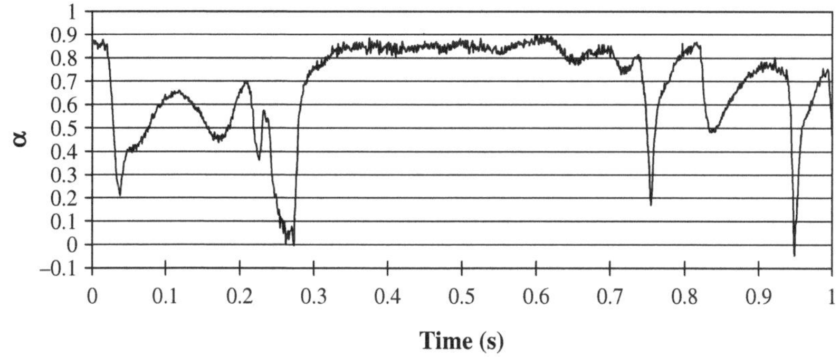

Figure 72. Void fraction time trace at μ-*g* for flow 12

Transitional Flow Regime at μ-g, $\alpha_{\text{ave}} = 0.790$
Flow 13 (F1P33): $V_{\text{SL}} = 0.16\,\text{m/s}$, $V_{\text{SG}} = 5.96\,\text{m/s}$

Figure 73. Flow image at μ-*g* for flow 13. The μ-*g* flow shows a smooth film interrupted by a frothy slug region

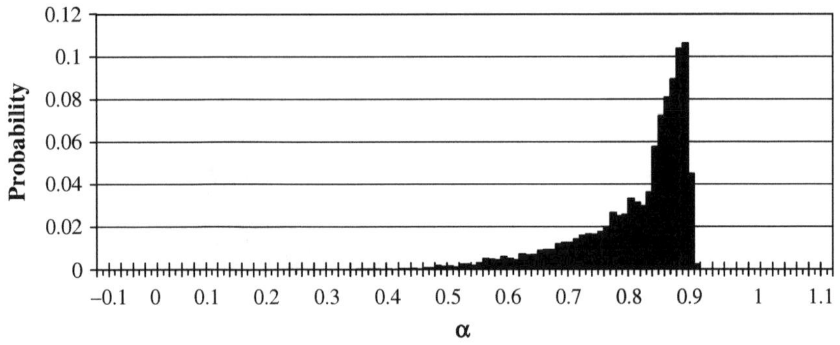

Figure 74. Void fraction PDF at μ-*g* for flow 13

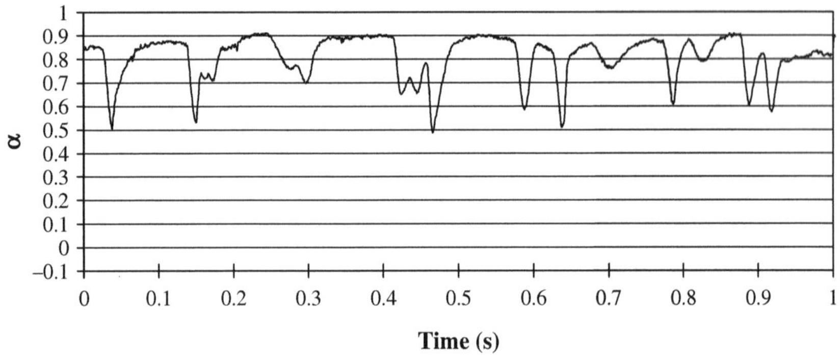

Figure 75. Void fraction time trace at μ-*g* for flow 13

Transitional Flow Regime at 1-*g*, $\alpha_{ave} = 0.739$
Flow 13 (G1P29): $V_{SL} = 0.16\,m/s$, $V_{SG} = 5.90\,m/s$

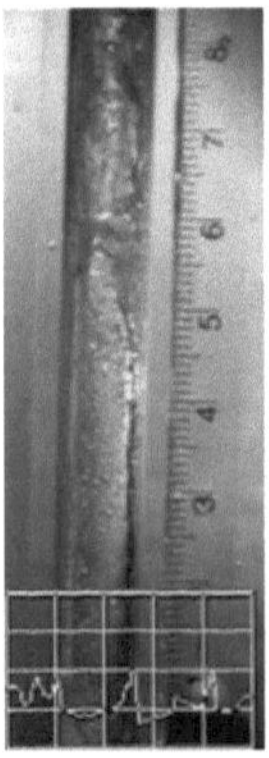

Figure 76. Flow image at 1-*g* for flow 13. The 1-*g* flow has a distinct annular region containing disbursed bubbles, and a short slug which appears to bridge the tube

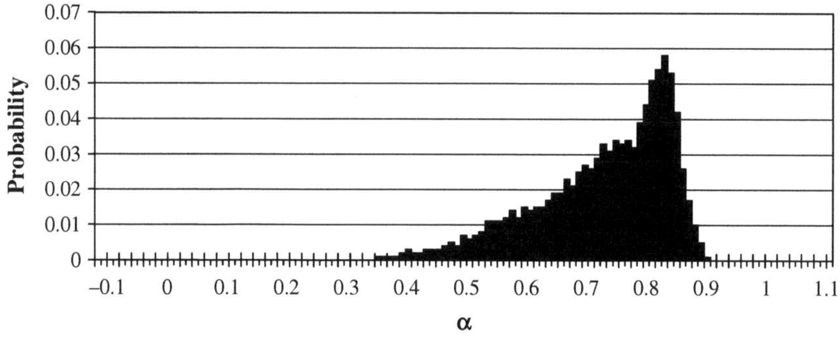

Figure 77. Void fraction PDF at 1-*g* for flow 13

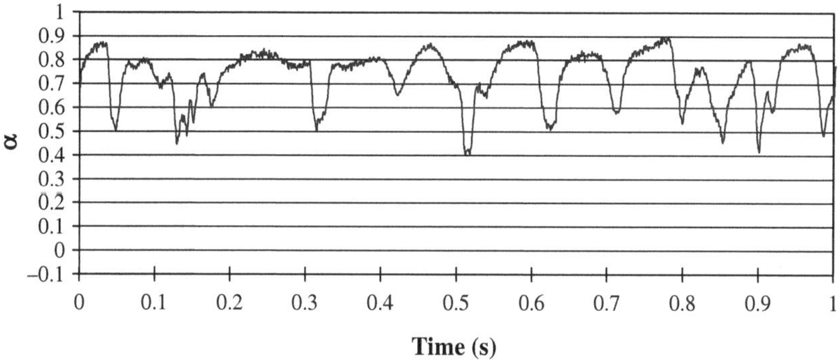

Figure 78. Void fraction time trace at 1-*g* for flow 13

Transitional Flow Regime at μ-*g*, $\alpha_{ave} = 0.798$
Flow 14 (96F2P19): $V_{SL} = 0.20\,m/s$, $V_{SG} = 8.93\,m/s$

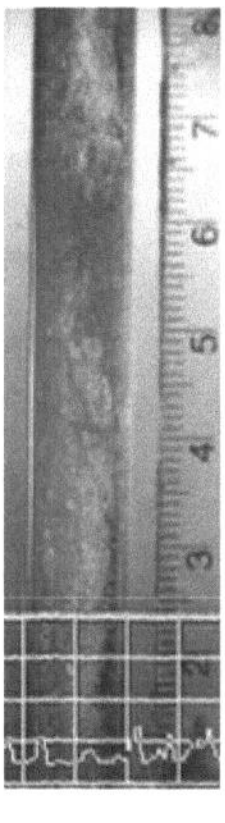

Figure 79. Flow image at μ-*g* for flow 14. The μ-*g* flow contains a broad frothy slug that appears to bridge the tube

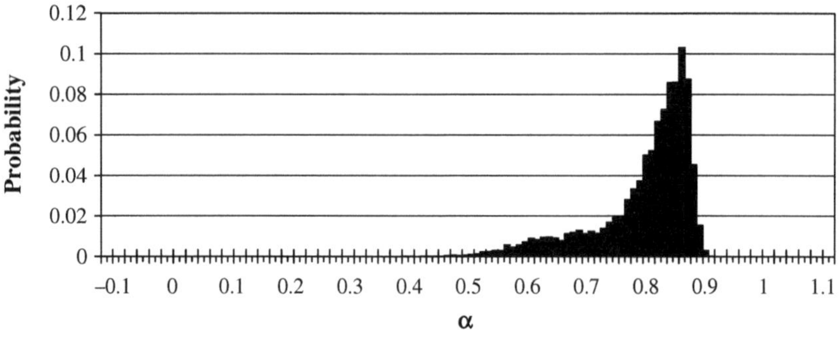

Figure 80. Void fraction PDF at μ-*g* for flow 14

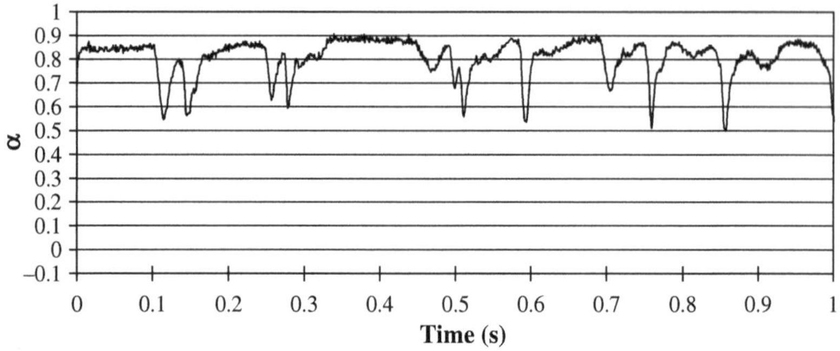

Figure 81. Void fraction time trace at μ-*g* for flow 14

Transitional Flow Regime at 1-g, $\alpha_{ave} = 0.772$
Flow 14 (96G2P17): $V_{SL} = 0.20\,m/s$, $V_{SG} = 8.90\,m/s$

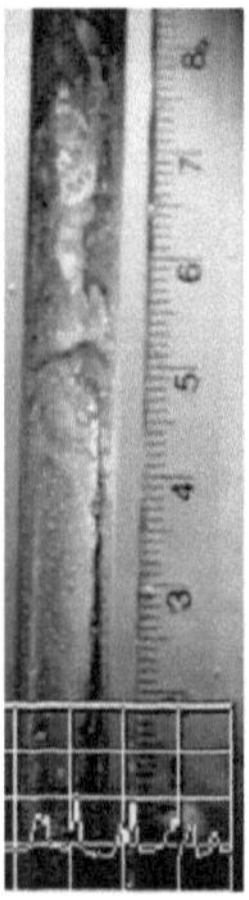

Figure 82. Flow image at 1-*g* for flow 14. The 1-*g* flow contains a very irregular pattern caused by the passage of a wave and falling film

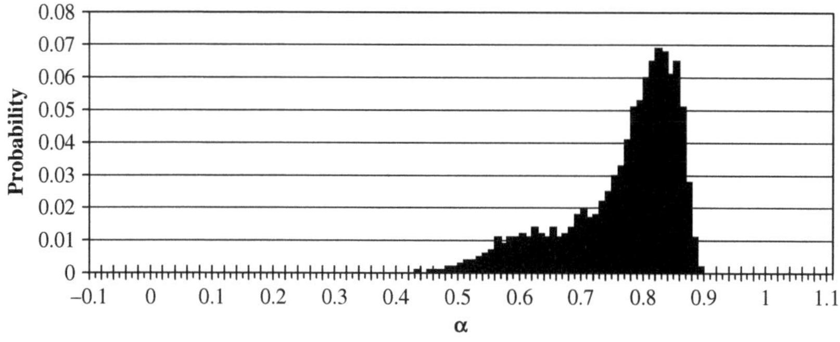

Figure 83. Void fraction PDF at 1-g for flow 14

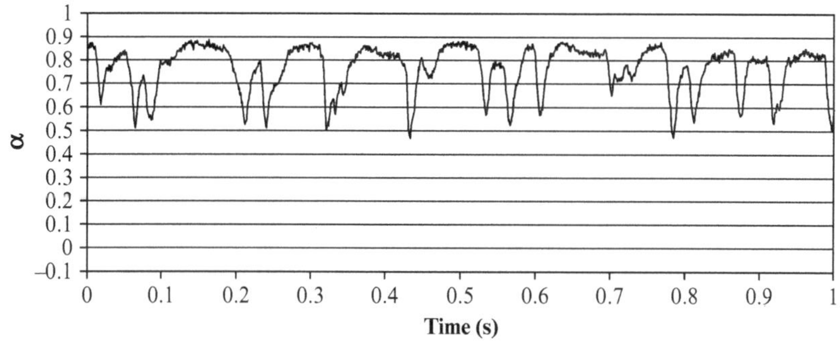

Figure 84. Void fraction time trace at 1-g for flow 14

Annular Flow Regime at μ-g, $\alpha_{\text{ave}} = 0.809$
Flow 15 (96F2P28): $V_{\text{SL}} = 0.31\,\text{m/s}$, $V_{\text{SG}} = 13.92\,\text{m/s}$

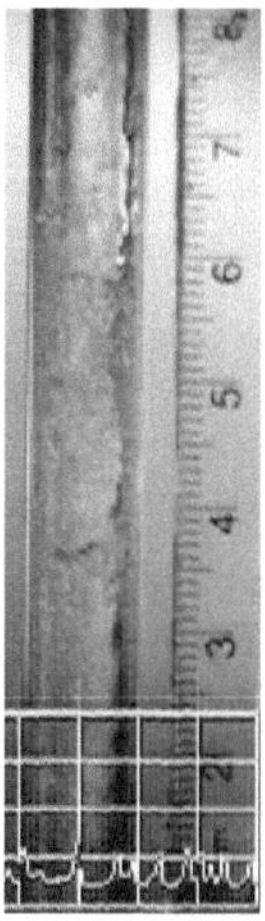

Figure 85. Flow image at μ-g for flow 15. The μ-g flow shows a frothy region in the center of the image, possibly the passing of a wave

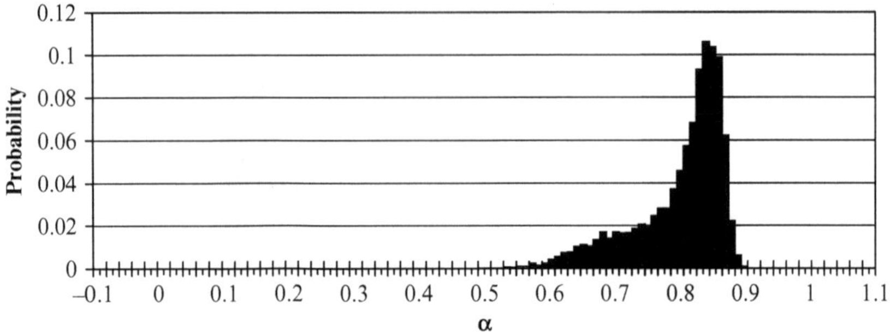

Figure 86. Void fraction PDF at μ-*g* for flow 15

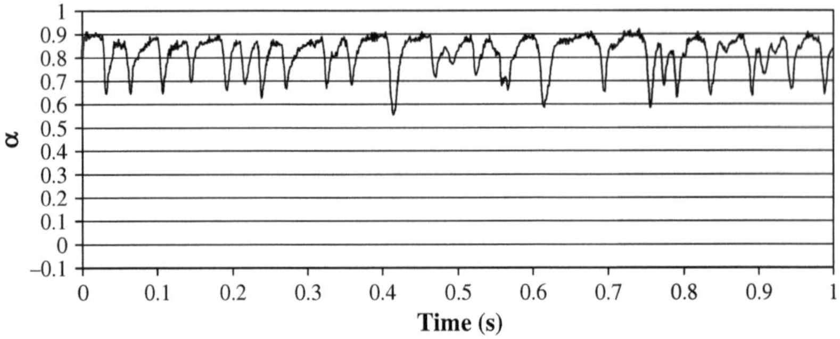

Figure 87. Void fraction time trace at μ-*g* for flow 15

Annular Flow Regime at 1-g, $\alpha_{\text{ave}} = 0.810$
Flow 15 (G2P26): $V_{\text{SL}} = 0.31\,\text{m/s}$, $V_{\text{SG}} = 13.82\,\text{m/s}$

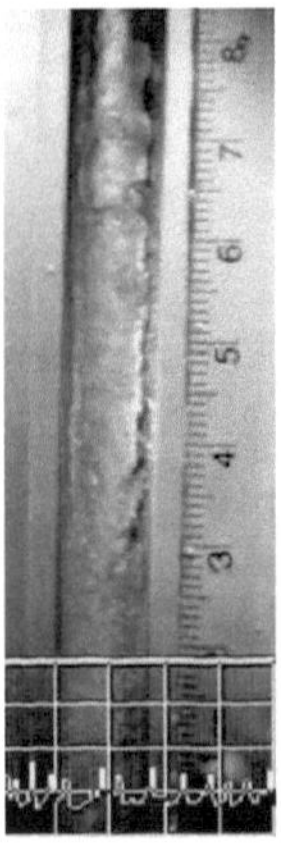

Figure 88. Flow image at 1-*g* for flow 15. The 1-*g* flow shows a similar frothy region at the bottom of the image and a notably irregular annular region at the top

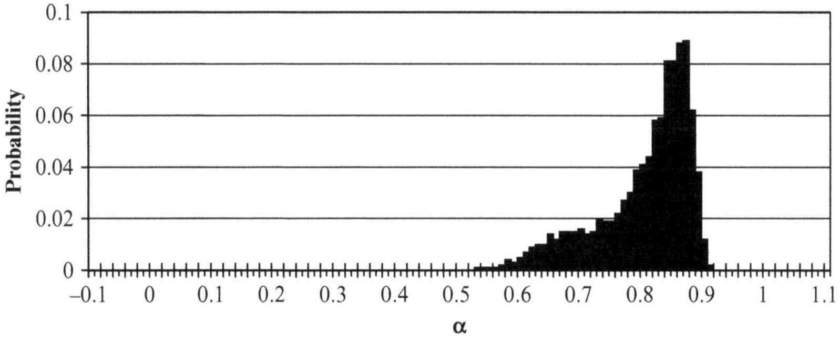

Figure 89. Void fraction PDF at 1-*g* for flow 15

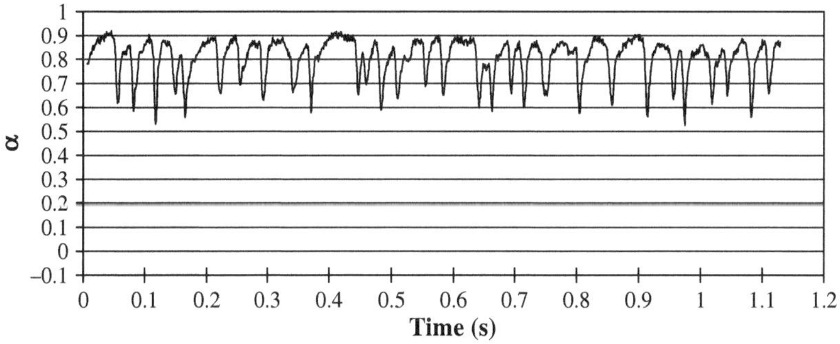

Figure 90. Void fraction time trace at 1-*g* for flow 15

Transitional Flow Regime at μ-*g*, $\alpha_{ave} = 0.826$
Flow 16 (96F1P24): $V_{SL} = 0.09\,m/s$, $V_{SG} = 7.94\,m/s$

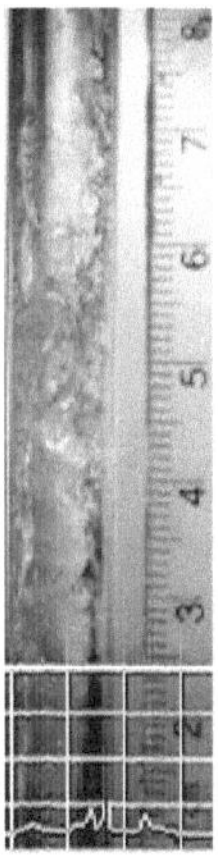

Figure 91. Flow image at μ-*g* for flow 16. The μ-*g* flow is irregular on both sides of the film and a frothy disturbance wave passes in the center of the image

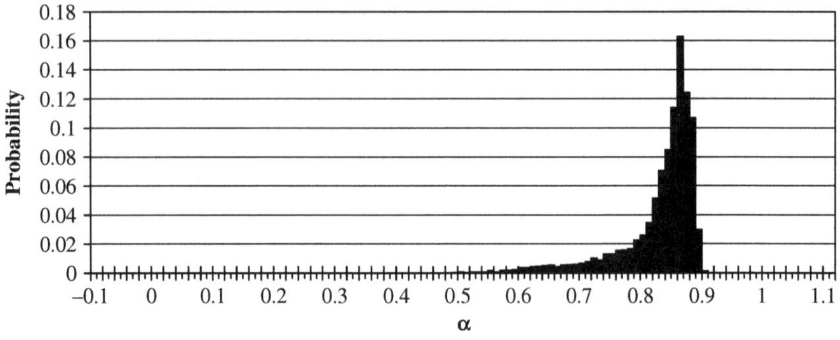

Figure 92. Void fraction PDF at μ-g for flow 16

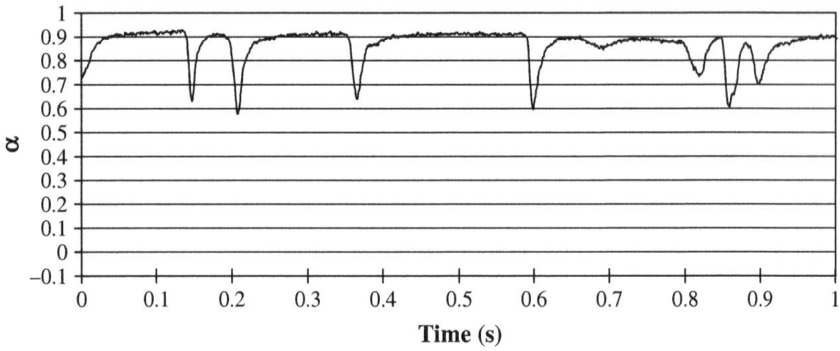

Figure 93. Void fraction time trace at μ-g for flow 16

Transitional Flow Regime at 1-g, $\alpha_{ave} = 0.806$
Flow 16 (96G1P22): $V_{SL} = 0.09\,m/s$, $V_{SG} = 7.90\,m/s$

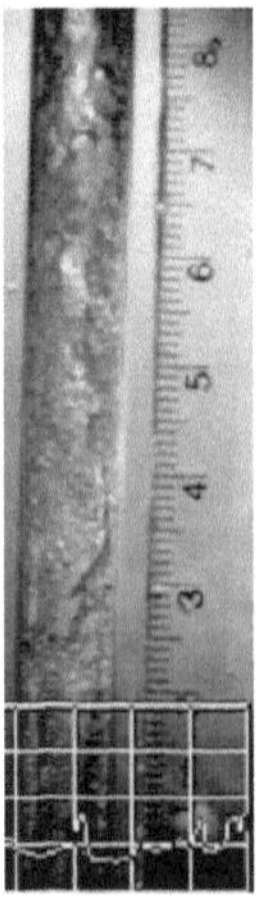

Figure 94. Flow image at 1-g for flow 16. The 1-g flow shows a much longer disturbance that fills the entire viewing section and varies between frothy and semi-annular regions

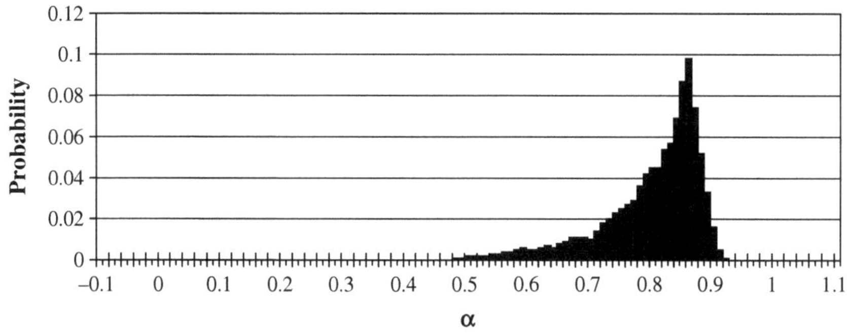

Figure 95. Void fraction PDF at 1-*g* for flow 16

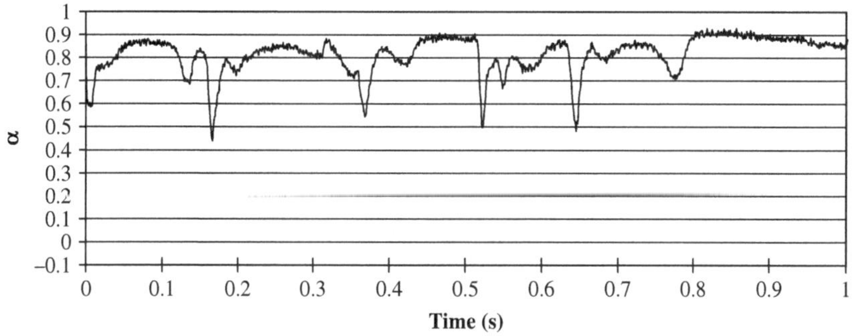

Figure 96. Void fraction time trace at 1-*g* for flow 16

Annular Flow Regime at μ-*g*, $\alpha_{ave} = 0.828$
Flow 17 (96F2P13): $V_{SL} = 0.21$ m/s, $V_{SG} = 13.95$ m/s

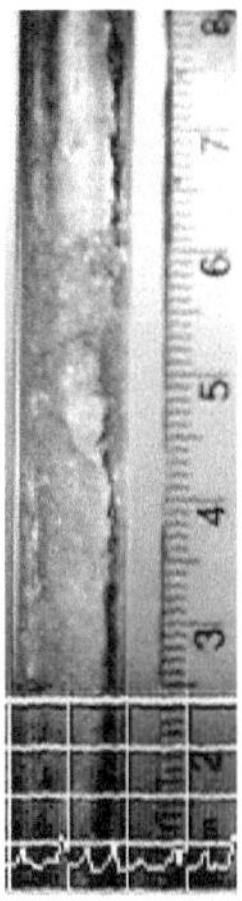

Figure 97. Flow image at μ-*g* for flow 17. The μ-*g* flow shows a frothy disturbance in the center of the image, this may be due to the passing of a wave

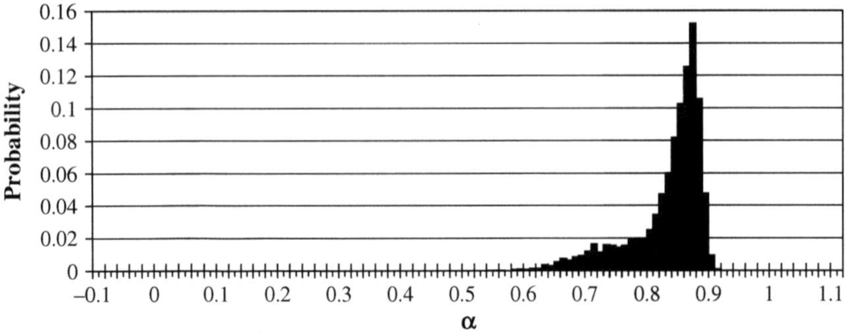

Figure 98. Void fraction PDF at μ-g for flow 17

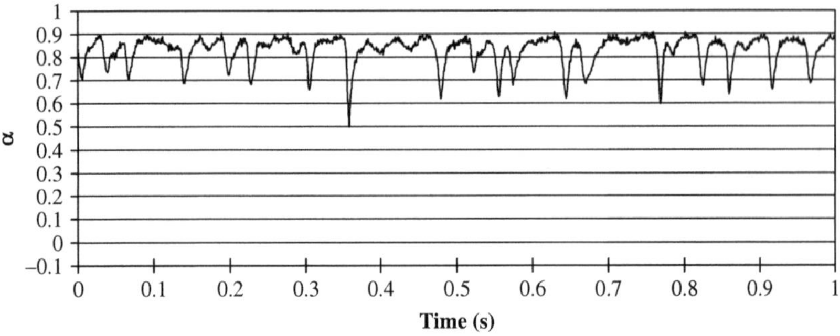

Figure 99. Void fraction time trace at μ-g for flow 17

Annular Flow Regime at 1-g, $\alpha_{ave} = 0.825$
Flow 17 (G2P11): $V_{SL} = 0.21\,m/s$, $V_{SG} = 13.87\,m/s$

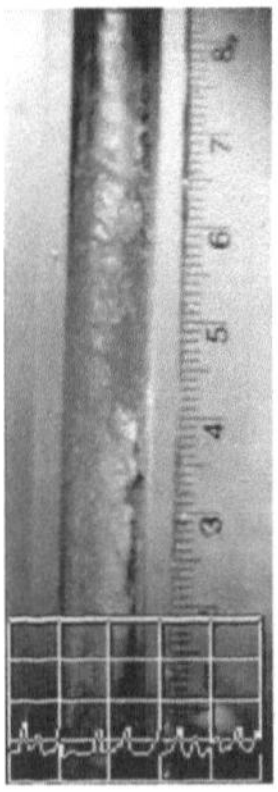

Figure 100. Flow image at 1-g for flow 17. The 1-g flow shows a much larger frothy disturbance and possibly bridging. Annular flow is present at the top of the image

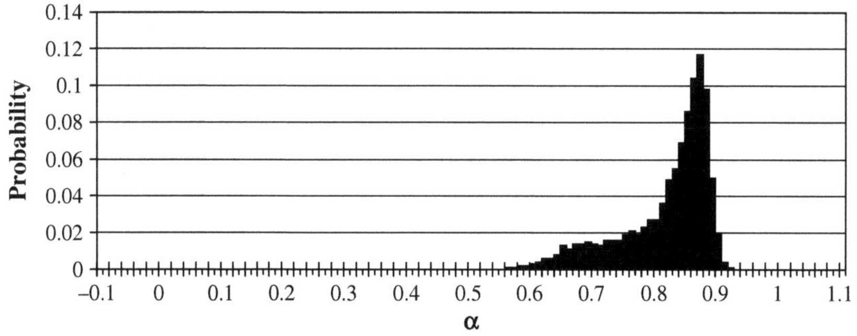

Figure 101. Void fraction PDF at 1-*g* for flow 17

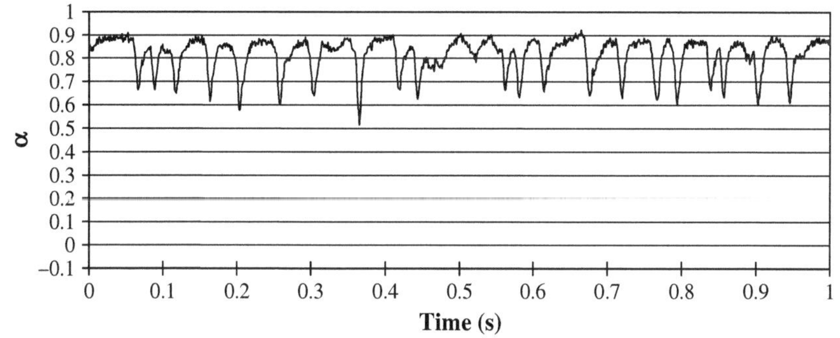

Figure 102. Void fraction time trace at 1-*g* for flow 17

Annular Flow Regime at μ-*g*, $\alpha_{ave} = 0.830$
Flow 18 (96F2P34): $V_{SL} = 0.11\,\mathrm{m/s}$, $V_{SG} = 9.92\,\mathrm{m/s}$

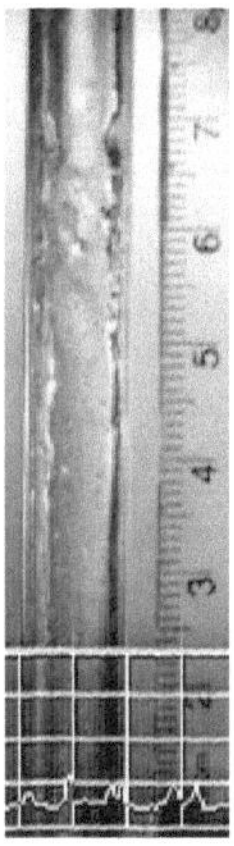

Figure 103. Flow image at μ-*g* for flow 18. The μ-*g* flow shows a smooth film interrupted by a wave in the center of the image

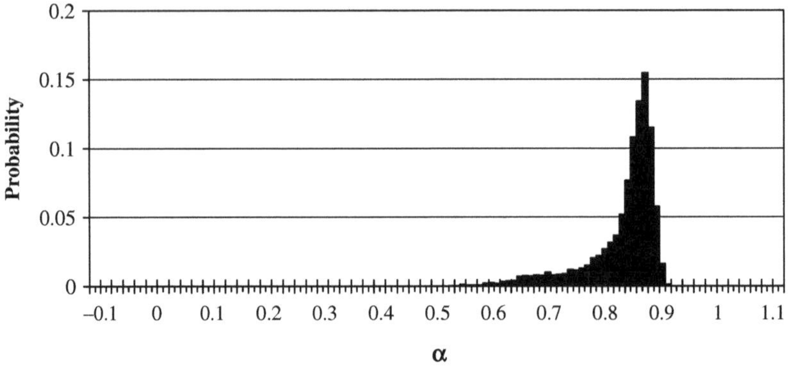

Figure 104. Void fraction PDF at μ-*g* for flow 18

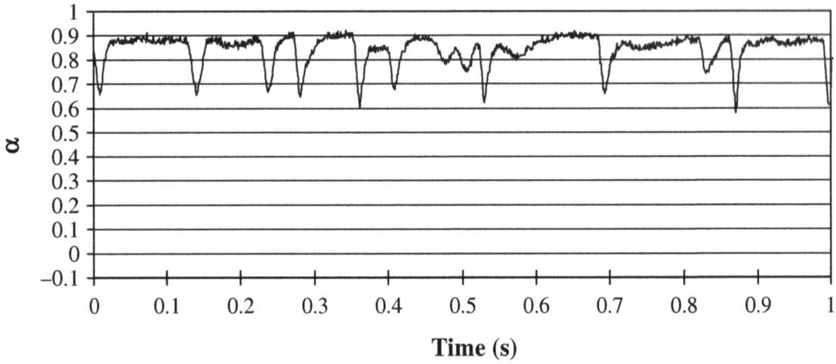

Figure 105. Void fraction time trace at μ-*g* for flow 18

Annular Flow Regime at 1-*g*, $\alpha_{\mathrm{ave}} = 0.811$
Flow 18 (96G2P31): $V_{\mathrm{SL}} = 0.11\,\mathrm{m/s}$, $V_{\mathrm{SG}} = 9.86\,\mathrm{m/s}$

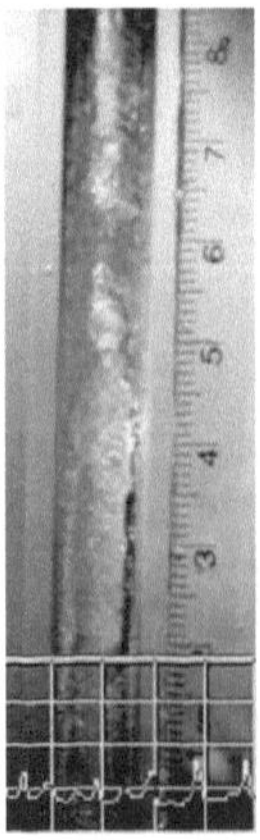

Figure 106. Flow image at 1-*g* for flow 18. The 1-*g* flow shows a more irregular film interrupted by decay of the annular flow and a frothy region with possible bridging

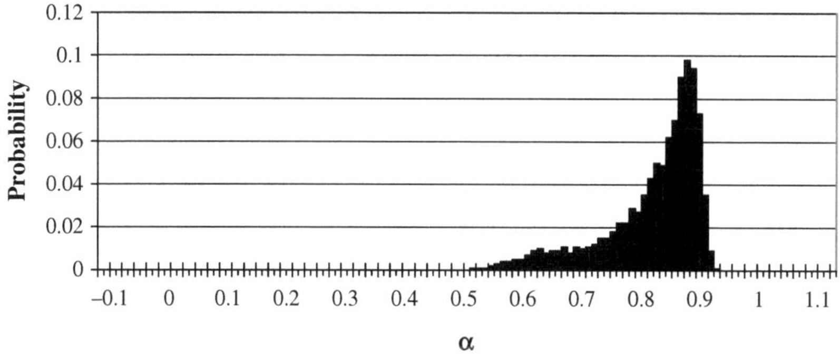

Figure 107. Void fraction PDF at 1-g for flow 18

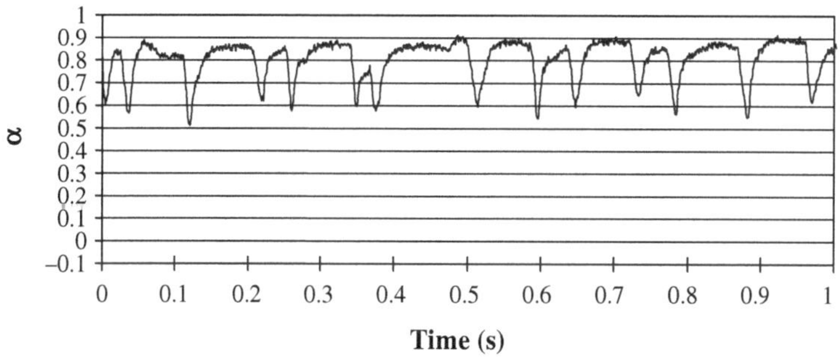

Figure 108. Void fraction time trace at 1-g for flow 18

Annular Flow Regime at μ-g, $\alpha_{ave} = 0.855$
Flow 19 (96F1P3): $V_{SL} = 0.05\,m/s$, $V_{SG} = 6.92\,m/s$

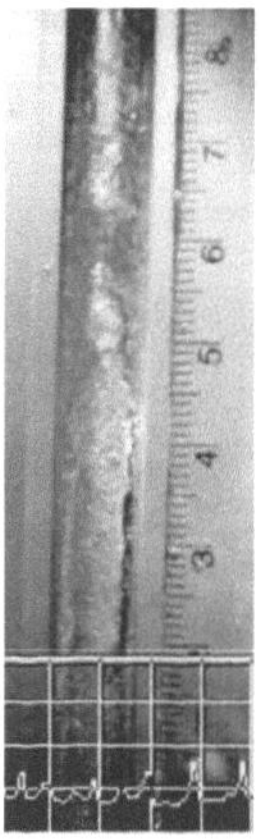

Figure 109. Flow image at μ-g for flow 19. The μ-g flow is annular with no complete breaks in the annular film. The smooth film is interrupted by the passage of a wave at the center of the image

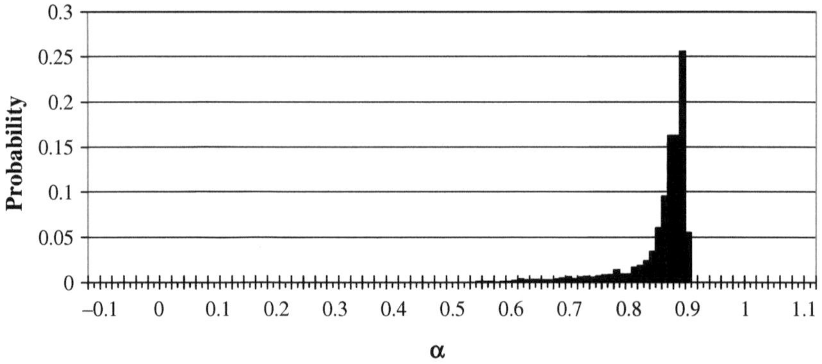

Figure 110. Void fraction PDF at μ-g for flow 19

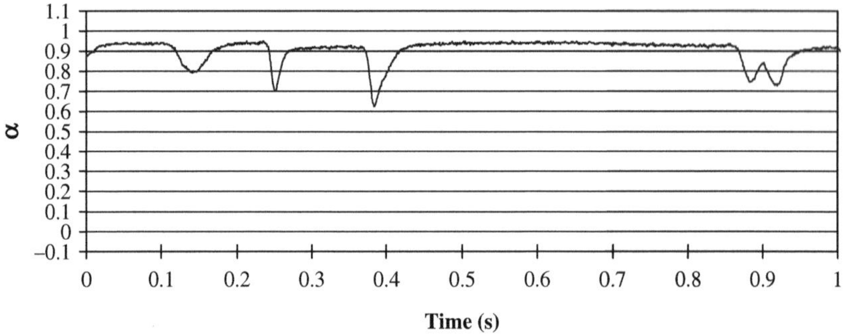

Figure 111. Void fraction time trace at μ-g for flow 19

Annular Flow Regime at 1-g, $\alpha_{ave} = 0.791$
Flow 19 (G1P3): $V_{SL} = 0.05\,\mathrm{m/s}$, $V_{SG} = 6.85\,\mathrm{m/s}$

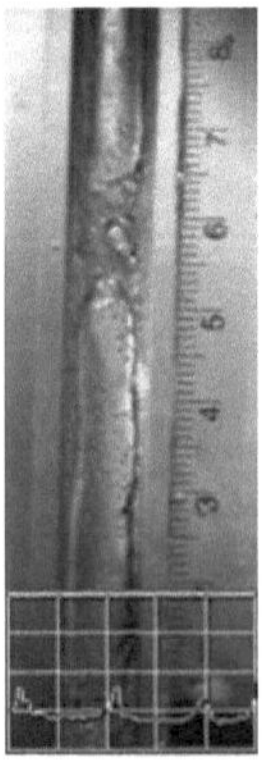

Figure 112. Flow image at 1-g for flow 19. The 1-g flow is interrupted by a frothy slug. It is difficult to determine if bridging is complete due to the presence of bubbles in the slug

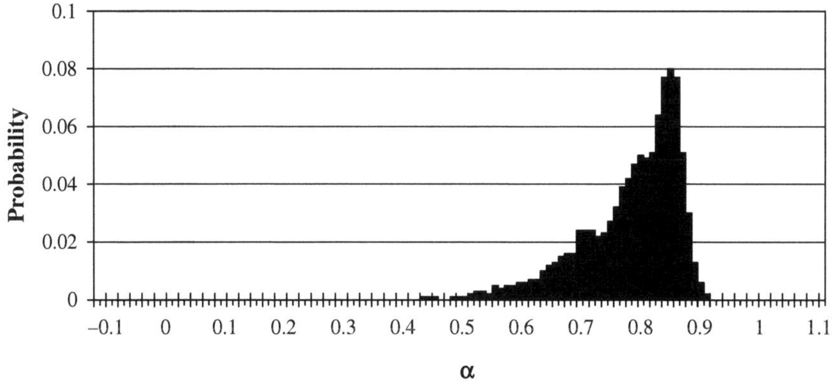

Figure 113. Void fraction PDF at 1-g for flow 19

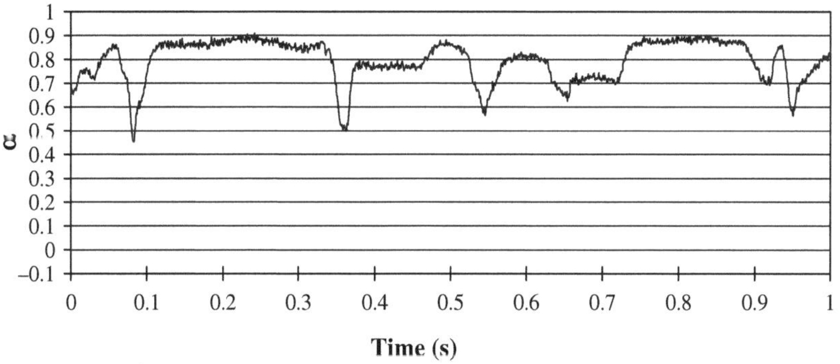

Figure 114. Void fraction time trace at 1-g for flow 19

Annular Flow Regime at μ-g, $\alpha_{\text{ave}} = 0.885$

Flow 20 (F1P8): $V_{\text{SL}} = 0.05\,\text{m/s}$, $V_{\text{SG}} = 16.95\,\text{m/s}$

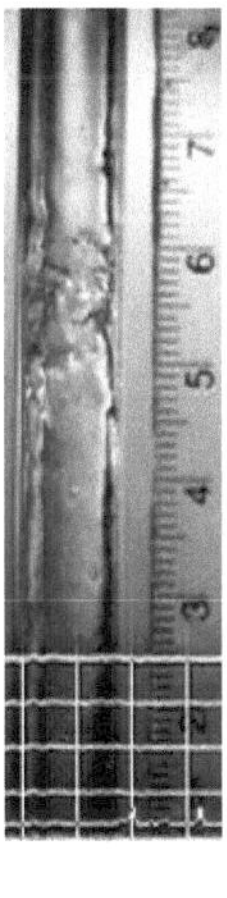

Figure 115. Flow image at μ-g for flow 20. The μ-g flow is annular and the passage of a wave is shown at the center of the image

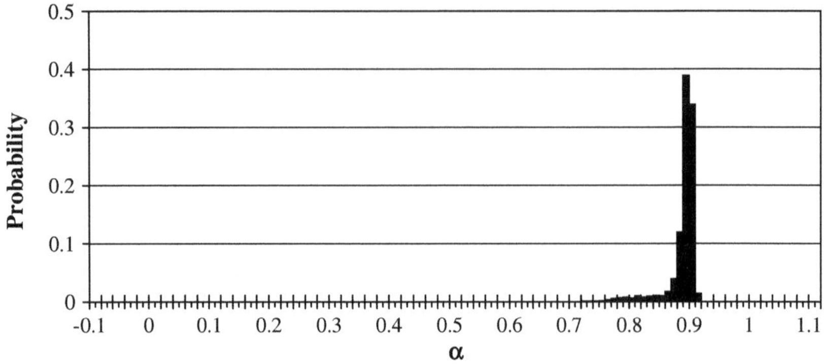

Figure 116. Void fraction PDF at μ-g for flow 20

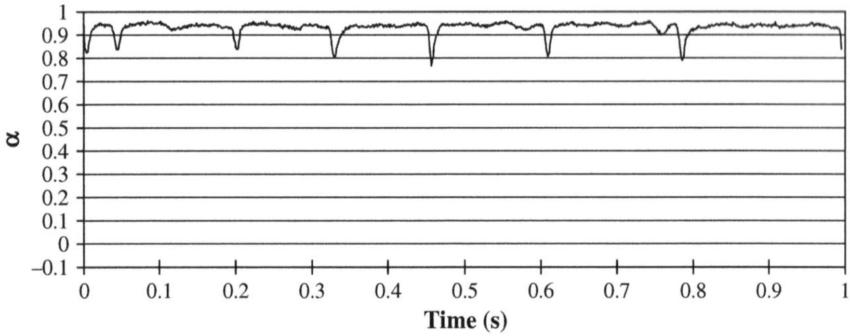

Figure 117. Void fraction time trace at μ-g for flow 20

Annular Flow Regime at 1-g, $\alpha_{ave} = 0.886$
Flow 20 (G1P8): $V_{SL} = 0.05\,m/s$, $V_{SG} = 16.86\,m/s$

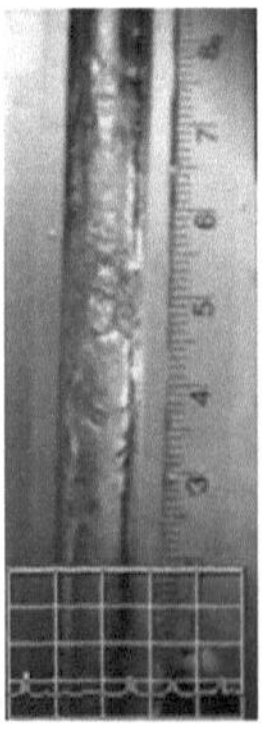

Figure 118. Flow image at 1-g for flow 20. The 1-g flow shows a significant wave disturbance

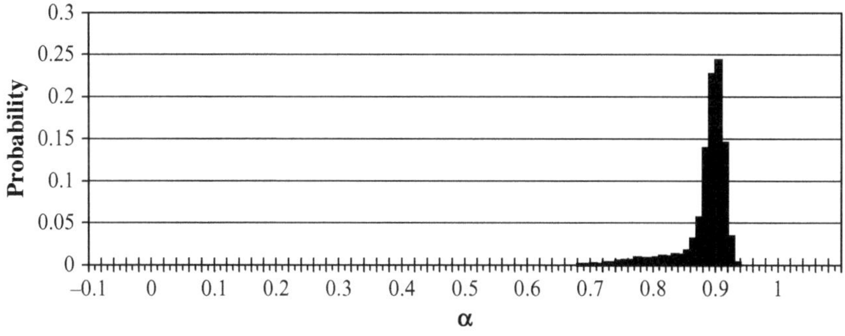

Figure 119. Void fraction PDF at 1-g for flow 20

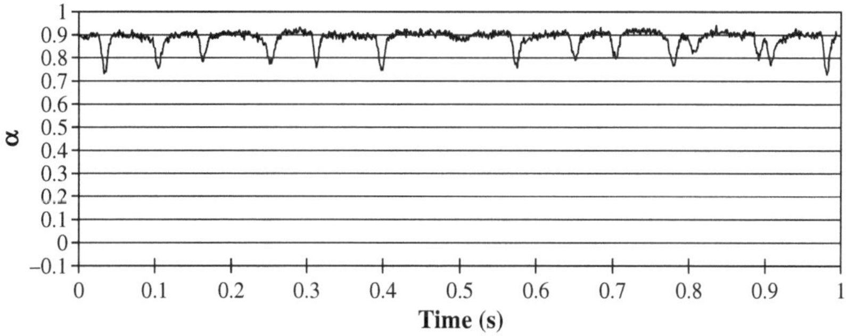

Figure 120. Void fraction time trace at 1-g for flow 20

Annular Flow Regime at μ-g, $\alpha_{ave} = 0.889$
Flow 21 (F1P10): $V_{SL} = 0.05\,m/s$, $V_{SG} = 20.94\,m/s$

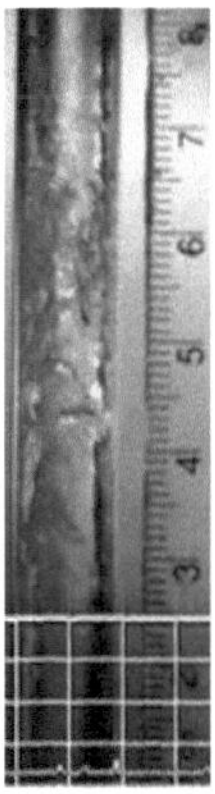

Figure 121. Flow image at μ-g for flow 21. The μ-g flow shows a disturbance at the center of the image. This could be due to the passing of a wave

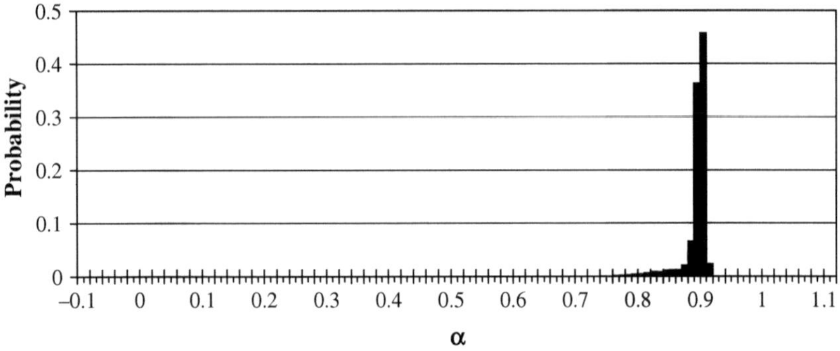

Figure 122. Void fraction PDF at μ-g for flow 21

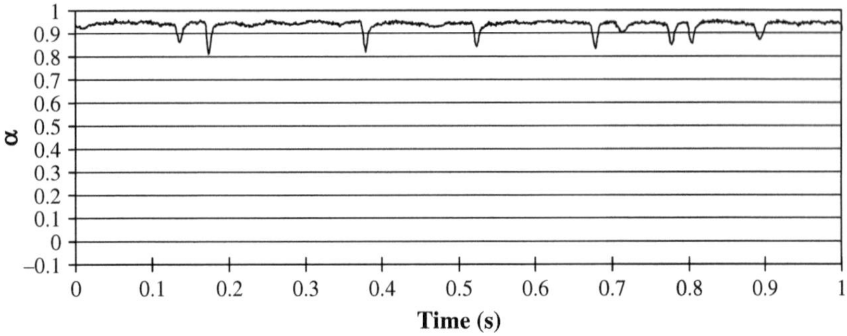

Figure 123. Void fraction time trace at μ-g for flow 21

Annular Flow Regime at 1-g, $\alpha_{ave} = 0.900$
Flow 21 (G1P10): $V_{SL} = 0.05\,\mathrm{m/s}$, $V_{SG} = 20.91\,\mathrm{m/s}$

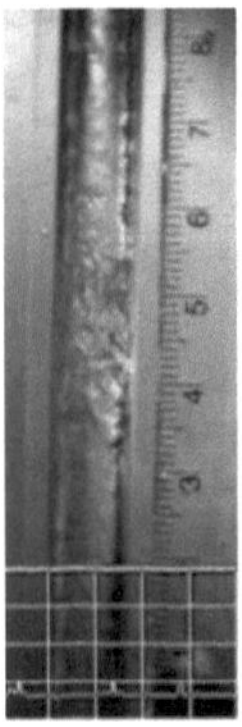

Figure 124. Flow image at 1-g for flow 21. Similarly, the 1-g flow also shows a disturbance at the center of the image, possibly the passing of a wave

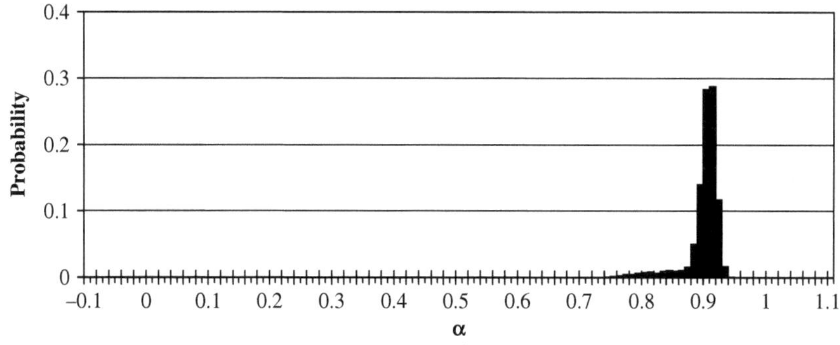

Figure 125. Void fraction PDF at 1-*g* for flow 21

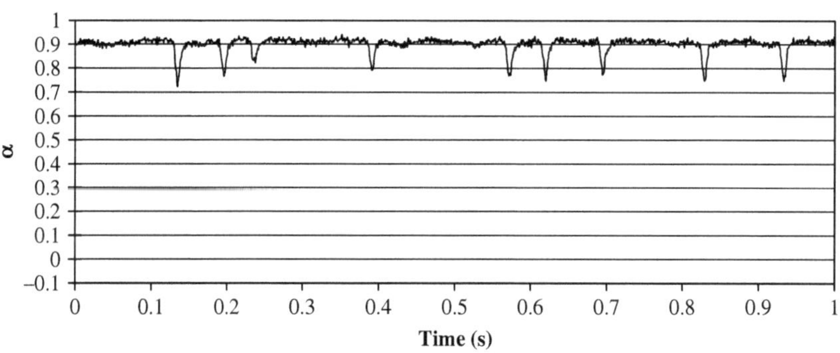

Figure 126. Void fraction time trace at 1-*g* for flow 21

REFERENCES

Antar BN, Nuotio-Antar VS (1993) Fundamentals of low gravity fluid dynamics and heat transfer. CRC Press. Florida, USA.

ASHRAE Handbook (1981) — Fundamentals, American society of heating, refrigerating and air-conditioning engineers, Inc.

Aung W (1987) Mixed convection in internal flow In: Kakaç S, Shah RK, Aung W (eds) Handbook of single-phase convective heat transfer. , Wiley-Interscience, New York, pp 15.10–15.43.

Barnea D, Taitel Y (1986) Flow pattern transition in two-phase gas-liquid flows. Encyclopedia of fluid mechanics, Vol. 3. Gulf Publishing Company pp 403–474.

Bousman WS, Mcquillen JB (1994) Characterization of annular two-phase gas-liquid flows in micro-gravity. Second microgravity luid physics conference, Cleveland, OH, June 21–23, pp 5.1–5.6.

Bousman WS (1995) Studies of two-phase gas-liquid flow in microgravity, NASA Contractor Report 195434.

Bugg JD, Mack K, Rezkallah KS (1998) A numerical model of Taylor bubbles rising through stagnant liquids in vertical tubes, Int. J. Multi-phase flow 24(Z):271–281.

Chen JC (1963) A correlation for boiling heat transfer to saturated fluids. Transactions of the ASME - Journal of Heat Transfer 63-HT-34.

Chen I, Downing R, Keshock E, Al-Sharif M (1991) Measurements and correlation of two-phase pressure drop under microgravity conditions. J Thermophysics 5: 514–523.

Chisholm D (1967) A theoretical basis for the Lockhart-Martinelli correlation for two-phase flow. Int J Heat Mass Transfer 10: 1767–1778.

Chisholm D (1983) Two-phase flow in pipelines and heat exchangers. George Godwin, London and New York.

Colin C, Fabre JA, Dukler AE (1991) Gas-liquid flow at microgravity conditions – I, dispersed bubble and slug flow. Int J Multiphase Flow 17: 533–544.

Collier JG (1972) Convective boiling and condensation, 1st Edn. McGraw-Hill, New York, pp 387–395.

Delil AAM (1991) Thermal gravitational modelling and scaling of two-phase heat transport systems similarity considerations and useful equations predictions versus experimental results. Proceedings of the First European Symposium on Fluids in Space, ESA SP-353, Ajaccio, France, November 18–22, pp 579–599.

Dukler AE, Wicks M, Cleveland RD (1964a) Frictional pressure drops in two-phase flow: A. a comparison of existing correlations for pressure loss and holdup. AIChE J 10: 38–43.

Dukler AE, Wicks M, Cleveland RD (1964b) Frictional pressure drops in two-phase flow: B. an approach through similarity analysis. AIChE J 10: 44–51.

Dukler AE, Taitel Y (1986) Flow pattern transitions in gas-liquid systems: Measurement and modeling, Multiphase Science and Technology, Hemisphere Publishing Corporation, Vol. 2. Hemisphere Publishing Corporation, pp 1–94.

Dukler AE, Fabre JA, McQuillen JB, Vernon R (1988) Gas-liquid flow at microgravity conditions: Flow pattern and their transitions. Int J Multiphase Flow 14: 389–400.

Elkow K (1996) Void fraction measurement and analysis at normal gravity and microgravity conditions, M.Sc. Thesis, Saskatoon: University of Saskatchewan.

Feldmanis CJ (1966) Pressure and temperature changes in closed loop forced convection boiling and condensing processes under zero gravity conditions Institute of Environmental Sciences' 1966 Annual Technical Meeting Proceedings.

Friedel L (1979) Improved friction pressure drop correlations for horizontal and vertical two-phase pipe flow. European Two-Phase Flow Group Meeting. Paper E2, Ispra.

Glycerine Producers Association (1973) Physical properties of glycerine and its solutions, pp 1–27.

Govier GW, Aziz K (1972) The flow of complex mixtures through pipes, Van Nostrand-Reinhold, New York, p 503.

Greger R, Rath HJ (1995) Measurement of the thermal conductivity of fluids with low viscosity under reduced gravity conditions using the transient hot-wire technique. International Journal of Heat and Mass Transfer 38(6): 1105–1110.

Hamacher H, Fitton B, Kingdon J (1987) The environment of earth-orbiting systems. In: Walter, HU (ed) Fluid science and material science in space, Springer Verlag, pp 1–50.

Heppner DB, King CD, Littles JW (1975) Zero-gravity experiments in two-phase fluids flow patterns. Presented at the ICES Conference, San Francisco, California, ASME Paper No. TS-ENAS-24.

Hirt CW, and Nicholas BD (1981) volume of fluids (VOF) method for the dynamics of free boundaries, J.Comp.phys.39:201–225.

Huckerby CS Rezkallah KS (1992) Flow pattern observations in two-phase gas-liquid flow in a straight tube under normal and microgravity conditions, AIChE Proceedings 88: 139–147.

Incropera FP, DeWitt DP (1981) Fundamentals of heat transfer, John Wiley and Sons, New York, pp 761–792.

Kakaç S (1987) The effect of temperature-dependent fluid properties on convective heat transfer. In: Kakaç S, Shah RK, Aung, W (eds) Handbook of single-phase convective heat ransfer. Wiley-Interscience, New York, p. 18.6.

Kamp A, Colin C, Fabré J (1993) Bubbly flow in a pipe: Influence of gravity upon void and velocity distribution, 3rd World Conference on Experimental Heat Transfer, Fluid Mechanics, and Thermodynamics, Honolulu, HI, October 30–November 5.

Karri SBR, Mathur VK (1988) Two-phase flow pattern map predictions under microgravity. AIChE Jl, 34: 137–139.

Kirpich A, Kruger G, Matteo D, Stephen J (1990) SP-100 space reactor power system, Thermal-hydraulics for space power, propulsion, and thermal management system design, progress in astronautics and aeronautics 122: 29–39.

Kays WM Crawford ME (1980) Convective heat and mass transfer, 2nd Edn. McGraw-Hill, New York, pp 103–113, 260–270.

Kays W M London AL (1984) Compact heat exchangers, 3rd Edn. McGraw-Hill, New York, pp 80–101.

Kays WM and Perkins HC (1985) Forced convection, internal flow in ducts. In: Rohsenow WM, Hartnett JP, Ganiç E N (eds) Handbook of heat transfer fundamentals, 2nd Edn. McGraw-Hill, New York, pp 7.30, 7.134.

Knott RF, Anderson RN, Acrivos A, Peterson EE (1959) Experimental study of heat transfer to nitrogen-oil mixtures, Industrial and Engineering Chemistry 51: 1369–1372.

Kudirka AA (1964) Two-phase heat transfer with gas injection through a porous boundary surface, ANL-6862.

Lance M, Bataille J (1991) Turbulence in the liquid phase of a uniform bubbly air-water flow, J Fluid Mech 222: 95–118.

Lockhart RW Martinelli RC (1949) Proposed correlations of data for isothermal two-phase two-component flow in pipes, Chem Eng Prog 45: 39–48.

Merte H Jr, Clark J (1963) Boiling heat transfer with cryogenic fluids t standard, fractional, and zero-gravity Transactions of the ASME – Journal of Heat Transfer, 63-HT-28.

Ohta H, Fujiyama H, Inoue K, Yamada Y, Ishikura S, Yoshida S (1994) Microgravity flow boiling in a transparent tube 10th International Heat Transfer Conference, Brighton, England, No. 111.

Papell SS (1962) An instability effect on two-phase heat transfer for subcooled water flowing under conditions of zero gravity, American Rocket Society 17th Annual Meeting and Space Flight Exposition.

Petukhov BS, Polyakov AF (1988) Heat transfer in turbulent mixed convection, Launder BE (ed) Hemisphere Publishing, New York, pp 147–153.

Reinharts TR, Best FR, Hill WS (1992) Definition of condensation two-phase flow behaviors for spacecraft design AIP Conference Proc. No. 246, MS El-Genk, MD Hoover (eds) American Institute of Physics, New York, 1: 1216–1225.

Rezkallah KS, Sims GE (1987) An examination of correlations of mean heat-transfer coefficients in two-phase, two-component flows in vertical tubes AIChE Symposium Series-Heat Transfer, Pittsburgh 83(257): 109–114.

Rezkallah KS, Sims GE (1989) New data on two-phase, two-component heat transfer and hydrodynamics in a vertical tube. Journal of Thermophysics and Heat Transfer 3(2): 213–219.

Rezkallah KS (1990) A comparison of existing flow-pattern predictions during forced-convective two-phase flow under microgravity conditions. Int J Multiphase Flow 16: 243–259.

Rezkallah KS (1996) Weber number based Flow-Pattern maps for liquid-gas flows at microgravity, Int J Multiphase Flow 22: 1265–1270.

Rite RW (1995) Heat transfer in gas-liquid flows through a vertical, circular tube under microgravity conditions, Ph.D. Thesis, Saskatoon: University of Saskatchewan.

Rite RW, Rezkallah KS (1994a) Heat transfer in two-phase flow through a circular tube at reduced gravity. Journal of Thermophysics and Heat Transfer 8(4): 702–708.

Rite RW, Rezkallah KS (1994b) The influence of liquid viscosity on heat transfer coefficients in gas-liquid flows under microgravity conditions. In: El-Genk MS, Hoover MD (eds) AIP Conference Proc 301, American Institute of Physics, New York, pp 1129–1136.

Rite RW, Rezkallah KS (1993) An investigation of transient effects on heat transfer measurements in two-phase gas-liquid flows under microgravity conditions. ASME Heat Transfer in Microgravity Systems – 1993, pp 49–57.

Serizawa A, Kataoka I, Michiyoshi I (1975a) Turbulence structure of air-water bubbly flow – II. Local properties. International Journal of Multiphase Flow 2: 235–246.

Serizawa A, Kataoka I, Michiyoshi I. (1975b.) Turbulence structure of air-water bubbly flow – III. Transport properties. International Journal of Multiphase Flow 2: 247–259.

Soliman HM (1986) The mist-annular transition during condensation and its influence on the heat transfer mechanism. International Journal of Multiphase Flow 12(2): 277–288.

Sridhar KR, Chao BT, Soo SL (1992) Pressure drop in fully developed, turbulent, liquid-vapor annular flows in zero gravity. AIAA Journal 30: 1016–1026.

Taitel Y, Dukler AE (1976) A model for predicting flow regime transition in horizontal and near horizontal gas-liquid flow. AIChE J, 22: 47–55.

Taitel Y, Barnea D, Dukler AE (1980) Modeling flow pattern transitions for steady upward gas-liquid flow in vertical tubes. AIChE J, 26: 345–354.

Tennekes H, Lumley JL (1972) A first course in turbulence, The MIT Press, Cambridge, p 7.

Usiskin CM, Siegel R (1961) An experimental study of boiling in reduced and zero gravity fields. Journal of Heat Transfer – Transactions of the ASME, Series C, 83: 243–251.

Vijay MM (1977) A study of heat transfer in two-phase two-component flow in a vertical tube, Ph.D. Thesis, University of Manitoba, Winnipeg, pp 234–240, 394–491.

Vijay MM, Savic P, Sims GE (1978) The Graetz problem for two-phase two-component gas-liquid flow in vertical tubes. 6th International Heat Transfer Conference, Toronto, Canada, 7–11 August 1978, Vol. 1, Washington DC: Hemisphere, pp 493–497.

Vijay MM, Aggour MA, Sims GE (1982) A correlation of mean heat-transfer coefficients for two-phase two-component flow in a vertical tube. Proceedings of the 7th International Heat Transfer Conference, Washington DC, Vol. 5, pp 367–372.

Wallis GB (1966) Pressure gradients for air flowing along porous tubes with uniform extraction at the walls. Institution of Mechanical Engineers Proceedings, Vol. 180, No. 1, pp 27–35.

Wallis GB (1969) *One-dimensional two-phase flow*, McGraw-Hill Book Company, New York.

Weisman J, Duncan D, Gibson J, Crawford T (1979) Effects of fluid properties and pipe diameter on two-phase flow patterns in horizontal lines. Int. J Multiphase Flow 5: 437–462.

Weisman J, Kang SY (1981) Flow pattern transitions in vertical and upwardly inclined lines. Int J Multiphase Flow 7: 271–291.

Whalley PB (1987) *Boiling, condensation and gas-liquid flow*, Oxford University Press, N.Y., USA.

Zhao L, Rezkallah KS (1993a). Gas-liquid flow patterns at microgravity conditions. Int J Multiphase Flow 19: 751–763.

Zhao L, Rezkallah KS (1993b) The transient effects on the two-phase flow heat transfer at microgravity. 29th National Heat Transfer Conference, Paper No. 93-3905, Atlanta, GA, August 8–11.

Zhao L, Rezkallah KS (1995a) Pressure drop in gas-liquid flow at microgravity conditions. Int J Multiphase Flow 21: 837–849.

Zhao L, Rezkallah KS (1995b) A flow pattern map for two-phase liquid-gas flows under reduced gravity conditions. Advances in Space Research 16(7): (7)133–(7)136.

INDEX

MIX
Papier aus verantwortungsvollen Quellen
Paper from responsible sources
FSC® C105338

If you have any concerns about our products,
you can contact us on
ProductSafety@springernature.com

In case Publisher is established outside the EU,
the EU authorized representative is:
Springer Nature Customer Service Center GmbH
Europaplatz 3, 69115 Heidelberg, Germany

Printed by Libri Plureos GmbH
in Hamburg, Germany